TONGASS TIMBER

TONGASS TIMBER

A History of Logging & Timber Utilization in Southeast Alaska

James Mackovjak

FOREST HISTORY SOCIETY

The Forest History Society is a nonprofit educational and research institution dedicated to the advancement of historical understanding of human interaction with the forest environment. It was established in 1946. Interpretations and conclusions in FHS publications are those of the authors; the institution takes responsibility for the selection of topics, competency of the authors, and their freedom of inquiry.

This edition was published with support from the Kendall Foundation, Mike Blackwell, the SB Foundation, the Alaska Historical Society, the Alaska Humanities Forum, and the Lynn W. Day Endowment for Forest History Publications.

Printed in the United States of America

Forest History Society
701 William Vickers Avenue
Durham, North Carolina 27701
(919) 682-9319
www.foresthistory.org

Design by Zubigraphics, Inc.

On the cover: U.S. Forest Service Ranger J. M. Wyckoff and Alaska District Assistant Forester Frank Heintzleman cruising Sitka spruce timber at Bond Bay, Behm Canal, in 1930. Heintzleman worked tirelessly for more than three decades to establish pulp mills in Southeast Alaska. (USDA Forest Service, Alaska Regional Office, Juneau)

Library of Congress Cataloging-in-Publication Data
Mackovjak, James R.
Tongass timber : a history of logging and timber utilization in southeast Alaska / James Mackovjak.
p. cm.
Summary: "Tongass timber traces the history of logging and timber utilization in southeast Alaska from Native Indian use to Russian settlement and through the establishment and closing of the pulp mills. This background, shown in great detail, reveals the forces that influence the present choices about forest management in southeast Alaska"--Provided by publisher.
Includes bibliographical references and index.
ISBN 978-0-89030-074-9 (pbk. : alk. paper)
1. Logging--Alaska--Tongass National Forest--History. 2. Timber--Alaska--Tongass National Forest--History. 3. Lumber trade--Alaska--Tongass National Forest--History. 4. Pulp mills--Alaska--Tongass National Forest--History. 5. Forest policy--Alaska--Tongass National Forest--History. 6. United States. Forest Service--History. 7. Tongass National Forest (Alaska)--History. 8. Tongass National Forest (Alaska)--Management. 9. Tongass National Forest (Alaska)--Environmental conditions. I. Title.
SD538.2.A4M234 2010
338.1'749809798--dc22
                    2010027611

*This book is dedicated to the many librarians and archivists
who assisted me with this project. Their commitment
and helpfulness were a joy to behold.*

Contents

Foreword

Receiving my academic degrees and earning a practical education in economics were totally separate experiences carried out in two different worlds—the Academy and the Real World. The backbone of academic economics at the time (the late 1930s) was based upon an oversimplification of reality—it excluded any forces or factors that did not fit into its clean, simple version of cause-and-effect Newtonian physics. When I complained about this state of affairs to one of my mentors, he advised me that the study of neo-classical economics was more interesting than playing bridge but not any more useful. His advice was to play the game, get the degree, then find an island for my laboratory and explore the non-economic forces that were really driving its economic change. World War II sent me to Alaska and this became my "island."

In this book, James Mackovjak digs deeply into his "island"—the small, well-defined geographic unit of Southeast Alaska. The result is an in-depth history of Southeast Alaska's forest region from pre-Western contact through the Russian-American period to the establishment of pulp mills at Ketchikan (1954) and Sitka (1960). This book's saga is a fascinating case study of the dynamics of change in this bit of the real world. The forces of change were a complex mix of geopolitics, globalization, fluctuating heavy government subsidies, increasing environmental regulation, and technological changes, which do not fit neatly together in a traditional neoclassical economic statistical format. What this book provides is the human story behind the economics. This background, shown in great detail, reveals the forces that influence the present.

The research is impressive in its coverage and includes contemporary views of boosters, politicians, and the press, giving a sense of the spirit of the times. The chronological narrative flows through the rises and falls of timber with a detailed footnoting of sources. The appendices contain detailed information about the technology used to harvest timber in Southeast Alaska's forest, and the mills, large and small, that processed the timber.

If this book had been available when I started my work in and about Alaska it would have been a much-used reference kept close at hand.

—George Rogers

George Rogers received his training in economics at the University of California (Berkeley) and Harvard University. He came to Alaska in 1945 and served as an advisor to a number of governors. Among them was Frank Heintzleman, who championed the development of a wood pulp industry in Southeast Alaska. Rogers was also a consultant to Alaska's Constitutional Convention during the winter of 1955–56. His principal interest has been what he terms "real world" economics, in which dynamic forces of change—often political, bureaucratic, or technical—shape the economic landscape. He used Alaska as his "laboratory" to explore these forces. Internationally, he has worked with the Sorbonne and Cambridge University as well as served as special advisor to the Canadian Privy Council. The author of a number of books on northern economies, including Alaska in Transition: The Southeast Region *(Baltimore: Johns Hopkins Press, 1960), Rogers resides in Juneau with his wife, Jean.*

Preface

Viewed from the deck of the *Ol'ga*, the Russian-American Company ship transporting Alexandr Baranov to New Archangel (Sitka) in 1804, Southeast Alaska presented itself as rugged and mountainous. More often than not, clouds shrouded the landscape, dulled its colors, and gave it a soft, almost mystical appearance. Fed by a generous amount of rain and snow, numerous streams coursed to the ocean, and myriad channels wound between the many islands and into the Coast Range beyond. The lowlands and lower mountains were mostly covered with a vast coniferous forest, and some of the trees were magnificent. Handsome Sitka spruce trees sometimes reached more than 200 feet into the sky from bases that exceeded eight feet in diameter; some Alaska-cedars had been growing since biblical times.

The forest Alexandr Baranov saw represented millennia of growth, decay, and rejuvenation. The cultures of the Tlingit and Haida people that had occupied Southeast Alaska for some 10,000 years were oriented more toward the region's rich marine resources—particularly fish—than those of the land, and their impact on the forest was minimal. Trees were used by the region's aboriginal inhabitants for homes and household goods, for firewood, for canoes, and for their iconic and highly-visible totem poles. Baranov had at his disposal a vast forest that had never heard the ring of a timberman's axe.

But Baranov had little interest in Southeast Alaska's verdant forest wealth. His charge, as an officer of the Russian-American Company, was to establish a trading post and to harvest sea otters for their pelts, which had a widespread market in both Europe and Asia. Timber would be useful to build a fort, to construct houses and other buildings, to repair ships, and to burn as firewood. Only after the sea otters had been hunted to near extinction did selling lumber begin to interest the Russians, and even then only to a limited extent. Nineteenth-century hand-powered technology for moving logs confined logging to the shoreline areas, and the difficulty of shipping wood vast distances to markets in Chile, China, and other Pacific destinations made any timber enterprise marginal at best.

The large-scale cutting of Southeast Alaska's forest would not take place until a century and a half after Baranov built the Russian-American Company's headquarters on the island that came to bear his name. This harvesting would supply the two large pulp mills constructed at Ketchikan and Sitka in the 1950s.

This book records the history and development of the logging, lumbering, and pulp manufacturing industries in Southeast Alaska from 1804, when the Russian-American Company felled the first trees to construct its Sitka stockade, through 1960, when the second of two modern pulp mills became fully established in the region. The U.S. Forest Service's long effort to establish a wood pulp industry is chronicled in detail. The text is followed by two appendices. Appendix A, "The Sawmills of Southeast Alaska," contains brief descriptions of nearly every sawmill that operated in Southeast Alaska prior to 1960. Appendix B describes the evolution of the technology that was used to log in the region. The appendices are followed by a glossary of timber-related terms.

The information presented was gleaned from a wide variety of sources. In promoting the utilization of Southeast Alaska's timber, the Forest Service conducted numerous studies and prepared reports and articles, many of which contained useful historical information as well as insights into the agency's philosophy. As well, reports submitted by Alaska's territorial governors provided details and an understanding of the official Alaska perspective on the use of Southeast Alaska's timber resource. Much information came from trade publications such as *The Timberman* and *Pulp & Paper*, and professional journals such as the *Journal of Forestry*. Southeast Alaska's newspapers provided a local perspective and details regarding mills and logging operations, while publications such as the *New York Times* and *Washington Post* furnished a national perspective on the development of Southeast Alaska's forest. In addition, the author obtained information from several individuals who were directly involved in the industry.

Acknowledgments

I would first like to thank my late friend and neighbor, Ken Youmans, in whose collection of historical Alaska papers I first discovered the potential for writing this history. The project would not have been possible without the generous financial support of Mike Blackwell, the SB Foundation, the Kendall Foundation, the Alaska Historical Society, and the Alaska Humanities Forum. I cannot begin to express my appreciation for the almost blind faith they put in me and their recognition of the value of the project.

Alaska historian Robert De Armond, Alaska economist George Rogers, and veteran Pacific Northwest forester William Hagenstein provided inspiration, guidance, and a valued connection to Alaska and the timber industry's past. It is an honor to build on some of their work. Joe Mehrkens, former U.S. Forest Service economist in Alaska, was among those who reviewed early drafts of the manuscript. Jim LaBau, a former U.S. Forest Service forester and currently with the University of Alaska (Fairbanks), generously gave his time to critically review and provide insight on technical aspects of the draft manuscript. Frank Norris, former historian for the National Park Service in Alaska, and Sally Atwater, former editor of the *Journal of Forestry*, with Steven Anderson, president of the Forest History Society, edited the manuscript. My neighbor, Bill Eichenlaub, and my son, Seth Mackovjak, prepared several maps.

I would also like to thank those at the U.S. Forest Service Alaska Regional Office and the Alaska Forestry Sciences Laboratory who assisted me, especially Mark Shultz, Ray Massey, Sarah Stiles, Bob Price, Shiela Spores, and Paul Hennon.

Among the numerous other individuals who assisted and supported me in various ways were, in no particular order: Greg Streveler, Judy Brakel, John Sisk, Cheryl Oakes, William Brown, Richard VanOrder, Louis Blackerby, Jim Calvin, John Daly, Wes Tyler, Tracy Churchill, Rob and Koren Bosworth, Ralph Dale, Elizabeth Bluemink, Don Nelson, Marinke Van Gelder, Brixie Crabtree, Adam Greenwald, Hal Salwasser, Paul Barnes, Tom Waldo, Narda Pierce, Kristen Griffin, Mike Salee, Don Muller, Larry Edwards, Wayne Weihing, Alf Skaflestead, Fred Hosford, Robert Thorstenson, Joe Smith, John Schnabel, Kirk Dahlstrom, Greg Head, Florian Sever, Jim Clark, Buck Lindekugel, Laura Vidic, Paul Rushmore, and Richard Van Cleave.

It was my pleasure to gather information at the Alaska State Archives, Clausen Museum (Petersburg), Wrangell Museum, Tongass Historical Museum (Ketchikan), Juneau-Douglas City Museum, Library of Congress, Southeast Alaska Conservation Council, Klondike Gold Rush National Historical Park, Federal Records Center (Anchorage), National Archive (College Park, Md.), Knight Library (University of Oregon), Valley Library (Oregon State University), Suzallo Library (University of Washington), Doe Library (University of California, Berkeley), Consortium Library, University of Alaska/Alaska Pacific University (Anchorage), Bancroft Library, and the Rasmuson Library (University of Alaska, Fairbanks). This project would not have been possible without the considerate help of the folks who staff these institutions and organizations, and those who had the foresight to establish the collections. I would especially like to thank the staff at the Alaska State Library (Juneau) and the Historical Section, in particular, for the enjoyable and productive work environment they always provided to me.

Finally, I would like to again thank Steven Anderson and Cheryl Oakes and their colleagues at the Forest History Society for their willingness to take on and shepherd this project to completion.

CHAPTER ONE

Southeast Alaska and Its Forest

PHYSICALLY, SOUTHEAST ALASKA (THE "PANHANDLE") IS COMPRISED OF HUNDREDS OF ISLANDS, KNOWN COLLECTIVELY AS THE ALEXANDER ARCHIPELAGO, AND A LARGER ADJACENT RUGGED STRIP OF MAIN-LAND THAT RUNS TO THE CREST OF THE COAST RANGE. THE ALEXANDER ARCHIPELAGO WAS NAMED BY THE U.S. GOVERNMENT IN HONOR OF RUSSIA'S TSAR ALEXANDER II. IT WAS UNDER ALEXANDER II'S REIGN THAT RUSSIA SOLD ALASKA TO THE UNITED STATES.

The region has extremely varied geology and forms a rough arc along the eastern Gulf of Alaska from Dixon Entrance (55° N. lat.) in the south to Yakutat Bay (60° N. lat.) in the north. It is more than 500 miles long and, in places, 125 miles wide. Southeast Alaska comprises about 35,560 square miles, roughly the same area as the state of Indiana. About 40 percent of Southeast Alaska's acreage is located on some 1,100 islands, 67 of which have an area of more than four square miles. Of these, 15 exceed 100 square miles in area, and six—Prince of Wales, Chichagof, Admiralty, Baranof, Revillagigedo, and Kupreanof—exceed 1,000 square miles. Prince of Wales is the largest island, with an area of 2,770 square miles. The islands and the mainland together have nearly 10,000 miles of shoreline. Approximately 85 to 90 percent of the shoreline is characterized by precipitous rock faces.[1]

Southeast Alaska's climate is governed largely by the "Aleutian Low," a large, semi-stationary North Pacific weather system that is most intense during the winter months. Offshoots of this system travel southeastward, and amplify the onshore flow of relatively warm, moisture-laden air brought against the region's mountainous coast by the westerly winds that prevail globally at this latitude. Upon striking the coast, this air is forced to rise, causing it to cool and its moisture to be condensed into rain and snow.[2] The result is a maritime climate in which three conditions predominate: mild temperatures (cool summers and moderate winters), high, fairly well-distributed precipitation, and prolonged cloudiness. The annual precipitation at sea level in Southeast Alaska ranges from fewer than 50 inches in upper Lynn Canal to more than 200 inches at some locations to the south. Precipitation in higher

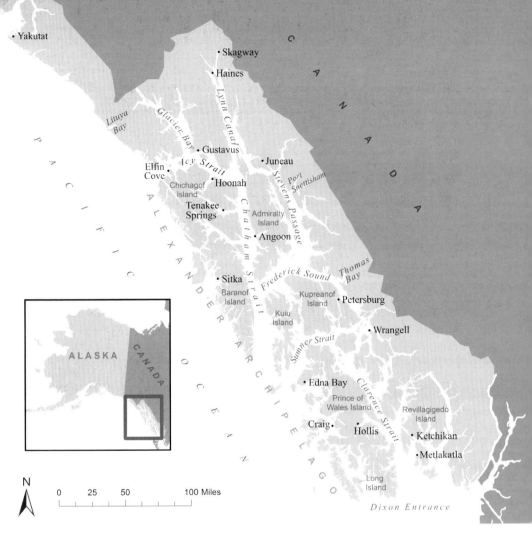

FIGURE 1. Southeast Alaska (Bill Eichenlaub)

elevations tends to be considerably higher. This ample precipitation, combined with high humidity and relatively mild temperatures, is conducive to luxuriant coniferous forest growth—provided adequate drainage is present.

About 41 percent of Southeast Alaska is rock or covered with ice, 25 percent is muskeg and scrub forest, and 34 percent is what can be considered commercial forest, which grows to an elevation of about 2,000 feet. Commercial forest lands comprise about 6.3 million acres of southeast Alaska's land base, of which 45 percent produce 8–20 thousand board feet (MBF) per acre, another 36 percent produce 20–30 MBF per acre, and 19 percent produce more than

30 MBF per acre. But because lands that produce less than 20 MBF per acre are virtually never logged, the remaining 3.5 million acres of forest provide the basis for the region's timber and wood products industry. The commercial forest is comprised almost entirely of four coniferous species: western hemlock (*Tsuga heterophylla*), Sitka spruce (*Picea sitchensis*), western redcedar (*Thuja plicata*), and Alaska-cedar (*Chamaecyparis nootkatensis*). By species, about 73 percent of the pre-industrial stand of commercial timber in Southeast Alaska was western hemlock, 21 percent was Sitka spruce, 3 percent was western redcedar, and 3 percent was Alaska-cedar. A fifth species, Mountain hemlock (*Tsuga mertensiana*) occurs at higher elevations and has occasionally been logged, but is unimportant commercially.[3]

The forest of Southeast Alaska is a large component of the region's resource base. Stands of timber, however, are variable in quantity and quality, with extensive areas of good forest unusual, and the very best material located in tracts of 20 to 100 acres.[4] Muskeg tends to develop in poorly drained areas.

Wind is a major factor in natural tree mortality in Southeast Alaska. Storms regularly roll off the Gulf of Alaska, particularly in the fall, and associated strong winds often topple trees, at times over large areas. A major storm caused extensive windthrow damage around Juneau in about 1884.[5]

The classic old-growth forest of Southeast Alaska is different from that of the Pacific Northwest. In Southeast Alaska, high precipitation without a summer drought period generally precludes the catastrophic fires that periodically plague the Pacific Northwest, and the classic old-growth forest is best described as a "climax association of all-aged timber." In an area as small as one-tenth of an acre, trees may vary in age from a few years to 1,000 years. Tree growth can sometimes be inhibited due to a variety of natural factors, and it is not uncommon to find trees two to four inches in diameter that are more than a century old.[6]

Due to harsher climatic conditions and a shorter growing season, trees in Southeast Alaska generally grow more slowly than, and do not attain as great a size as, their same-species counterparts in the Pacific Northwest. The slower growth affects wood characteristics, one of which is a tighter annular ring pattern, a quality considered highly desirable in specialty applications such as sounding boards for musical instruments. Additionally, trees in Southeast Alaska live about twice as long as those in the Pacific Northwest.[7]

An estimated 5,000 rivers and streams in Southeast Alaska—representing a total of approximately 20,000 miles—support populations of salmon. Though

FIGURE 2. Most of Southeast
Alaska's forest is comprised
of a mixture of Sitka spruce
and western hemlock.
(NARA-Anchorage, RG 95)

salmon are not usually considered to be
creatures of the forest, this is certainly the
case in Southeast Alaska, where each year
millions of salmon spawn and die in forest
rivers, streams, and lakes. In doing so, salmon
contribute to the forest a substantial portion of their nutrient content, much
of which is ultimately transported well beyond the margins of spawning areas
by the various animals that feed on them, including bears, otters, eagles,
insects, etc. The fertilizer effect of the transfer of marine-derived nutrients into
the forest ecosystem of Southeast Alaska has only recently been recognized
as a significant factor in forest productivity.[8]

Low-elevation alluvial fans are an important landform type for big-
tree/high-volume forests in Southeast Alaska. More important, however, are
areas underlain by carbonate bedrock such as limestone and marble. These
areas have a distinctive topography characterized by landforms such as caves
and sinkholes, and are known as karst areas. The carbonates buffer the forest
soil's natural acidity, which fosters forest growth by increasing the availability
of nutrients.[9]

Southeast Alaska possesses some of the best developed karst areas in
the world, particularly on Prince of Wales Island and some of the neighboring

islands. The largest and highest quality trees in the region were found almost invariably on karst areas that were sheltered from the prevailing heavy winds.[10] Illustrating the karst factor, J. M. Wyckoff, the Forest Service's Ketchikan district ranger in 1930, wrote that "the largest and best spruce trees are found on the west coast of Prince of Wales Island and adjacent islands, a region of limestone formation. Areas have been logged in this section yielding more than 100,000 board feet per acre and averaging 5,000 feet per tree."[11] In 1927, logger J. R. Reynolds described some particularly fine timber at Ham Cove, on Dall Island:

> The timber runs 75,000 per acre, 90 per cent spruce, 10 per cent hemlock. Most of the trees are three and one-half to six feet on the stump with merchantable lengths up to 160 feet. The largest timber scales as high as 15,000 feet to the single stick. The ground is partly broken up limestone and marble formation; the elevation 100 to 350 feet.[12]

Government estimates of the volume of merchantable timber in Southeast Alaska increased as a long progression of timber surveys were completed, from 70 billion board feet (1917) to 74 billion board feet (1924), to 78.5 billion board feet (1928). By 1954, the merchantable timber volume had increased to 137 billion board feet, and by 1975 had swelled to 166 billion board feet.[13]

PRINCIPAL COMMERCIAL TIMBER SPECIES

Western hemlock (*Tsuga heterophylla*) Western hemlock is the predominant tree in Southeast Alaska's forest because of its ability to grow in relatively poorly-drained areas, and its tolerance of shade.

A typical mature western hemlock on a good site in Southeast Alaska is three to four feet in diameter and 100 to 140 feet tall. Because the lower branches are shed readily, the trunk tends to be free of large knots. Little decay is found in western hemlock trees fewer than 100 years old.[14]

The root system of the western hemlock is shallow and wide-spreading, though in wind-protected locations it can be almost superficial. Even sound trees can be very susceptible to blowdown should exposure to high winds occur, such as the case might be at the edge of an existing blowdown or clearcut.

As compared to the other coniferous trees in southeast Alaska, Western hemlock is not particularly long lived. Specimens sometimes reach an age of

FIGURE 3. Western hemlock forest near Ketchikan about 1909. Trees range from 12 to 24 inches in diameter and are from 75 to 100 feet tall. (Alaska State Library, USFS Collection)

500 years, though some have been found in Southeast Alaska in the 625- to 725-year-old range.[15]

There is little similarity between the wood of the western hemlock and that of the eastern hemlock (*Tsuga canadensis*). Eastern hemlock was despised by lumbermen, its wood weak with coarse and often crooked grain and little resistance to decay. As well, it is difficult to work, splinters badly, and is prone to curling and twisting. Held as nearly worthless for many years, utilization of eastern hemlock began only when the supply of more desirable trees had been largely depleted. The species remained a hard sell, however, and the lumber industry in 1914 was forced to contend with a widely-published newspaper article titled "Avoid the Wood Called Hemlock."[16]

As a commercial species, western hemlock suffered due to its association with its eastern relative. As logging progressed from the East Coast through the Great Lakes states and on to the Pacific Northwest, the eastern hemlock's poor reputation accompanied it, and impaired the marketability of its western namesake. Early-day timber cruisers blocking out timberlands on the West Coast for eastern buyers contemptuously referred to western hemlock as "weed trees." To those familiar with the positive qualities of western hemlock, however, the lack of appreciation for it was due to the general "ignorance of the timber dealers, architects and wood consumers." They claimed that the species had been "sinned against" when given the name hemlock, for it was not an "Orphan Annie," but the "Cinderella of western woods." In 1926, Aldo Leopold, the assistant director of the Forest Products Laboratory who later became a leader in the American conservation movement, ranked western hemlock as one of North America's most valuable trees.[17]

Green western hemlock is very heavy due to high water content. Because of the wood's weight, there was a bias against it among loggers and millmen, particularly in the early years of Southeast Alaska's lumber industry, when much material was moved by hand. In 1902, Governor John Brady noted that "loggers and mill men...do not like to handle [western hemlock], on account of its great heft."[18] Loggers sometimes faced an additional problem: western hemlock logs stored in booms sometimes became completely waterlogged and sank.

In 1931, western hemlock was described as "moderately light in weight, moderately weak in bending and compression, moderately stiff, moderately hard, moderately low in shock resistance, and has a moderately large shrinkage." Among the many desirable qualities of this light-colored wood are ease of working, resistance to splitting, lightness combined with considerable strength and toughness, and the ability to take a handsome finish. It is not easily scratched, has excellent wearing qualities, and takes paint and stains well.[19]

The Tlingits' use of western hemlock was fairly limited; primarily they used the species' sapwood to make canoe paddles.[20] In modern society, western hemlock is used for framing lumber, and is particularly suitable for flooring, molding, paneling, and all inside finish lumber. Clear western hemlock found considerable use as ladder stock. As late as the early 1920s, lumbermen in Southeast Alaska continued to generally avoid western hemlock. But in 1927, the Juneau Lumber Mills began advertising western hemlock, claiming that "used once [it] will be favored for flooring, finishing, siding, studding, framing, boards and boxes, plank, dock timbers."[21] Western hemlock was also used for railroad ties. Overall demand, however, remained low for the wood until the 1940s, but in more recent years the market for western hemlock has expanded.

One local need met by the tough, somewhat teredo-resistant western hemlock was for piling. Southeast Alaska's cities and towns are largely built along the water, and a substantial number of the region's buildings are constructed on piles, as are its wharves and docks. Additionally, the once-ubiquitous salmon (pile) traps required large quantities of piling.

The pulpwood-of-choice of early West Coast manufactures of mechanical (groundwood) and sulfite pulp was Sitka spruce. The species was also widely used for lumber, and the demand for it became so great that there was a very real concern that the resource would soon be exhausted. Efforts were made to find a less costly and more available substitute. Western hemlock was found to be very suitable, and about 1930, the species had become the region's pulpwood-of-choice.[22]

Consideration of Southeast Alaska's vast resource of western hemlock primarily as raw material for the manufacture of pulp began early. In 1913, B. E. Hoffman, wrote in *Forestry Quarterly* that "the western hemlock of southeastern Alaska is expected to reach its highest commercial importance in the manufacture of paper pulp."[23]

Sitka spruce (*Picea sitchensis*) The vigorous and handsome Sitka spruce was given its scientific name by Russian naturalists in 1827, after its scientific discovery on Baranof Island, which was then known as Sitka Island. The "spruce from Sitka" was an item traded for food by the Russians colonists at New Archangel (Sitka) to the Spaniards and Mexicans in California, and thus the species received its common name. The name is very appropriate: *sitka* is a Tlingit word that is said to mean "by the sea."[24] The Sitka spruce is strictly adapted to the coastal environment and very seldom found far from the ocean.

"Sitka spruce is, in every way, the finest tree that grows in Alaska," wrote the Forest Service's Frank Heintzleman in 1923. "It reaches the enormous size of 10 feet in diameter and 250 feet in height, and its clean, symmetrical trunk extending up through the dense stand of smaller hemlock presents the appearance of a huge dark granite shaft."[25] Almost 40 years later, in 1962, the Sitka spruce was designated as Alaska's state tree, based on the simple fact that it was considered to be Alaska's most valuable tree.

In 1903, Republican Congressman James Tawney of Minnesota, a person long identified with timber interests, visited Southeast Alaska with a group of timbermen. They visited Shakan, at the north end of Kosciusko Island, where the group

> entered the forest to find spruce trees measuring from two to six feet in diameter, and from sixty to 100 feet of trunk entirely free from limbs. Two of the lumbermen in the party, after we had gotten into the forest, within a couple of

FIGURE 4. Loggers on springboards felling a large Sitka spruce, circa 1918. The wooden wedge helps keep the cut open to prevent their saw from binding. Note complete lack of safety equipment. (NARA-Anchorage, RG 95)

hundred yards radius counted twenty spruce trees measuring four feet and over in diameter, and having sixty to 100 feet of clear trunk.

Tawney returned from his trip "an enthusiast over the resources and the future of Alaska and particularly with respect to the timber of that territory."[26]

The Sitka spruce is the largest of the spruce trees. It has a capacity for rapid growth that makes it well-adapted to take advantage of sunlight-providing disturbances such as the relatively frequent blowdowns that occur in Southeast Alaska. The shallow root system of the Sitka spruce can be very extensive, particularly in trees growing in open locations.

Pure stands of Sitka spruce seldom exceed 80 acres, and tend to be ten acres or less. Historically, these stands were quite common and frequently contained as much as 100 MBF per acre. A very exceptional 5-acre stand of Sitka spruce at Clover Passage (Revillagigedo Island) logged by the Alaska Spruce Log Program during World War II contained an estimated 1 million board feet. Pure stands of Sitka spruce are often located on new ground: areas of extensive, contemporaneous disturbance such alluvial fans, alluvial terraces along valley floors, raised outer coast beaches or areas where land is exposed by retreating glaciers. More commonly, Sitka spruce is found in a varying mixture with western hemlock with the best stands occurring over karst or on alluvial fans.[27] Sitka spruce trees seven feet in diameter and 200 feet high were once common in Southeast Alaska. William Weigle, supervisor of the Tongass National Forest, described some timber being logged circa 1916: "Many spruce trees are being cut at the present time on the West Coast of Prince of Wales Island and adjacent islands which range from six to eight feet in diameter above the butt swells, and are from 150 to 200 feet tall. One tree just scaled contained 24,000 feet of merchantable timber."[28] Another observation of particularly fine timber was by the Forest Service's J. M. Wyckoff in 1924:

> The finest stands of spruce on the Tongass National Forest are found in that section known as the "West Coast," embracing Prince of Wales Island and adjacent westerly islands…. Mills prefer the class of logs produced from this section for the high per cent of clears they yield. Trees from this section yielding from 5 to 12 M. bd. ft. are common. Some have been cut scaling 18 to 20 M. bd. ft. The Forest Service sold two areas in 1924 situated on Long Island and Dall Island, with an estimated stand of 45,000,000 bd. ft. This timber is in compact bodies and will average better than 4 M. bd. ft. per tree.[29]

A typical mature Sitka spruce on a good site in Southeast Alaska is about five feet in diameter and 160 feet high. The Forest Service estimated in 1940 that about 65 percent of the Sitka spruce trees in Southeast Alaska had a diameter greater than 24 inches, with the average tree scaling about 2,500 board feet. On a good site where competition is minimal, Sitka spruce can grow very rapidly and add as much as a half-inch of radial growth in a year. Those in competitive situations grow far more slowly. Some of a group of 500- to 520-year-old Sitka spruce trees examined by the Forest Service on Kosciusko Island in 1946 were found to have added only an inch of radial growth in the previous 100 years.[30]

In 1909, when Assistant Forester Royal Kellogg was in Alaska to collect material for his Forest Service publication on Alaska's forests, part of his work involved the documentation of great age in Sitka spruce. At Portage Bay (between Petersburg and Juneau) on a section of a log 54 inches in diameter taken 25 feet above the ground Kellogg counted 600 rings. Another log 54 inches in diameter eight feet above the ground had 525 rings. Kellogg noted both trees to be entirely sound. In his 1931 study of the mechanical properties of Alaska woods, the Forest Service's L. J. Markwardt recorded a Sitka spruce from Southeast Alaska that was 757 years old. Later observations by others have reported patriarchs over 900 years old.[31]

What was thought by Frank Heintzleman to be the largest individual Sitka spruce tree in the region measured 14½ feet in diameter six feet above the ground. Heintzleman did not divulge the location of the tree. Slightly smaller was a 600-year-old tree cut on Sukoi Island, near Petersburg, in 1909. It had a circumference of 44 feet at the base, and was 217 feet tall. Five men worked five days to fell the tree, the largest log from which was shipped to the Alaska-Yukon-Pacific Exposition in Seattle.[32]

The wood of the Sitka spruce is creamy white in color and has no distinct odor. It has been described as "moderately light in weight, moderately weak in bending and compressive strength, moderately stiff, moderately soft, moderately low in resistance to shock and has a moderately large shrinkage." In addition, it was described by the Forest Service as "easy to kiln-dry, good in ability to stay in place, easy to work, low in nail-holding ability, very easy to glue, and very low in resistance to decay." What makes Sitka spruce very special, however, is that it has the highest strength-to-weight ratio of any wood.[33]

Southeast Alaska's Native population used Sitka spruce in many ways, perhaps the most prominent of which was for the beautiful seagoing canoes that were carved from single logs. Tlingits in the northern part of the region built houses of Sitka spruce logs or hand-split planks. Sitka spruce was often used to construct the famed bentwood boxes that were used for all sorts of storage and even coffins. General utensils and tools were made of Sitka spruce, as were ceremonial hats and masks.[34]

In more recent years, many uses have been developed for Sitka spruce lumber. Construction-grade material has been used to frame buildings, and finer material has been used as sash for doors and windows. It has been used to make ironing boards and clothes drying racks for the household. Sitka spruce has been present on farms as windmill slats and poultry brooders, and has been used for playground equipment, and in trunks and veneers. Light and strong, it has been used to make ladders. It is excellent material for the manufacture of paddles and oars, and has been widely used for the manufac-

FIGURE 5. Loggers felling an exceptionally large Sitka spruce near Natoma Point, on Long Island, in 1941. This tree scaled 45 thousand board feet. (USDA Forest Service, Alaska Regional Office, Juneau)

ture of packing cases, butter containers, baskets, and fruit and vegetable containers. Finally, it has been used in the construction of caskets.[35]

There are two uses, however, for which Sitka spruce has no equal: the manufacture of wooden airplanes and stringed musical instruments.

For early airplane fuselage and wing structures, the wood-of-choice was Sitka spruce because of its superior strength-to-weight ratio, its elasticity and ability to withstand sudden strain and shock, and the fact that it could be obtained in clear, straight-grained pieces of large size and uniform texture with few hidden defects. Sitka spruce, proclaimed in *American Forestry* in 1918 to be "an aristocrat among woods," was used in the wing construction of 77 out of 84 commercial airplanes displayed at the All-American Air Show in 1928.[36] Brigadier General Brice P. Disque, the head of the Army's Spruce Production Division during World War I, praised Sitka spruce:

> And there must be wood. Wood that should be both tough and light. Wood that should not splinter from the impact of a bullet. Wood that should be stout and strong but not brittle. Wood that should be straight-grained and free from flaw. A perfect wood in brief.
>
> Sitka spruce…is the wood which is found to have the necessary qualities of toughness and lightness which make it the airplane wood par excellence.[37]

In more modern times, laminated Sitka spruce was used to construct the nose cones of Poseidon and Trident I submarine-launched ballistic missiles. The nose cones are designed to ablate (vaporize) as the missile reenters the atmosphere to dissipate heat that would otherwise damage the payload. Sitka spruce was the chosen material because it is strong and light and burns evenly.[38]

Makers of fine pianos, guitars, and other stringed musical instruments have long recognized clear, vertical-grained Sitka spruce as the premier music wood where resonance is the primary consideration. It is the uniformity and organization of the Sitka spruce's wood fibers that make the wood so special. Sitka spruce is used in pianos built by Steinway & Sons as well as in high-end acoustic guitars.

The utilization of Sitka spruce for the manufacture of wood pulp began around 1907. The specie's suitability for this use was quickly recognized, and it soon became the pulpwood-of-choice on the Pacific Coast. Its suitability is attributed in part to the fact that Sitka spruce wood contains a very high amount of cellulose, but is relatively free of resins, gums, and tannins.[39] Sitka spruce was also used to make dissolving pulp from which rayon, cellophane, and some plastics are made.

Western redcedar (*Thuja plicata*) Slow-growing, large, and long-lived, the western redcedar is not a true cedar. In Southeast Alaska, the species is at the northern edge of its range, and rarely occurs in the northern half of the region. The western redcedar attains its best growth in Southeast Alaska at elevations below 500 feet and is most prevalent on gentle slopes. Though it is able to tolerate poorly-drained soils, it grows best on well-drained sites.

On a good site in Southeast Alaska, mature western redcedars reach a height of 100 feet and a diameter of four to six feet. An exceptional Southeast Alaska specimen was reported to be 158 feet tall and have a diameter of 9.5 feet. The trunks in older trees taper rapidly and are heavily root-buttressed, which, in combination with a well-developed root system, make the species relatively windfirm. Western redcedar trees 800 years old have been reported in Southeast Alaska.[40]

The color of the wood of western redcedar is a dull brown tinged with a purplish red, with the thin sapwood being nearly white. Western redcedar is light in weight, moderately soft, moderately weak when used as a beam or post, low in shock resistance, and gives off an odor that is agreeable and pleasant to humans but repels moths and other insects. The wood is easy to work, dries easily, and once dry, shrinks and swells only slightly. Western redcedar takes paint and stains well, provided it is dry.[41]

Because it is light in weight, easy to work, and decay resistant, western redcedar was preferred by Southeast Alaska's Native people for a number of important uses. The most famous traditional use of western redcedar in Southeast Alaska was for the totem poles that continue to grace the region.

As well, western redcedar was the preferred material for Native canoes throughout Southeast Alaska and British Columbia, and most particularly by the Haida. Where available, western redcedar logs or hewn timbers were also used by Southeast Alaska's Natives for house construction. Western redcedar was also used for ceremonial masks and household items.

In more recent years, western redcedar is considered to be the finest shingle wood in the United States, primarily because of its lightness and resistance to decay. The manufacture of shingles for the regional market was a small but fairly stable business in Southeast Alaska, and the region's shingle manufacturers were proud that the quality of their shingles was considered equal in all respects to that of those manufactured in the Pacific Northwest and British Columbia. Three shingle mills operated in Southeast Alaska in 1924, producing some 5.2 million shingles that sold locally for about $4 per thousand.[42]

FIGURE 5. Western redcedar near Ketchikan, 1909. (Alaska State Library, USFS Collection)

Western redcedar is also much used in other applications where exposure to weather is a factor: house siding, porches, porch columns and railing, gutters, lattice, fences, window boxes, greenhouses, and lawn furniture. For interior construction uses, western redcedar is used as ceiling and wall paneling, and because it repels moths, it is used in clothing closets.

Western redcedar was used as a component of wood pulp in British Columbia, but was not used in Southeast Alaska's pulp mills.[43] With little demand for the material, most western redcedar trees encountered in early pulp mill-era clearcut logging operations were felled but left behind.[44] To foster more complete utilization of this species, the Forest Service began to permit the export of unprocessed western redcedar logs, and the Ketchikan Pulp Company assembled Davis rafts of western redcedar for transport to the Puget Sound area.[45] The value of western redcedar in Alaska increased as the supply in the Pacific Northwest became depleted, and by about 1970 the value of good quality western redcedar logs in Southeast Alaska was exceeded only by that of Alaska-cedar.[46]

Alaska-cedar (*Chamaecyparis nootkatensis*) Commonly known as "yellow cedar" because its wood turns sulfur-yellow when wet, Alaska-cedar, like western redcedar, is not a true cedar. The species grows throughout Southeast Alaska, and in the southern area of the region it generally reaches its best development at elevations between 500 and 1,200 feet, while in the northern area its best development occurs near sea level. Mature Alaska-cedar trees can be 48 inches or more in diameter and 100 feet high, though Frank Heintzleman considered the best quality sawlogs to be those about 24 inches in diameter. Commercial stands of Alaska-cedar usually occur in patches of fewer than 20 acres, with timber volumes ranging from 5,000 to 25,000 board feet per acre. These stands are often located at the heads of valleys and ravines. A particularly fine stand of Alaska-cedar, noted in 1943, was located at Cape Fanshaw, near Petersburg. The stand was said to contain several million board feet, with trees up to seven feet in diameter (dbh).[47]

The Alaska-cedar is a survivor. It is often free of fungus attacks, and is able to tolerate marginal sites, such as along the edges of muskegs, where competition is minimal. Alaska-cedar usually grows very slowly, with 50 to 60 growth rings per inch common and 360 growth rings per inch having been recorded. Alaska-cedar is able to outlive its competition, with sound trees 700 to 1,200 years old being fairly common. Some veterans have been estimated to be 3,500 to 4,000 years old.[48]

Alaska-cedar is in decline in Southeast Alaska. The cause is thought to be natural, with extensive mortality of the species having been noted in undisturbed areas before 1900. The decline may be related to a climatic warming trend that has resulted in less snow, which leaves the trees' roots more exposed. This is critical in the springtime, when they are going through a de-hardening process during which their shallow roots are very vulnerable to injury by freezing episodes.[49]

Some of the mechanical characteristics attributed to Alaska-cedar are that it is moderately light, moderately hard, fairly strong, and brittle. Alaska-cedar is also repels moths and termites, and experiments have been done in which rodents chose starvation over gnawing through Alaska-cedar to freedom. In marine environments, Alaska-cedar is considered to be practically immune to the borings of the notorious teredo.

FIGURE 6. Grove of Alaska-cedar on Pleasant Island, in northern Southeast Alaska.
(James Mackovjak)

About the only criticism of Alaska-cedar was that there was not enough of it on which to base a substantial industry.

Alaska-cedar carves extremely well. Native Alaskans in Southeast Alaska carved Alaska-cedar into various ceremonial and functional items such as masks and bowls. Alaska-cedar was esteemed by Southeast Alaska's Native people for the manufacture of canoe paddles, of which the largest canoes required 30 or 40.[50]

The maritime value of Alaska-cedar was recognized by the Russian colonists at Sitka, who used it in the construction of the hulls of some twenty vessels built there. The *New York Times* reported in 1884 that Congress was once urged, unsuccessfully, to declare Prince of Wales Island a government reservation to preserve Alaska-cedar for its own use as ship timber and piling. In his 1870 book, *Alaska and Its Resources*, William H. Dall stated that Alaska-cedar was the most valuable wood on the Pacific Coast. John Muir thought Alaska-cedar to be Alaska's most important commercial timber tree.[51]

As a cabinet wood, Alaska-cedar compares favorably with any species in North America. Alaska-cedar came into temporary favor for interior work in the early 1900s and was used in the construction of cabinets, shelving, molding, and flooring.[52] As well, it was used for musical instruments and household electronics, such as radios and phonographs. Alaska-cedar is considered a "fixed dimension" wood, and once seasoned and finished, does not tend to shrink, expand, or warp. Because of this characteristic it was used for the manufacture of measuring and engineering instruments as well as printers and pattern makers' gauges. Early boat builders in Southeast Alaska often used Alaska-cedar for boat ribs or for similar applications.

There was little demand for common Alaska-cedar lumber in Southeast Alaska prior to the 1920s, but in 1922 a major effort arose for its anticipated use in conjunction with expected pulpwood operations.[53] Because of the small size of the timber stands and the fact that the best material is located a considerable distance from saltwater, the cost of logging Alaska-cedar has been relatively high.

As was the situation with western redcedar, Alaska-cedar trees encountered in early pulp mill-era clearcut logging operations were felled but often left behind.[54] To foster more complete utilization, the Forest Service permitted the export of unprocessed Alaska-cedar logs to Japan, and by about 1970 the species had become the most valuable in the region.

CHAPTER 2

The Russians Arrive in Southeast Alaska

THE RUSSIAN-AMERICAN COMPANY WAS CHARTERED BY TSAR PAVEL IN 1799. THE COMPANY WAS GIVEN A MONOPOLY ON HUNTING AND TRADING IN ALASKA, AS WELL AS THE RESPONSIBILITY FOR THE ADMINISTRATION OF THE REGION. NEW ARCHANGEL (SITKA) WAS FOUNDED BY ALEXANDR BARANOV IN 1804 AND BECAME THE COLONY'S ADMINISTRATIVE CENTER IN 1808. ACCORDING TO FERDINAND WRANGELL, ONE OF THE RUSSIAN-AMERICAN COMPANY'S LATER GOVERNORS, THE PRINCIPAL REASON FOR ESTABLISHING A SETTLEMENT AT NEW ARCHANGEL WAS THE GREAT NUMBER OF SEA OTTERS THAT INHABITED THE AREA. SEA OTTER PELTS WERE VALUABLE IN EUROPE AND ASIA, WHERE THEY WERE USED IN THE MANUFACTURE OF COATS AND HATS. ANOTHER IMPORTANT CONSIDERATION NOTED BY WRANGELL WAS THAT THE AREA'S FOREST WOULD YIELD THE "BEST TIMBER FOR BUILDING."[1] AS A PART OF ITS CHARTER, TSAR PAVEL GAVE THE RUSSIAN-AMERICAN COMPANY PERMISSION TO CUT TIMBER FROM ALASKA'S FORESTS.[2]

Firewood was consumed at Sitka in great quantities, and logs were used in various construction projects, including buildings and houses. Mikhail Tebenkov, an explorer and later colonial governor, wrote that some of the logs used for construction of a warehouse in 1826 were up to 154 feet long, and absolutely straight. Finnish workmen, expert with the broadaxe and adze, were employed by the Russian-American Company to construct many of the buildings at New Archangel. Logs were hewed square, and dovetailed and pinned at the corners.[3]

The principal uses of lumber at New Archangel were for the construction of buildings, houses, furniture, ships, boats, and barrels. By 1860, there were more than 100 wooden houses at New Archangel. Lumber—most of it intended for similar purposes—was also shipped to Russian settlements throughout Alaska, including St. Michael, on Norton Sound.[4]

The construction and repair of both sail and steam-powered vessels, which extensively used Alaska-cedar, was an important component of the

Russian-American Company's operations, particularly after the sea otter had been hunted to near extinction. Barrels were used mostly for packing salted fish for export. Boards, some of them thin slats, were provided to Aleut sea otter hunters on various islands to be used for making kayak frames. Efforts were also made to export logs and lumber. The manufacture of charcoal at New Archangel provided another considerable enterprise.

The lack of technology for moving or sawing large logs, however, limited the size of the timber the Russian-American Company was able to utilize. In 1867, a newspaper reporter from Alta, California, noted, "A stream of moderate size flows though the town [of New Archangel], on which is a sawmill; yet there is less use of boards here than any other civilized town I have visited."[5]

LOGGING

When the colonists arrived at New Archangel in the fall of 1804, one of their first projects was to build a stockade, for which about a thousand trees were felled. Harvesting continued as the community grew and its enterprises increased. Reports of the amount of timber cut are incomplete, but in 1820 some 350 Sitka spruce trees and 130 Alaska-cedar trees were cut for logs and lumber.[6] Even at this very early stage of industrial development, timber was rapidly becoming scarce at New Archangel, and loggers were forced to go farther and farther afield. "Although the rugged shores are covered with trees as thickly as wild animals are covered with fur," remarked Kyrill T. Khlebnikov, chief of the countinghouse at the Russian-American Company's main office in New Archangel from 1818 until 1832, "the closest ones have been the ones first cut, and over a period of time this has been a growing problem."[7]

Efforts were increasingly directed inland. As early as the spring of 1826, logging operations were being conducted as far as a verst (.66 mile) from saltwater, and in 1846 the Russian-American Company cut 1,725 logs in Krestof Sound, about 18 miles by water from New Archangel. In 1849, some 2,600 logs were delivered to New Archangel. An 1861 historical survey of the Russian-American Company estimated that between approximately 1850 and 1860, about 9,500 logs were cut around the port. By the time the United States purchased Alaska in 1867, most logging took place 20 to 33 miles from New Archangel. To obtain Alaska-cedar, so valuable in vessel construction, the Russian-American Company's loggers were forced to travel to locations in Peril Strait, some 60 miles to the north.[8]

Trees were felled using only iron axes, and the transporting of logs to saltwater was done by brute manpower. The process was described by Kyrill Khlebnikov:

> In the fall 20 men are sent out for one or two months to cut trees. In one day two men can cut as many as five trees, making logs from six to 12 sazhens [42 to 84 feet] long and from six to ten vershoks [10.5 to 17.5 inches] thick. They prepare from 300 to 400 such logs; when all have been prepared, 40 or 50 men are sent out to bring them to the shore. The shoreline is quite irregular, hilly and covered with fallen trees, which makes it extremely difficult to snake the lumber out. They cannot drag more than ten such logs to the shore on some days, but usually they manage from ten to fifteen. Bad weather makes this more difficult. By May all the timber is brought to the harbor. Generally they use rowboats to tow, each one towing from 10 to 16 logs; but sometimes they tie 30 or 40 logs into a raft, and pull with two or three boats.[9]

The company's steam-powered *Baranov*, with 30 horsepower, was said to be scarcely able to tow more than 17 logs, even in the vicinity of New Archangel, where tidal currents are minimal.[10]

Beginning in the fall of 1854, in part to deter Tlingit attacks, labor for the Russian-American Company's logging and lumber operations was drawn from the lower ranks of the Russian army's Fourteenth Siberian Line Battalion. Between 1854 and 1856, some 200 soldiers from this battalion arrived at New Archangel to begin a seven-year term of service. They remained until 1866. The Russian-American Company paid the men—who became temporary employees—about twice what they would have received in the army, but they were paid only once each year.[11]

An 1862 report noted that logging was done mainly by soldiers, who were deployed in groups of 15 to 20 to work for a month or more at a time, all the while living in earthen dugouts. They received food from New Archangel and were rationed one cup of vodka per day. Despite the hardships, there were many volunteers for this duty, and some soldiers worked in the forest for months at a time, with only infrequent visits to New Archangel.[12] After the transfer of Alaska to the United States, logs 40 feet long and roughly hewn into 36- or 42-inch squares were found more than a mile from the beach. It was later presumed that very large pieces of material such as these would usually be hauled out over the winter snow on sleds.[13]

Prior to 1833, when the first sawmill began operating at New Archangel, logs were cut into lumber using the standard hand-powered technology of the day: a pit saw, with which logs were whipsawn into boards by two-man teams. A log was rolled atop a scaffold-like wooden structure, and one man worked his end of the saw from above while straddling the log. His partner (usually wearing a big hat to protect him from a constant rain of sawdust) worked his end from below, often in a pit. It has been said two angels could not saw their first log in this manner without getting into an argument. In 1826, 14 to 18 men were employed full-time to supply New Archangel with lumber.[14] A good team could cut 150 board feet of lumber per day, making total daily production on the order of 1,000 to 1,400 board feet.[15]

THE FIRST SAWMILLS

In 1831, Ferdinand Wrangell, then chief manager of the Russian-American Company, wrote to his superiors in St. Petersburg about the desirability of establishing a water-powered sawmill at New Archangel to supply lumber for the settlement as well as for export to California and Chile. Wrangell requested that a sawyer, a model of a sawmill, and whatever cast iron components needed for the construction of a mill be sent to New Archangel.[16]

The site selected for the sawmill was at Ozerskoi redoubt, a supplementary fortification about 12 water miles south of New Archangel, near the outlet of what is now known as Redoubt Lake. Construction began in the fall of 1832, and on August 29, 1833, the mill produced the first lumber in Alaska to be sawn with anything other than manpower.[17] The type of saw mostly in use at that time was the muley saw (sometimes called a jig saw), in which a single, heavy vertical blade was made to move up and down by a wooden beam attached to a crankshaft connected to a water wheel. Such saws cut only on the down stroke. Another type of saw coming into use at that time was the gang sash saw. Considered an improvement over the muley saw, it operated in a similar manner but cut with multiple blades that traveled in guides inside a frame (sash) to simultaneously make multiple cuts. When new and operating a single blade, the saw at Ozerskoi redoubt cut through wood approximately a foot thick at the rate of about 25 feet per hour—5 inches per minute.[18]

During the summers of 1841 and 1842, the sawmill was ordered to run day and night to meet the expected demand for lumber.[19] Over time, however, the quantity and quality of lumber produced at this sawmill left much to be desired. In May 1843, Adolph Etolin, then the Russian-American Company's chief manager at New Archangel, described the problem:

> The water-powered sawmill at Ozerskoi redoubt had begun to operate ever worse over time, with the boards coming out crooked and in insignificant quantity, so…finally, in order to satisfy demands for boards at [New Archangel's] port alone, one constantly had to have up to six pairs of sawyers for hand sawing.[20]

Much of the timber to supply the mill was cut around Redoubt Lake and floated downstream. Transporting materials by water to and from this somewhat remote location was difficult—and sometimes dangerous. Once the nearby timber had been cut, the Russian-American Company decided to build a new and better sawmill closer to New Archangel. The machinery at the Ozerskoi redoubt sawmill was dismantled in approximately 1845.[21]

In 1842, a water-powered sawmill was constructed near New Archangel's port, at the mouth of Malyshevka Creek, which flows from what is now known as Swan Lake. Water was routed by flume from the lake to a 10-horsepower overshot wheel at the sawmill. The mill was initially equipped with a gang sash saw, which could operate with as many as 20 blades. Some of the mill's machinery was American made. The Malyshevka Creek sawmill began cutting lumber on January 1, 1843, and with a single saw, it cut at the rate of 50 feet per hour, providing the timber being cut was no more than 13 inches thick. This was about twice as fast as the Ozerskoi redoubt sawmill.[22] Oscar Levy, a Canadian businessman, visited the mill in 1867 and observed its operation:

> Four men were attending the mill; at least I should judge that was their business. When I entered they were all asleep, waiting for the saw, an upright, to get through a log 13 feet long. By a close calculation I found the saw was wearing itself through the log about a sixteenth part of an inch on each stroke and was making two strokes a minute. At that rate it would take the saw 15 hours and 36 minutes to make one cut. The engineer in charge told me they could take a heavier cut, but to do so would stop the mill.[23]

Some 10,000 boards of various dimensions were cut at New Archangel in 1843, and a similar amount in 1844. It is likely most of them were cut at the Malyshevka Creek sawmill. A shed for storing boards was built near the mill in the summer of 1845.[24]

Partly in anticipation of exporting lumber to the Hawaiian Islands and to California to a lesser extent, the Russian-American Company began construction of a third sawmill in 1845. It was built over the Medvetcha River (now known as Sawmill Creek), at Silver Bay, about five miles east of New Archangel. Water for the sawmill was impounded behind a log dam that was 70 feet long, 15 feet high, and 42 feet wide, and a 30-horsepower waterwheel-powered a gang sash saw with 20 blades.[25]

The Medvetcha River mill began sawing lumber on August 7, 1847, but because of "imperfections in its arrangement," it ceased operating in the spring of 1852—the same year that the sawmill was almost completely refurbished. Details are scarce, but there are reports that the Medvetcha River sawmill and part of the dam were destroyed in 1852 by Tlingits, who also may have killed some of the workmen. It was apparently difficult to keep workmen at the mill because of frequent attacks by Tlingits. Moreover, the Medvetcha River location was problematic for another reason: in low-water conditions, logs could not be floated to the mill. By 1862, this sawmill was rarely used.[26]

During the second half of the 1840s, the sawmills at New Archangel annually produced some 6,000 boards of various dimensions. In 1846, when the Malyshevka Creek sawmill was the only sawmill operating at New Archangel, as many as 3,900 boards were cut.[27]

Major renovations that began in 1854 tripled the size of the Malyshevka Creek sawmill. To augment the gang sash saw, a circular saw was installed, as well as a 5-horsepower steam engine that could be used jointly with waterpower to operate the circular saw. In 1862, the mill was reported to be equipped with circular saws ranging from 14 inches to four feet in diameter. The Russian-American Company had at great expense been importing wooden shingles from California, and with some improvisation, one of the circular saws was used to cut up to 2,000 shingles per day. A planer and equipment for manufacturing cornices and door and window frames were installed in 1859, as was a shingle machine, which produced 155,000 shingles in 1861. The sawmill's planer was put to good use: two-thirds of some 9,000 boards of various sizes sawn at New Archangel in 1862 were planed.[28] Oscar Levy, the Canadian businessman who visited New Archangel in 1867, described the quality of the lumber: "My first impression was that it had been smoothed by rats' teeth."[29] Sawdust from the Malyshevka Creek mill was used to pack ice cut from nearby Swan Lake for shipment to San Francisco during the early 1850s, and the mill continued in operation after Alaska was transferred to the United States in 1867.[30]

In 1852, the Russian-American Company began the construction of a floating sawmill that was used to cut lumber at locations some distance from New Archangel. The barge upon which it was constructed was 85 feet long and had a beam of 32 feet. After many delays, the floating mill was launched in June 1855. A 30-horsepower steam engine powered what was probably a gang sash saw. It was said to be capable of sawing one hundred 21-foot-long logs in 24 hours. From 40 to 50 Russian-American Company employees plus a number of Tlingits (employed on a more casual basis) were expected to be engaged in providing logs and operating this sawmill.[31]

While the floating sawmill was reported to be in good repair in 1862, at the time of the transfer of Alaska to the United States, a "Floating Steam Sawing Shop," considered public property, was reported to be aground at Sitka. The sawmill that the U.S. Army operated during the 1869–1870 period may have been the floating sawmill.[32]

EARLY WOOD PRODUCTS

Charcoal, which generates a higher heat than wood, was needed by local blacksmiths, metalworkers, and coppersmiths. About 1830, a Russian at New Archangel remarked on the cycle of toil: "Work here involves making axes to cut trees to make charcoal, which is used to make axes."[33] Charcoal was also used in heating and cooking stoves, in samovars (urns used to heat water for tea), and for smoking rats out of ships. Wood was cut and laid in a pile about 30 feet in diameter, covered with dirt, and set afire. The pile was then allowed to burn for as many as 15 days.[34] Sites of Russian colonial charcoal manufacturing operations are evidenced today at several locations near Sitka by the remains of charcoal pits surrounded by even-aged stands of second-growth timber, with the 189-acre site at Mount Verstovia possibly being the oldest industrial-scale clearcut in western North America.

Shipbuilding at New Archangel began in 1805, and the 8-horsepower sidewheel tugboat *Mur* (37 feet long, with a beam of 9 feet), built about 1839, was likely the first steam vessel—boiler and engine included—to be entirely constructed on North America's western coast. At least two additional steam-powered vessels, the 60-horsepower *Imperator Nikolai I* and the 30-horse-power *Baranov*, were constructed at New Archangel, in addition to the sailing schooners *Kvikhpak* and *Chil'kat* and the brig *Promysel*. An American ship, renamed the *Lady Wrangell*, was re-timbered at New Archangel.[35]

Oar-powered vessels were also constructed. Framed with Alaska-cedar, these were usually sold outside the settlement or used in trade for other goods. Some, however, were used by the Russians, as was the case with the "16-oar military cutter" completed in 1852.[36]

Ferdinand Wrangell reported that "we use a variety of cypress [Alaska-cedar] called *dushnoe dervo* [fragrant wood] for the ship's ribs.... We use larch [western hemlock] for the covering."[37] Another report stated that all the steamers constructed at New Archangel were "said to have been built" of Alaska-cedar. One vessel was reported to have had a new deck of Alaska-cedar.[38]

Shipbuilding at New Archangel proved to be a marginal enterprise. According to a 19th-century historian,

> It was found that vessels could be purchased from foreign-ers, and especially from Americans, to better advantage than they could be built in the colonies, and it is probable that the managers would have saved money if no attempt at shipbuilding had been made in Russian America, except perhaps for inter-colonial traffic.[39]

Used American ships constructed of oak and pine were considered to last five times as long as ships built at New Archangel.[40]

Probably the very first export of timber from Southeast Alaska was in 1829, when the trading schooner *Alabama* visited New Archangel and accepted 410 Sitka spruce timbers in partial exchange for salt and other material.[41] No sawmill had yet been built at New Archangel; the timbers had been sawed by hand or hewed.

For the Russian-American Company, the lumber trade was little more than an experiment. In search of trade opportunities and often food, the company's ships traveled the Pacific coast as far south as Chile, to the Hawaiian Islands, and to the Orient. Merchant ships departing New Archangel would speculatively carry a few Alaska-cedar logs on deck. On ships carrying lightweight cargoes of fur, logs may have been transported in the lower hold as a sort of tradable ballast. Alaska-cedar was said to have been prized in China, where it was used to make trunks and chests.[42] Ships trading from

Sitka commonly left for that nation and others with a few Alaska-cedar logs in their cargo.

A cargo of lumber was reportedly transported to Chile and traded for foodstuffs in 1839, and there were said to be in Macao and Canton (China) structures constructed of Sitka spruce lumber cut at New Archangel. On an 1853–1854 trading voyage to California and Hawaii, the company ship *Kodiak* carried some 40,000 square feet of lumber. The portion sold in Honolulu reportedly fetched $60 per 1,000 square feet (not necessarily board feet).[43] Ship masts were also sent abroad for sale, again as an experiment. Pavel Golovin, a Russian Navy captain who was sent to assess Russia's American colonies in 1861, characterized the amount of lumber sent out for sale from Sitka as being "rather insignificant," but apparently not because of any deficiency in the material: "On Sitka," he wrote, "it is not unusual to see tree trunks 90 feet high and two and a half feet in diameter at the crown."[44]

In 1854, the San Francisco-based American-Russian Commercial Company, predecessor to the Alaska Commercial Company, entered into a 20-year contract (later renegotiated) with the Russian-American Company for a supply of ice, lumber (particularly Alaska-cedar), coal, and fish. Only the ice was profitable.[45]

THE TRANSFER

When the United States took possession of Alaska, Secretary of State William H. Seward specified that private dwellings and facilities such as shops and sawmills would remain under the control of their owners.[46] The Russian-American Company's sawmill at Sitka's port at some point became the property of the American-Russian Commercial Company, which intended to use it to cut Alaska-cedar for export to San Francisco.[A] William Phillipson, the sawmill's manager, received the following instructions:

> You will run the mill so long as it shows a cash profit. Keep the stock in the yard at 30,000 feet. The present output of the mill should be 3,000 feet a day. The wages of Startsoff, the sawyer, are $1.75 a day; Corbitshoff, the machinist, $1.25 a day, and other hands $1 a day. You will pay no more than $3 per thousand for logs.

[A] The term "export" is often used in Alaska to refer to logs or lumber that is shipped from Alaska to either foreign nations or other states.

The usual rates for lumber are $20 a thousand all around, but should the government or the Greek-Russian church require any, charge them $30; they can stand it. You need not speak of the difference to anyone. Cedar lumber you will sell invariably for $35 a thousand, and for shingles charge $4 a thousand.[47]

The market for Alaska-cedar in San Francisco, however, proved to be less than hoped for, and the company changed its focus to supplying the local market.[48]

CHAPTER 3

Alaska's Timber Comes under U.S. Control

THE U.S. TIMBER INDUSTRY HAD ITS GENESIS IN THE NORTHEAST. BY 1850, THE REGION'S PRODUCTION BEGAN TO DECLINE STEEPLY AS THE EASILY LOGGED TIMBER BECAME EXHAUSTED. THE INDUSTRY THEN MIGRATED TO THE PINE FORESTS OF THE GREAT LAKES REGION. AROUND 1890, PRODUCTION THERE TAILED OFF AS THAT REGION'S ACCESSIBLE TIMBER WAS IN TURN DEPLETED, BUT THIS WAS MORE THAN OFFSET BY INCREASED LOGGING IN THE WEST AND THE SOUTH. FROM ABOUT 1850 TO 1890, THE WEST HAD ACCOUNTED FOR ONLY ABOUT FIVE PERCENT OF THE TOTAL U.S. LUMBER PRODUCTION, BUT BY 1920 FULLY ONE-THIRD OF THE U.S. TIMBER HARVEST CAME FROM THE WEST.[1]

For Secretary of State William H. Seward, Alaska's timber resources were of almost no consideration when the United States purchased Alaska from Russia. His desire to purchase Alaska was not so much for the territory's considerable resources, but as an early step in establishing a Pacific empire for the United States. It was commonly thought that if the United States acquired Alaska, British Columbia, which would then be sandwiched between Alaska and the Washington Territory, would eventually be acquired as well. Congress, in the meantime, had to figure out how pay for Alaska and its administration.

In the U.S. Congress, the main advocate for the purchase of Alaska was Massachusetts Senator Charles Sumner, a Republican. Sumner advocated for the purchase in large part because he believed it would contribute economically to the nation. Armed with a Smithsonian Institution report on Alaska, Sumner made a three-hour speech to Congress in which he spoke of "forests of pine and fir waiting for the axe." In an address marking the ceding of the territory by Russia, he predicted, "Russian America may be on the Pacific like Maine on the Atlantic, and the lumbermen of Sitka may vie with their hardy brethren of the East."[2] Sumner Strait, which borders some of the most productive timberlands in Southeast Alaska, was subsequently named in his honor.

The United States formally took possession of Alaska in October 1867. The Civil War had ended but two years earlier, and the nation was preoccupied with Reconstruction. What to do with Alaska ranked low in its priorities. The unorganized territory had no civil or criminal code and no system of courts. This period, during which no one there was legally entitled to settle, engage in commerce, marry, or even be pronounced dead, is sometimes referred to as Alaska's "era of no government," or the "era of total neglect."[3] The U.S. Army was initially given jurisdiction, a position it held until mid-1877, when it was withdrawn and the U.S. Collector of Customs put in charge by default. Jurisdiction of Alaska was handled by the U.S. Navy from 1879 until 1884, when Congress created the District of Alaska and provided for civil government with a governor appointed by the President. In 1906, Alaska was accorded a non-voting delegate to Congress, and six years later Congress formally created the Territory of Alaska. With territorial status, Alaska gained limited self-government by an elected legislature.

During the summer of 1869, Secretary of State Seward journeyed to Alaska. At Sitka, formerly New Archangel, he gave a speech in which he spoke of the region's timber: "No beam, or pillar, or spar, or mast or plank is ever required in either the land or the naval architecture of any civilized state greater in length and width than the trees which can be hewn down on the coasts of the islands and rivers here."[4] Seward likely based this statement on information collected by George Davidson of the U.S. Coast Survey, who had conducted a four-month geographical reconnaissance of Alaska's coast in 1867. Davidson's party had measured a felled Sitka spruce that was 180 feet long and four feet in diameter at the base, and it encountered standing trees that were more than six feet in diameter and branchless for over 50 feet.[5]

The distance of Southeast Alaska from major timber markets, however, was a problem. An 1870 report by Army Major E. H. Ludington stated, "The timber in Alaska is of a superior quality, and would be a source of revenue if near market, but the expense of transporting it is so great as to preclude the idea of its being commercially valuable."[6]

The following year, a U.S. Army report on Alaska was submitted to Congress; in it, Major John C. Tidball addressed the timber resources of Southeast Alaska:

> There are but two species of trees, viz, the Sitka spruce and yellow cedar, which have only value in an economical or commercial point of view. The first of these, the spruce,

constitutes the principal growth; it grows tall and straight, and of excellent size for lumbering purposes…[but] in comparison with the timber of Washington Territory and Oregon, it is vastly inferior. The other, the cedar, possesses some rare and excellent qualities… It possesses the quality of lightness, toughness, ease of workmanship, and great durability. It is susceptible of high polish, and is beautiful as a cabinet wood. It possesses a strong aromatic odor, which is said to be a preventive against moths and other insects destructive to furs or woolens; hence chests made from it are in great favor.…

The excessive ruggedness of the country…interposes almost insuperable barriers to the procuring of timber of good quality in large quantities… It may therefore be safely inferred that as long as the forests of British Columbia, Washington Territory, and Oregon hold equal to the demand, the lumber trade will not come so much further to obtain an inferior article, at much greater expense. This applies to the spruce; as to the cedar, the qualities before mentioned may make it before long a valuable article of trade.[7]

An 1876 report made much the same point: "When the forests of Oregon and Washington are gone, Alaska will be our permanent supply," wrote Captain J. W. White of the U.S. Revenue-Marine Service.[8]

A Treasury Department report that same year, however, painted a gloomier picture. Special Agent Henry W. Elliot portrayed Southeast Alaska as being covered by "a dense jungle of spruce and fir [western hemlock], cedar and shrubbery, so thick, dark, and damp that it is traversed only by the expenditure of great physical energy." The spruce and hemlock of this "forest-jungle," he wrote, were "so heavily charged with resin" that they could be used for "nothing but the roughest work."[9]

In the fall of 1880, the *New York Times* interviewed Lester H. Beardslee, who was stationed at Sitka as commander of the U.S. Navy warship *James-town*. Beardslee offered a kinder assessment of the quality of Southeast Alaska's timber than did Tidball and Elliot, but like Tidball and White, he tied its future utilization to the exhaustion of supplies in Washington and Oregon. Beardslee noted:

The mountains which skirt the coast of Sitka are covered with a dense growth of evergreen timber, such as larch, spruce, hemlock, pine, and yellow cedars. Within five miles of Sitka are acres of spruce and hemlock, from which logs 4 feet in thickness can be cut, and which will furnish from 4,000 to 5,000 feet of lumber to each tree. This timber I consider one of the greatest resources of Alaska in the future. The time must eventually come when the mills on Puget Sound, which are turning out lumber at the rate of 1,000,000 feet a day will exhaust the supply from the forests of Oregon and Washington Territory, and then this great Alaskan resource is coming into play.[10]

In 1880, Ivan Petroff, Special Agent of the Census, reported on the population, industry, and resources of Alaska. Regarding timber, Petroff gave an evaluation similar to Tidball's, stating that spruce trees "have not grown in proportion to their age as fast as they would have done were they growing in a more congenial latitude to the south, such as Puget Sound or Oregon," and spruce boards were "ineligible for nice finishing work." Petroff's praise for Alaska-cedar was tempered: "The best timber of Alaska is the yellow cedar, which in itself is of great intrinsic value; but this cedar is not the dominant timber by any means; it is the exception to the rule." Even the territorial governor, A. P. Swineford, made no claims regarding the value of Alaska's timber; he called it "neither so good as it has been represented by enthusiasts, on the one hand, nor yet nearly so worthless as has been claimed by Alaska's interested detractors."[11]

One early outdoor writer recognized the potential: "The timber forests of Alaska are a standing testimony to the value of the 'Seward Purchase,'" wrote Charles Hallock in 1886, "which even the most obstreperous objectors could not deny." But deny it they did. The timber resources of Southeast Alaska were considered—if at all—only as a reserve supply that could help bridge the gap in production between the exhaustion of virgin timber supplies and the maturation of second-growth stands in the contiguous 48 states. A 1907 Forest Service publication titled *The Timber Supply of the United States* made no mention of the Alaska resource.[12]

TIMBER TRESPASS

Until 1884, the U.S. Army, the Collector of Customs, and the U.S. Navy each, in turn, had jurisdiction over Alaska. None of these entities showed an interest in regulating the use of Alaska's timber. They were focused on maintaining a semblance of order, and the cutting of public timber was overlooked.

Early Congressional legislation provided some noncommercial access to government timber. A loosely-worded act passed in 1878 permitted "citizens of the United States and other persons" to fell and remove timber from government "mineral lands" for "building, agricultural, mining, or other domestic [household] purposes."[13] Southeast Alaska in its entirety was essentially "mineral lands," defined as places where claims had been filed under the 1872 Mining Act. Mineral claims in the region predate the act: what was likely the first lode claim in Alaska, the Copper Queen, was staked near Kasaan on Prince of Wales Island in 1867, the same year Alaska was purchased from Russia.[14]

There was no charge for timber cut under the 1878 legislation, but also no specific provision by which sawmill interests could legally obtain government timber. Though timber was cut to supply the sawmills at Sitka, Klawock, and Shakan, doing so was technically illegal.

One early effort to obtain a major tract of land in Southeast Alaska was documented in 1879 by Department of the Treasury agent William Gouverneur Morris. His *Report upon the Customs District, Public Service, and Resources of the Territory of Alaska* (which offered little new information on Southeast Alaska's forest resources) described an unsuccessful effort in 1877 by the Alaska Ship-building and Lumber Company of San Francisco to purchase over a period of ten years from the government up to 100,000 acres on Kuiu Island and adjacent islands at a price of $1.25 per acre. In exchange for the right to do so, the company would construct on Kuiu Island at least one 1,200-ton ship and thereafter maintain a shipyard and "vigorously prosecute shipbuilding." Morris considered the effort, which was unsuccessful, an "astounding proposition, a preposterous guise" by "a party of San Francisco capitalists, who are endeavoring to gobble up [Kuiu Island]."[15]

The passage of the Alaska Organic Act on May 17, 1884, provided for civil government in Alaska but contained no timber-related provisions.[16] Essentially repeating the wording of the 1878 legislation, a federal administrator at that time counseled Alaskans that government timber could be cut "for their own use in the vicinity where obtained, for building, agricultural, mining and other domestic purposes."[17] No provision was made for the purchase of government timber by mills engaged in the commercial sale of lumber. Trespass remained the standard operating procedure and trespassing continued to be overlooked by the government until early in 1886, when federal officials in Alaska received the following instructions:

> You are hereby authorized and directed to take all proper measures to promptly discover depredations upon public timber and to immediately make complaint before a commissioner and cause the summary arrest of all persons engaged in such depredations. You will seize or cause the seizure of all timber cut or removed from lands in the territory of Alaska, whether found upon the land where cut, or in transit, or on ship-board, or elsewhere. You will not, however, interfere with bona fide residents who are cutting or removing timber from mineral lands for their own use for building, agriculture, mining and other domestic purposes. All other cutting or removal of any kind of timber from any lands in Alaska is hereby prohibited, and you will see that the restraining power of the courts and the penalties of the laws are invoked and enforced in all such cases.
>
> You will particularly guard against the export of any timber or lumber of any character from any portion of the territory whether cut or obtained under pretext of purchase from Indians or otherwise.[18]

Several months later, the illegal export of lumber was brought to the attention of a federal marshal, who learned at Juneau that two schooners had sailed for San Francisco with aggregate cargoes of about 500,000 feet of Alaska-cedar lumber cut on Prince of Wales Island. The value of Alaska-cedar lumber in San Francisco at that time was $80 per thousand board feet (MBF), presumably for a high-grade material. The vessel *Gem* had departed from Klawock, and the *San Buenaventura* from Shakan. The marshal took a passenger steamship south and from Portland telegraphed authorities in Washington, D.C., who in turn notified agents in San Francisco. The ships and their cargo were seized upon arrival. William Thonegal, captain of the *Gem*, was later tried in court and found guilty of cutting 500,000 feet of lumber on government land in Alaska. He was fined $3,145.[19]

The effect of this restrictive government policy on timber and land was noted by Governor A. P. Swineford in his 1887 report to the Secretary of the Interior:

> The dearth of mills can readily be accounted for. The general land laws not being in force in the Territory, it is impossible to secure title to timber lands, and the manufacturers of lumber, not being permitted to export any part of their cut, do not care to comply with the stringent regulations promulgated by the Department in relation to the cutting of timber on public lands, by which they are compelled to rely solely upon a market altogether too limited to justify the undertaking of even a single enterprise of the kind on a large scale.[20]

The restrictions were ridiculed. "Much embarrassment is occasioned by the present [Alaska] timber law, which forbids any white person to use timber from the public lands even for domestic or local purposes," wrote John W. Noble, Secretary of the Interior, in 1889. Governor Lyman Knapp, Swineford's successor, likewise observed in 1890 that "it seems absurd that with timber one of the most prominent features of this country, it should be necessary to import lumber to build the houses needed to protect the people from the inclemency of the weather."[21]

The first U.S. timber agent in Alaska was Charles Gee. A prominent Republican from Virginia, he resigned his office in the state's senate in 1889 to be appointed special timber agent for the District of Alaska in the Department of the Interior's General Land Office. Gee was sent to Sitka, where his job was to work with the U.S. district attorney to suppress timber depredations and to handle the examination, appraisal, and disposal of government timber. Sitka's newspaper, *The Alaskan*, termed it "a rather unpleasant duty to perform among us." During a five-month inspection trip in 1890, Gee visited every sawmill in Southeast Alaska, as well as several in British Columbia that were reportedly milling timber illegally cut in Alaska, and then reported 13 cases of timber trespass:

- Sitka Milling Company [John Brady], Sitka, 213,000 feet;
- Presbyterian Mission, Sitka, 35 logs;
- Theodore Haltern, Sitka, 1,700 feet;
- Lake Mountain Mining Company, Sitka, 150,000 feet;

- Alaska M & M Company, Douglas, 5,014,000 feet;
- Eastern M & M Company, Douglas, 300,000 feet;
- Alaska Fish Oil & Guano Company, Killisnoo, 2,200 cords;
- Silver Bow Basin Mining Company, Juneau, 80,000 feet;
- Willson & Sylvester, Wrangell, 1,000,000 feet plus 3,000 logs;
- Alaska Trading & Lumber Company, Shakan, 800,000 feet;
- North Pacific Trading & Packing Company, Klawock, 218,000 feet;
- Edward Cobb, Shakan, 25,000 feet; and
- William Duncan, Metlakatla, 3,000,000 feet.[22]

All of the cases were eventually settled. The most common form of settlement was a claim of "innocent trespass," in which timbermen who had knowingly cut timber on government lands, if caught, pleaded not guilty and paid whatever fines were due. Innocent trespass was tacitly accepted, particularly given the lack of government administrative personnel.

Of those charged by Gee, at least one individual, John Brady, chose to argue his case in court rather than pay a fine. The trial, in which the government sought damages of $500, based on a value of $0.25 for each tree Brady had cut, was held in Sitka in September 1891. After hearing the evidence, the jury deliberated for more than an hour and returned a verdict in favor of the government. Total damages, however, were set at only $1.[A/23] That same year lumber prices at the Haltern Sawmill in Sitka, per 1,000 board feet, were: "Rough lumber $16; clear, without knots $20; yellow cedar, without knots $50; Third class boards $10."[24]

Beginning to address the trespass issue, in 1889 Secretary Noble recommended to Congress that "arrangements should be made for the purchase of wood upon the public lands" in Alaska. His recommendation was likely in response to Alaska's Governor Knapp, who, in his official 1889 report to Noble, suggested that provision be made to allow timbermen to purchase timber lands from the government. According to Knapp, such an arrangement would afford a timberman "at least a feeling of self-respect and a sense of safety which he can not have when he knows he is subject to penalties of a broken law and

---

[A] John G. Brady first arrived in Sitka in 1878. During his college years, he worked for two summers at sawmills in Pennsylvania. He graduated from Yale in 1874, and was later ordained into the Presbyterian ministry. His work as a missionary in Alaska gradually evolved into a lumbering venture. Brady knew lumbering from the logging of timber to the manufacture and marketing of lumber, and his timber trespass had likely turned him a tidy profit.

is dependent upon the possible caprices of official favor." Congress did not act, and, in his 1891 report to the Secretary of the Interior, Knapp described the continuing problem faced by timbermen seeking access to government timber: "The lumber business has been harassed by the unfortunate conditions of land titles and most of the lumber used has been imported from the States. Those who endeavored to supply the demand for it from the Territory are now involved in suits for timber depredations."[25]

Though timber issues were a component of the Forest Reserve Act of March 3, 1891, the legislation failed to offer a procedure for legally obtaining government timber to be sold in commerce. It simply reiterated what officials had been telling Alaskans:

> In the District of Alaska…in any criminal prosecution or civil action by the United States for a trespass on such public timber lands or to recover timber or lumber cut thereon, it shall be a defense if the defendant shall show that the said timber was cut or removed from the timber lands for use in such State or Territory by a resident thereof for agricultural, mining, manufacturing, or domestic purposes and has not been transported out of same.[26]

REGULATED CUTTING

The subsequent May 5, 1891, Rules and Regulations Governing the Use of Timber on Public Domain, however, provided for the legal, regulated cutting of commercial timber for which there was a public necessity, without charge for the timber. The rules and regulations, as they pertained to Alaska, stated,

> Section 6: Persons, firms, or corporations…who desire to procure permission to cut or remove timber from public lands for purposes of sale or traffic, or to manufacture same into lumber or other timber product as an article of merchandise…must first submit an application therefore, in writing, to the Secretary of the Interior…describing the lands by natural boundaries and the estimated number of acres therein. They must also define the character of the land and the kinds of trees or timber growing thereon, giving an estimate as to the quantity of each kind, stating which particular kind or kinds they desire authority to cut

or remove, and the specific purpose or purposes for which the timber or the product thereof is required. The application must be sworn to, and witnessed by not less than four reliable and responsible citizens of the state, district or territory in which the land is situated, and who reside in the locality of the particular land described.

Section 7: The petitioner, or petitioners, should also submit with the application such evidence as can be procured, to conclusively show that the preservation of the trees or timber on the land described is not required for the public good; but that its use as lumber or other product, and for the purposes named in the application, is a public necessity. Upon receipt of the application, with accompanying papers, it will be duly considered, and if deemed for the public interest, the desired permission will be granted, subject to such restrictions and limitations as may be deemed necessary; but if it shall appear that the cutting of timber in the locality described in the application will be detrimental to the public interests, or infringe upon the rights and privileges of the settlers in that locality, the application will be rejected....

Sawmill owners, lumber dealers and others, who in any manner "cause or procure" timber to be cut or removed from any public lands in violation of law or these rules and regulations; whether directly by men in their employ, or individuals who actually cut or remove such timber, and are alike liable to criminal prosecution.[27]

Pursuant to the May 1891 regulations, the first government license to cut timber in Alaska was issued to the Willson & Sylvester's Fort Wrangell Mills at Wrangell in the spring of 1892.[28]

Some continued to favor the Department of the Interior's traditional and more direct method of providing access to public timber, and advocated a liberal policy for the granting of public timberlands to private interests. In his 1897 annual report, Secretary of the Interior Cornelius Bliss optimistically stated, "The early disposal of these [Alaska] timber tracts is a matter of great concern [i.e., interest] to the people, and if authorized by law they would at

FIGURE 7. The Willson & Sylvester Mill Company sawmill at Wrangell, circa 1905. This mill was constructed in 1888 mostly to manufacture wooden cases for canned salmon. (Wrangell Museum)

once enter into the lumbering business and in the near future could build up a very profitable trade with Japan and China."[29]

The policy and process for obtaining government timber in Alaska continued to be refined, with provision for the sale of government timber first made by Department of the Interior in 1898. The revised rules and regulations governing the sale of timber on public lands in the District of Alaska went into effect on April 1 of that year. Section 4 read as follows:

In order, however, that the native timber of Alaska may be placed upon the home market for all legitimate purposes of trade, to such a reasonable extent as shall meet existing emergencies in the matter of the demand therefore, sales of timber on public land in Alaska may be directed by the department from time to time.

While such sales of timber are optional, the Secretary of the Interior may exercise his discretion at all times as to the necessity or advisability of any sale, petitions from responsible persons for the sale of timber in particular localities will be received by this department for consideration.

For a prospective buyer, the actual process for obtaining government timber was somewhat daunting. A description of the location of the timber and the character of the land upon which it stood was required, as was a statement attesting "whether or not the removal of the timber would injuriously affect the public interests." And it was necessary to cruise the timber:

> Of live timber, state the different kinds and estimate the quantity of each kind of trees per acre. Estimate the average diameter of each kind of timber, and estimate the number of trees of each kind per acre above the average diameter. State the number of trees of each kind above the average diameter is desired to have offered for sale, with an estimate of the number of feet, board measure, therein, and an estimate of the value of the timber as it stands.

An estimate was also required of the amount, by species, of dead timber, and the cause of its death. Once this was completed, the government official in charge could, if deemed necessary, have the timber examined and appraised. Public notice was then given that the timber was up for bid.[30] Under this program the government would for the first time receive revenue for the sale of public timber in Alaska.

The Department of the Interior's April 1, 1898, regulations were soon made law by Congress in legislation known as the Act of May 14, 1898. The legislation read,

> That the Secretary of the Interior, under such rules and regulations as he may prescribe, may cause to be appraised the timber or any part thereof upon public lands in the District of Alaska, and may from time to time sell so much thereof as he may deem proper for not less than the appraised value thereof, in such quantities to each purchaser as he shall prescribe, to be used in the District of Alaska but not for export therefrom. And such sales shall at all times be limited to actual necessities for consumption in the District from year to year.[31]

This legislation had little effect on timber trespass. In 1903, when a government inspector (almost certainly William Langille) reviewed operations on the Alexander Archipelago Forest Reserve, which had been created in Southeast Alaska in 1902 (discussed at length in the following chapter) with little provision for administration, he reported,

> Nearly, if not quite all, of the timber consumed by the mills in Alaska is cut [without permit] from public lands, and of course the persons who cut it are trespassers, unless it is taken for individual use. This has time out of mind been the custom, the people of Alaska having felt that their isolated condition and the demands of the country justified them in doing so, since any of the methods heretofore provided for the sale of Alaskan timber have been considered by the Alaskans to be extensive and tedious. I am informed by the register and receiver that not a single application to purchase has ever been presented under the act of May 14, 1898.[32]

F. E. Olmsted, who examined the Alexander Archipelago Forest Reserve in 1906, arrived at the same conclusion. So far as Olmsted could discover, "not a stick of timber in Alaska has ever been sold under [the Act of May 14, 1898's] provisions." As he explained, the standard procedure for obtaining government timber was straightforward:

> The logger or sawmill man goes wherever he pleases and cuts whatever he wants, without permission from anyone and without notifying any official of his doings. Once a year each mill is visited by the [General Land Office] Special Agent, who inquires as to the amount of its cut. An innocent trespass case is then made out against the owner who settles (if he does settle) on the basis of the lumber sawed… In southeast Alaska the General Land Office collects 20 cents per thousand for saw timber, one-half cent per foot for piling, and 25 cents per cord for wood.[33]

The governor's report of 1901 gives different figures: the "mills take out a license and pay the government 10 cents for each 1,000 feet cut." Based on individual mill production figures given by historian Robert DeArmond, about

8.5 million board feet (MMBF) of timber was cut in Southeast Alaska for the year 1900.[34]

Very early in 1902, the General Land Office issued regulations providing for the free use of timber generally to "settlers, farmers, prospectors, and others residing within or in the neighborhood of a forest reserve."[35] It also issued the following regulations pertaining to the purchase of timber on the forest reserves:

GENERAL

1. Timber will be sold, both live and dead, wherever the removal of such material will be beneficial, or at least not detrimental to the forest reserves.

2. In the disposition of this material the local demand will have preference, and in localities where this demand is so great that all available timber is likely to be needed, applications involving the export of the material to distant points will be refused.

HOW THE TIMBER MAY BE PURCHASED

1. The applicant who wishes to purchase timber will apply in person or in writing to the supervisor of the reserve, stating—

   a. How much timber he wishes to buy;
   b. The kind of material desired;
   c. Where the timber is located.

2. As soon as practicable the supervisor or his assistant will go over the ground with the applicant, and determine whether the timber may be sold, under what conditions and at what price.

3. After an agreement is reached, the applicant should sign a definite application, prepared on the regular form, with the assistance of the forest officer.

4. After this the forest officer marks out the blocks or area where the timber may be cut, maps it, and estimates the amount of timber on the whole, and also the particular kind applied for. He also makes a general forest description of the tract, block or quarter section.

5. Then the application, together with the forest officer's description and recommendation, is sent to the Department [of the Interior] at Washington.

6. If approved, the timber will be advertised in a local paper for 30 days (60 days in California). This advertisement will be waived only in cases where the amount involved in the sale is of $100 stumpage value [see Glossary], or less.

7. Bids on timber will then be in order. These bids, together with a deposit (insuring good faith of the bidder), would be sent by the bidders to the receiver of the local land office, and the bid will be forwarded from that office to the department.

8. At the end of 30 days (60 days in California), the timber will be awarded to the highest bidder; and if the applicant is the successful bidder, the deposit is credited on the sale; if not, the money will be refunded.

9. When the timber is awarded the applicant will sign a contract containing the specifications contained in the original application, as to manner of cutting, scaling and cleaning up, etc., and if the case seems to justify it, he will be asked to give a bond, usually in an amount of double the value of the timber, to secure the proper fulfillment of the contract.

10. Cutting may then begin.[36]

The requirement for approval from Washington, D.C., made the procedure in Alaska particularly time-consuming, and many loggers chose to take their chances with trespass. John Brady, who himself had earlier been convicted of trespass and was now Alaska's fifth governor, appointed by President William McKinley, explained: "We all feel that we have been trespassers but feel that it has been as necessary to use the timber for living purposes as it has been to use water."[37] When William Langille, who had been appointed supervisor of the Alexander Archipelago Forest Reserve, inspected his charge in 1905, he reported,

Twenty-two timber trespass cases have been acted on, and in nearly every case the trespassers have readily acceded to a demand for stumpage, all protesting their innocence of any intent to violate the reserve laws, but acted on the assumption that they were entitled to timber for domestic purposes.[38]

Enforcement was almost nil. It was the job of the General Land Office's agent at Sitka to investigate and report timber depredations. Langille called the agent's activities in the last years before the 1905 establishment of the Forest Service "perfunctory."[39]

An example of what seemed not to have been standard policy occurred in 1903, when Representative James Tawney (R-Minnesota), on behalf of the Alaska Fish & Lumber Company, requested from the General Land Office a 15-year permit to cut timber on Prince of Wales and adjacent islands. Alaska Fish & Lumber was owned primarily by the congressman's constituents in Mankato, Minnesota. Gifford Pinchot (then head of the Bureau of Forestry in the Department of Agriculture but providing consultation to the Department of the Interior), after conversations with William Langille and Representative Tawney, judged that a permit of five years' duration could be issued, with a stumpage of $0.50 per MBF for sawlogs and $0.10 per MBF for hemlock piling or mining timber. Alaska Fish & Lumber cut only a small amount of lumber in 1905 and was reported to be bankrupt in 1906.[40]

A summary of the timber and wood business transacted by the government during 1905 in the Alexander Archipelago Forest Reserve included revenues of $372.68 for timber sales, and $894.90 from settlement of timber trespass cases.[41]

In addition to the innocent trespass occurring in Southeast Alaska, there were also reports of "wanton depredations" by Canadian Indians logging illegally along the Alaska side of Portland Canal and at Fillmore Inlet. Their logs were towed to a mill at Port Simpson, British Columbia, a distance of approximately 20 miles. One raft of illegally cut timber reportedly contained about 450,000 board feet, and in 1906 F. E. Olmsted reported that "the revenue cutters have not as yet been able to catch the matter up."[42]

CHAPTER 4

The Forest Reserve Becomes a National Forest

THE MODUS OPERANDI OF TIMBERMEN IN THE LATE-NINETEENTH-CENTURY PACIFIC NORTHWEST WAS VERY SIMPLE: CUT OUT AND GET OUT. TIMBERLANDS AND LOGGING OPERATIONS WERE OFTEN HEAVILY MORTGAGED, AND THE OVERPRODUCTION OF TIMBER BY DEBT-RIDDEN OPERATORS WAS CHRONIC. THE RESULT WAS LOW TIMBER PRICES AND MARGINAL PROFITABILITY, HINDERING THE INDUSTRY'S ABILITY TO REFOREST CUTOVER LANDS, WHICH WERE USUALLY CONSIDERED NEARLY WORTHLESS AND OFTEN ABANDONED TO AVOID TAXES AND OTHER COSTS. WILLIAM GREELEY, THIRD CHIEF OF THE FOREST SERVICE, DESCRIBED THE PATTERN IN A 1928 ARTICLE:

> The logged-off land has been counted as no asset to forest industries, or as a residual asset only, to be junked like an abandoned sawmill… Our older forest regions have witnessed three or four hectic decades of logging, followed by the passing on of the camps and mills and the leaving of an immense residue of largely wrecked and unproductive land.[1]

Public concern over rapidly disappearing original forests and poor forest regeneration on cutover lands sparked demands for conservation on lands still under the control of the federal government. To this end, the Forest Reserve Act of 1891 was passed in early March as a rider on that year's agricultural appropriations bill. With this legislation Congress authorized the president to "set apart and reserve, in any State or Territory having public lands bearing forests, any part of the public lands wholly or in part covered with timber or undergrowth, whether of commercial value or not, as public reservations."[2] By the end of 1892, President Benjamin Harrison had created by proclamation 15 reserves containing more than 13 million acres, one of which was the Afognak Island Forest and Fish Culture Reserve, in the western Gulf of Alaska, just north of Kodiak Island. The purpose of the reserve was primarily to protect salmon habitat. In 1908, the Afognak Island Forest and Fish Culture Reserve was assimilated into the Chugach National Forest, which had been created in Southcentral Alaska the previous year.[3]

Though the 1891 legislation did not define the purposes of the reserves, they were subsequently viewed as preserves in which conservation would be accomplished by denying the public access. In fact, until the passage of South Dakota Senator Richard Pettigrew's amendment to the Sundry Civil Expenses Appropriations Act of June 4, 1897, it was technically illegal to set foot in a forest reserve.[4] This legislation opened the forest reserves to regulated use, including the authorization of the Secretary of the Interior to sell "dead, matured, or large growth of trees."[5] It is now generally referred to as the Organic Administration Act of the Forest Service, the Forest Management Act, or simply the 1897 Organic Act.

The forest reserves were administered by the Department of the Interior's General Land Office, but actual management was largely nonexistent as there was little authorization for on-the-ground personnel. Trespass—the cutting of timber without a permit—was a big issue, and some recommended that the Army be detailed to guard the reserves.[6] The Department of Agriculture's Bureau of Forestry, which had been established in 1881 and employed practically all the government's trained foresters, provided consultation to the General Land Office.

To a certain extent, western timber interests—already in possession of timberlands and suffering from chronically low prices due to overproduction—supported the establishment of the forest reserves as a means of reducing the timber supply, though there was the expectation of eventual access to timber on the reserves. A 1910 editorial in a trade journal reflected that "we should first economically manufacture the timber now in private hands, and when this is exhausted the National Forests can be drawn upon. This is true conservation, as *The Timberman* views it." Other interests fought the reserves, fearing their establishment would curtail the opportunities for the acquisition of public timber lands and prevent the cutting for local needs. Some opponents of the reserves in Alaska said the district held no timber worth conserving.[7]

In his first speech to Congress, in December 1901, President Theodore Roosevelt addressed access to the forest reserves, stating, "The fundamental idea of forestry is the perpetuation of the forests by use. Forest protection is not an end in itself; it is a means to increase and sustain the resources of the country and the industries which depend on them." Roosevelt, an ardent outdoorsman and conservationist, was pragmatic as well; he understood that the survival and expansion of the fledgling forest reserve system required provision for commercial activity. In 1902, when he established the Alexander Archipelago Forest Reserve in Southeast Alaska, there was at least some assurance that logging would be allowed.

THE ALEXANDER ARCHIPELAGO RESERVE

The boundaries of the Alexander Archipelago Forest Reserve were the result of work done at the request of President Roosevelt by retired Navy lieutenant George T. Emmons. Emmons had been sent to Southeast Alaska in 1882 to serve as executive officer aboard various naval ships assigned to patrol Southeast Alaska's waters and maintain peace and order among the Native population. During his military career and for 20 years after his retirement in 1899, Emmons spent considerable time each year exploring and collecting Native artwork and artifacts. He became an acknowledged expert on the region, and artwork and artifacts Emmons collected are today the foundation of major museum collections.

Emmons had been an acquaintance of Roosevelt's since about 1897 or 1898, when the future president was Assistant Secretary of the Navy. As president, Roosevelt gave consideration to the establishment of a forest reserve in Alaska early in his first term and turned to Emmons for advice. Like Roosevelt, Emmons was interested in conservation—and specifically, in protecting Southeast Alaska's forest from the kind of aggressive logging that was occurring around Puget Sound. To this end, Emmons in 1902 prepared for President Roosevelt a 16-page handwritten report titled "The Woodlands of Alaska." An informed and objective description of Alaska's woodlands, it made specific suggestions for areas to be included in a reserve. With the report was a government map on which Emmons had shaded in his suggested inclusions.

Emmons thought the most valuable timber was located on Southeast Alaska's islands, which, for the establishment of a reserve, had the additional benefit of easily recognizable boundaries. His selections, however, were tempered by the expectation that the General Land Office would oppose the inclusion of areas that would impede settlement by whites. Emmons' proposed reserve consisted of Prince of Wales Island—southeast Alaska's largest island and, in his words, "believed to be the most valuable of the wooded districts of Alaska, both in quantity and quality of its wood"—and

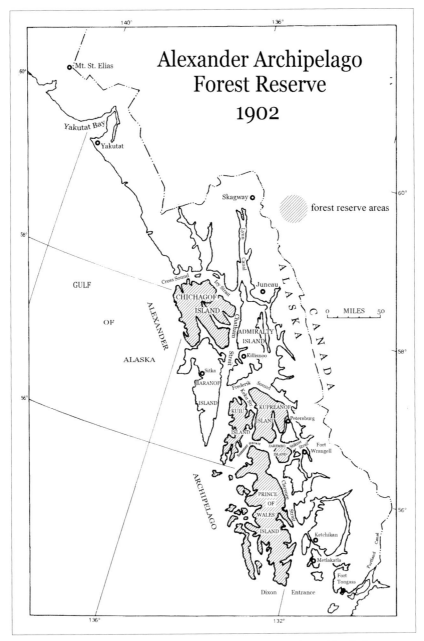

FIGURE 8. Alexander Archipelago Forest Reserve, 1902. (Adapted by Seth Mackovjak from David E. Conrad, "Creating the Nation's Largest Forest Reserve: Roosevelt, Emmons, and the Tongass National Forest," *Pacific Historical Review* 46, no. 1 [February 1977]: 65–83.)

its associated seaward islands, plus Zarembo, Kupreanof, and Kuiu islands, along with Chichagof Island and its associated seaward islands.[8] Despite its major stands of timber, Baranof Island was not selected because of the white population at Sitka. Heavily forested Admiralty Island was likewise not selected, perhaps because of its proximity to the large white population at Juneau.[9]

Roosevelt followed Emmons's suggestions exactly and used his map to determine the reserve's official boundaries. He proclaimed the Alexander Archipelago Forest Reserve, containing 4,506,240 acres, on August 20, 1902.[10]

The response of Alaska Governor John Brady, an experienced lumberman, to the creation of the reserve was highly critical. From Brady's 1902 report to the Secretary of the Interior:

> The President, unfortunately, has never had the opportunity to see Alaska like he has the arid region of the West. He has been guided by the arguments and advice of those who apparently knew all about the matter. In this instance the question may be asked, What is the public good? The reservations are surely not made to protect the valleys from freshets by the too sudden melting of snow at the sources of the streams. Anyone who knows the topography of the islands would not talk that way. While this is a valid argument for forest reserves elsewhere, it is not at all applicable to these islands. Is it then to preserve this timber, that the Government may derive much income from it? One of the fundamental principles of true forestry is that when a tree is full grown it should be removed and utilized. The fact is that the great bulk of the timber on these reserved islands has gotten its growth, and every year there is an incalculable waste in what falls and goes to decay. If this grown timber could be removed and converted into dwellings, mills, canneries, wharves, cross-ties, furniture, packing cases, etc., would the public good be jeopardized? On the contrary, would not all these things stand as a great sum in the asset of the wealth of the country? Would it not be part of sound public policy to encourage the people to use this timber that is now going to destruction?

Brady noted that the hardrock mines (mostly copper) on Prince of Wales Island would need considerable amounts of timber, and that clearcuts would quickly regenerate "and be ready for posterity." The proclamation creating the reserve "disturbs very many interests upon these islands, and especially upon Prince of Wales," he wrote.[11] Brady's report the following year continued his complaint:

> Too long the policy for Alaska has been either neglect or "you can't." On the other hand, we want a policy of encouragement. We have the good things of life here and we should seek the means to encourage the necessary things for life and happiness in Alaska is to have timber in all its special forms, and it should be the great aim of the [Interior] Department to encourage the proper utilization of it.[12]

In August 1904, Gifford Pinchot, as head of the federal Bureau of Forestry, wrote to Governor Brady explaining the federal government's policy for the management of forest reserves, noting that "forest reserves are not parks or game preserves, and the chief aim in their management is to provide for a practical, systematic, and conservative use of their various products." Regarding the Alexander Archipelago Forest Reserve, he wrote,

> The establishment of forest reserves in Alaska will effectually prevent any individual or corporation from monopolizing the timber supply, and will remove the danger of extensive exportation to the detriment of local industries. If the local timber supply becomes exhausted, mining, canning, and all other commercial enterprises will be seriously handicapped. Under forest reserve administration timber will not only be sold to meet local demand, but it will also be protected from destruction by fire where fire is dangerous. In short, the permanent success of the industries of Alaska can best be secured through the establishment of forest reserves. And, I may add, the prosperity of the Indians will follow regular employment in logging. What the Government intends to do in this respect is shown by what it has already done. When timber is ripe for cutting there is as much reason why it

should be cut from a forest reserve as from any other piece of land.[13]

Despite his assurances to Brady, Pinchot considered the General Land Office's administration of the forest reserves corrupt and incompetent, and he worked hard to have them transferred to the Department of Agriculture, which he considered a more appropriate home. As Overton Price, his colleague in the Bureau of Forestry, argued,

> Forestry is a component part of agriculture. Every source of wealth grown from the soil is in the sphere of the Department of Agriculture; hence the forest work rightly belongs to it. The production of timber is as naturally within the scope of the Department of Agriculture as is the production of field crops.[14]

Legislation by Congress in 1905 transferred the administration of the forest reserves from Interior to the Department of Agriculture's Bureau of Forestry. Roosevelt later that year changed the name of the Bureau of Forestry to the Forest Service, to better reflect the new agency's commitment to service.[15]

GIFFORD PINCHOT AND THE NEW SCIENCE OF FORESTRY

Theodore Roosevelt appointed Gifford Pinchot as the Forest Service's first chief. As well as being in charge of the Bureau of Forestry, Pinchot was this country's first native-born trained forester and Roosevelt's friend and advisor. Although he never set foot in Alaska, the policies Pinchot established had and continue to have a great effect on the region. Pinchot ardently promoted a "conservation-through-use" philosophy and was an advocate for the practice of forestry, for which he had a missionary zeal.

The word *forestry* has various definitions. Merriam-Webster currently defines it as "the science of developing, caring for, or cultivating forests." Theodore Roosevelt, however, defined it as "the preservation of forests by wise use." And Royal Kellogg, a representative of lumber and newsprint paper interests, gave it a utilitarian, human-centric definition: the "proper growing of trees for the use of man." But it was Gifford Pinchot who fundamentally defined forestry as practiced by the Forest Service: "forest management for continuous production." For Pinchot, forestry was part of a responsible utilitarian philosophy that favored the cutting of forests for the nation's immediate needs, but, unlike the reckless cut-and-abandon mentality that pervaded the age, in a manner that guaranteed a continuous supply of timber for future generations. Pinchot considered his job as "not to stop the ax, but to regulate its use."[16]

Pinchot's strategy was to build public support for national forests by promoting and facilitating their use, and he established the policies under which all the resources in the forests could be utilized, to quote Pinchot, "for the permanent good of the whole people."[17] Included was a realistic plan for the ordered cutting of timber under the management of trained professionals in the Forest Service. Logging, Pinchot and his associates wrote, was to be done carefully so "the stand is left in first-class condition for a second crop, and after that a third crop and any number of future crops."[18]

Pinchot was not shy about cutting forest reserve timber. That timber, he wrote in 1905, "is there to be *used*, now and in the future… The more it is used, the better."[19] His philosophy differed from those, such as John Muir, who were interested in protecting America's "noble primeval forests."[20] Basically all forest reserve timber for which there was an actual—rather than speculative—need, and that could be safely cut, was for sale. Pinchot had the firm support of President Roosevelt, who in a March 1903 speech to the fledgling Society of American Foresters outlined his administration's forest policy:

> First and foremost, you can never afford to forget for one moment what is the object of our forest policy. That object is not to preserve the forests because they are beautiful, though that is good in itself, nor because they are refuges for the wild creatures of the wilderness, though that, too, is good in itself; but the primary object of our forest policy, as the policy of the United States, is the making of prosperous homes. It is part of the traditional policy of home making in our country. Every other consideration comes as secondary.[21]

The forestry that was, and largely continues to be, practiced by the Forest Service on the Tongass National Forest was defined in 1933 as "that forestry practice which aims to realize through silvicultural treatment the nearest practical approach to the maximum productivity of a given site, building up in the shortest practical time as large an annual cut as is consistent with the productive capacity of the land."[22]

THE PEOPLE'S FOREST

The would-be beneficiaries of the switch to management by the Department of Agriculture were not impressed, according to a June 8, 1905, article in the *Alaska Sentinel* (Wrangell):

> Three or four weeks ago it was reported that loggers would have to pay 25 cents per 1,000 [board feet] stumpage for the purpose of plying their vocation. This was the price determined upon, but the land agent, Mr. Love, who is in this section, says that after looking over the ground and studying conditions he shall recommend that the stumpage be made 15 cents. That, even, is too much. Few loggers are making any money, and if those law makers who are responsible for these taxes could be on the ground and see what hardships the loggers are compelled to endure, in order to make a mere pittance, they would say "take the timber in welcome."[23]

William A. Langille, an accomplished outdoorsman who had prospected for gold in the Klondike and Nome, was hired by Gifford Pinchot in 1903 to travel to Southeast Alaska and report on the administrative needs of the newly-created forest reserve. On his return, Langille gave a talk about Southeast Alaska's forest at Pinchot's residence in Washington, D.C. In his opinion, a large proportion of the region's timber was rendered useless by rot that stemmed from "heart shake" and "frost cracks." He felt there should be very little restriction on logging in Southeast Alaska, apparently because of a desire to encourage people to settle in and develop the region.[24]

Langille was appointed forest supervisor in 1905 and initially made his headquarters at Wrangell. On assuming this position, Langille projected what a colleague termed an "abrupt, outspoken and occasionally mildly terrifying manner." According to the same colleague, Langille soon modified his tone and was "at the bottom…absolutely lenient and disposed to make every possible allowance to those doing business in the reserve." By 1907, Langille had moved his headquarters to Ketchikan, where the local *Mining Journal* wrote that "Mr. Langille has the confidence of the department and is a reliable officer and a good citizen."[25]

Yet complaints against the forest reserves persisted. In 1906, Pinchot dispatched his assistant forester, F. E. Olmsted, to Southeast Alaska to investigate.[26] Olmsted had trained as a forester in the United States and in Germany,

FIGURE 9. William Langille (left) at Native logging camp. (Alaska State Library, William Langille Collection)

and had gone to work for the Division of Forestry in 1900. His was the first detailed examination of Southeast Alaska's forests by a professionally trained forester.[27]

Olmsted traveled widely and interviewed many of the principal business people in Southeast Alaska. He found sentiment toward the forest reserve to be mixed and characterized it overall as "mildly unfriendly." The publishers of Wrangell's *Alaska Sentinel* vigorously opposed the reserve, and wrote in 1908 that "the whole forest reserve system of Southeastern Alaska is absolute nonsense, and is kept alive for no other purpose than to furnish jobs for a lot of pap-suckers whose accomplishments fit them for no higher position." On the other hand, Ketchikan's *Daily Miner*, which was published by A. P. Swineford, Alaska's governor from 1885 until 1889, supported it. H. Z. Burkhart, proprietor of the Ketchikan Power Company, which operated one of the largest sawmills in Alaska, favored the reserve because it offered some certainty regarding the cost and availability of timber and the conditions under which it could be purchased. Later, the *Daily Miner* would write that the Tongass

National Forest was created "at the request of the Ketchikan Power Company."[28]

Olmsted argued for the expansion of the forest reserve to include all of Southeast Alaska, "from Mt. Saint Elias to the Portland Canal...except for certain exclusions around the principal towns."[29] He foresaw little need for additional personnel to manage the land:

> There's no fire, no stock, no roads, no trails. Hence we need no Rangers. The Officer in Charge and two Deputy Supervisors can successfully manage the 16,000,000 acres of the proposed enlarged reserve, provided they have boats.[30]

In support of his position, he contrasted the Department of Agriculture's fledgling timber program on the forest reserve with that on the region's unreserved public lands, which were managed under regulations based on the Act of May 14, 1898, by the Department of the Interior. Olmsted wrote, "Here are public lands of precisely the same nature, lieing (sic) side by side, governed by two different policies with hardly a single point in common." He considered the Department of the Interior's handling of the timber business to be "absurd," given that it could not "sell a stick of timber without making a trespass case out of it."

Moreover, the prices charged for timber on the forest reserve were higher than those charged on adjacent unreserved government land. Sawtimber on the reserve was priced at $0.50 per MBF; similar timber on unreserved lands could be procured for a settlement of $0.20 per MBF.

Olmsted was interested in how the timber business in Southeast Alaska might be helped, so he proposed that the government "provide for machinery which will dispose of timber promptly and in a business-like way to those who apply for it." He made several recommendations, among them that certain rules requiring the exact designation of timber an operator was authorized to cut be "waived or very liberally construed." He suggested the publication of a special circular for Alaska to explain what the reserves were about, with a summary of the more important regulations. Olmsted also suggested the immediate advertising of 10 million board feet (MMBF) of timber in $500 lots.

So far as timber prices were concerned, Olmsted thought sawtimber—whether Sitka spruce, western hemlock, or western redcedar—was "beyond the shadow of a doubt" worth the established price of $0.50 per MBF, and that the half cent per linear foot price for piling was also fair. He thought the stumpage price of Alaska-cedar, of which some small lots of the best grades of lumber had sold for $70 to $80 per MBF, should be raised to $2 per MBF.[31]

As forest supervisor, Langille in 1907 was "authorized to sell not over 1,000,000 feet board measure of green or dead timber," and "Such other forest officers as the supervisor may designate are authorized to sell green and dead timber in amounts not exceeding $100 in value."[32]

THE TONGASS

In 1907, Congress decided that the forest reserves be renamed national forests, largely to dispel the notion that they were locked up, as implied by the word *reserve*. In 1924, Frank Heintzleman, then Alaska's assistant district forester, wrote the following:

> A National Forest is not a 'Reserve' in any sense and even the term "Forest Reserve," as applied to Federal timber-lands, was abandoned years ago. A "Reserve" implies something set aside for the future, whereas neither the timber nor any other resource of the National Forests is withdrawn from present use. The major purpose of the National Forests is the growing of timber for future needs, and not the reserving of the present stand for the future.[33]

A rider on the 1907 Agriculture Appropriations Bill deprived President Roosevelt of the authority to establish additional national forests in the states of Colorado, Idaho, Montana, Oregon, Washington, and Wyoming. The legislation required Roosevelt's signature to become law, but before he signed it he secretly signed 33 proclamations that created 16 million acres of hastily configured national forests in Colorado, Idaho, and other western states and Alaska. In Alaska, he created the 2,262,624-acre Tongass National Forest, which was located on the rugged mainland east of Ketchikan. Basically, it was bordered on the north by the Unuk River, on the east by the Canadian border, on the south by Portland Canal (through which runs the Canadian border), and on the west by Behm Canal and Revillagigedo Channel.[34]

The Tongass was named after a former Tlingit village on the south side of Tongass Island, itself named for a Native word meaning "sea lion people." The island lies at the southern tip of Southeast Alaska, about 35 miles east of Cape Fox. Fort Tongass, an army post, was established on the north side of Tongass Island in June 1868 and was maintained until September 1870. Also present was a customs house, which was operated until 1889. Tongass Island's population in 1890 was recorded as 255, but it declined as the Ketchikan area developed.[35]

In his 1907 report to the Secretary of the Interior, Alaska Governor Wilford Hoggatt expressed approval with the federal government's administration of the territory's forests. Hoggatt wrote: "The regulations governing the use of the timber and lands in these forests have been so modified during the past year as to allay the discontent growing out of the old regulations, and these forests are now so administered as to conserve the timber and provide for its most liberal use without hampering in any way the development of the country."[36]

The following year, 1908, President Roosevelt by executive order consolidated the Alexander Archipelago Forest Reserve into the Tongass National Forest, and in 1909 he proclaimed an additional 8,724,000 acres. Among the areas added were Admiralty, Baranof, and Revillagigedo islands, and nearly all the mainland from the Unuk River north to the Skagway River. Added as well was an area on the Gulf of Alaska between Dry Bay and Yakutat. Alaska Governor Walter Clark's comments in his annual report to the Secretary of the Interior that year echoed those of his predecessor: "The administration of these forests continues to conserve the timber and provide for its liberal use without hampering the development of the country, and is meeting with the general approval of the people of the Territory."[37]

The last major addition to the Tongass, approximately 1.1 million acres, was proclaimed by President Calvin Coolidge in 1925.[38] The addition was in the northern region of the forest, and included the Chilkat Peninsula and a substantial band of mainland along the north shore of Icy Strait and Cross Sound, as well as a narrower coastal band on the Gulf of Alaska, from Cape Spencer to Cape Fairweather.

The Tongass National Forest, with an area of 16,547,000 acres, was the nation's largest national forest, and it and Southeast Alaska were now nearly one and the same.[39]

The Tongass National Forest, as well as Southcentral Alaska's 4,960,000-acre Chugach National Forest, which had been proclaimed by President Roosevelt in 1907, were headed by a forest supervisor stationed at Ketchikan. Both forests were originally part of District 6, which was headquartered in Portland, Oregon. In 1921, in part due to the recognition of their importance as a source of wood pulp and in part to facilitate efficient administration,

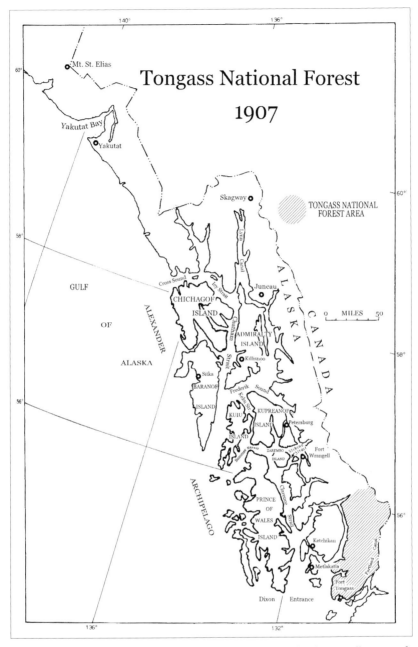

FIGURE 10. The original Tongass National Forest comprised only a small corner of Southeast Alaska. (See Figure 8 attribution)

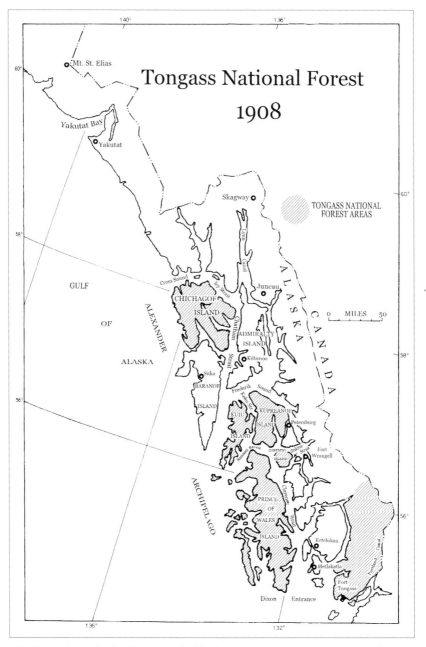

FIGURE 11. In 1908, the Alexander Archipelago Forest Reserve was consolidated into the Tongass National Forest. (See Figure 8 attribution)

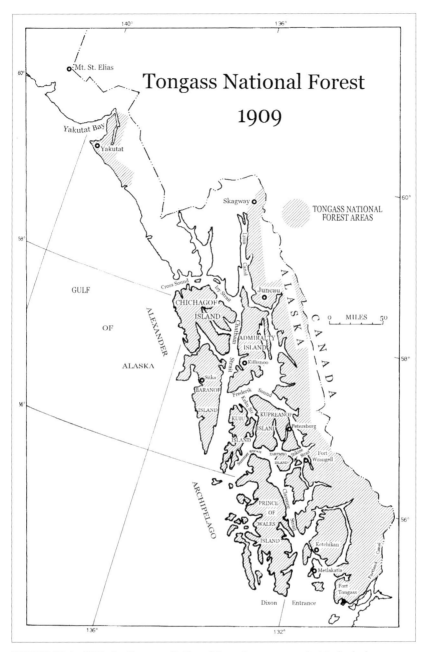

FIGURE 12. In 1909, the Tongass National Forest was expanded to include almost all of the region's islands, most of the mainland, and an area near Yakutat. (See Figure 8 attribution)

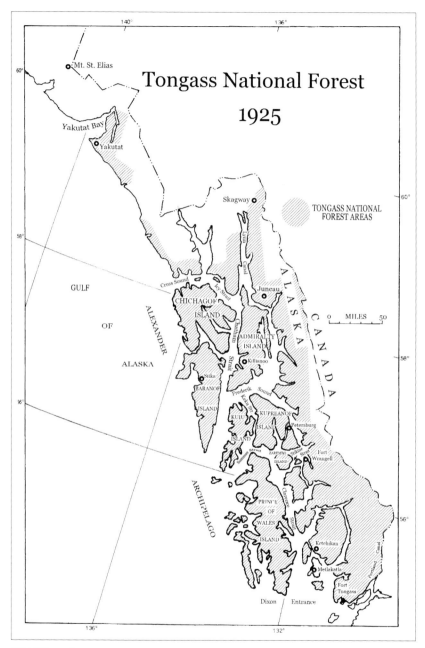

FIGURE 13. The last major addition to the Tongass, in 1925, added lands in the northern portion of the region. (See Figure 8 attribution)

Alaska's national forests were designated a separate District 8—the "Alaska District." The following year the district's headquarters were moved to Juneau. In May 1929, the Secretary of Agriculture approved a change in terminology in which all districts were renamed regions. District 8 became Region 8. Regions were renumbered in 1934, and Region 8 became Region 10.[40]

CHAPTER 5

Lumbermen Seek Consideration

A MODEST BUT REGIONALLY IMPORTANT LUMBERING INDUSTRY DEVELOPED IN SOUTHEAST ALASKA, MOSTLY TO MEET LOCAL DEMANDS. THIS INDUSTRY WAS ALMOST COMPLETELY DEPENDENT ON LIMITED STANDS OF LARGE-DIAMETER, OLD-GROWTH SITKA SPRUCE.

In 1902, the year that President Theodore Roosevelt established the Alexander Archipelago Forest Reserve, there were about 15 sawmills operating in Southeast Alaska. Mostly they were associated with mines or salmon canneries, had a capacity of less than 15 MBF per day, and operated seasonally or intermittently. Some cut fewer than 250 MBF per year. Most were powered by steam, but a few were water-powered. The region's most important sawmill at the time was the Willson & Sylvester Mill Company at Wrangell. The steam-powered mill had a capacity of about 35 MBF per day, which was relatively small in comparison to mills in Oregon and Washington that were cutting on the order of 125 MBF per day.[1]

The Willson & Sylvester Mill Company was established in 1888 primarily to manufacture wooden salmon cases for the region's growing salmon canning industry. It was the only sawmill operating in Wrangell in 1902 and employed 30 to 50 men. Almost as many more men were employed in logging operations. Production that year, the result of the usual operating season of eight or nine months, was 3.9 MMBF.

Juneau's first sawmill went into operation in the early 1880s. Of the several sawmills that operated in the Juneau area in 1902, the largest was the George E. James Company sawmill, which that year cut almost 2 MMBF.

In the Ketchikan area, three small sawmills were operating when the Alexander Archipelago Forest Reserve was established. The company that came to be Ketchikan Spruce Mills—ultimately the largest sawmill in the territory—was not one of them. Ketchikan Spruce mills had its genesis in 1903, when three area businessmen incorporated the Ketchikan Power Company. The firm engaged in two businesses: selling electricity and operating a sawmill. The sawmill's capacity in 1907 was about 40 MBF per day.

When Representative James Tawney (R-Minnesota) visited Southeast Alaska with a group of Northwest timber owners in 1903, he reported, "The saw mills in Alaska are of the most primitive kind. There is not a modern mill in the territory; in fact there are only three or four mills that make any pretension whatsoever of manufacturing lumber." Likewise, Royal Kellogg of the Forest Service characterized Southeast Alaska's sawmills in 1909 as "rather crude in character and of small capacity." His comments were later echoed by J. S. Baker, who wrote in *The Timberman* that the mills operating in Southeast Alaska ca. 1910–1915 were "crude, wasteful, and small in capacity."[2]

Writing in the same vein in a 1913 issue of *Forestry Quarterly*, B. E. Hoffman described Southeast Alaska's mills as "very inefficient, especially those that run box factories. In the first place they cannot produce enough lumber to supply the demand, and their methods are so far from perfect that they can afford to cut only the best timber at a profit." Hoffman added, "Mill methods are very wasteful, and a prodigality in the manner of working up material has been engendered by the easy way in which logs have been obtained." Many of the men employed by sawmills were inexperienced, he said, and no attempt was made to utilize the excessive waste.[3]

A fundamental problem with Southeast Alaska's early sawmills was their employment of circular saws instead of more efficient band saws. The circular saw was a great improvement over the old up-and-down muley saw, and sawmills were further improved with the introduction around 1870 of the double circular saw, in which two aligned circular saws, one above and slightly behind the other, were used to make a single cut. This arrangement made it possible to cut logs too large for any single saw. The main disadvantage with the circular saw was the thickness of the blade, usually ⅜ inch. Though more expensive to purchase and maintain, the band saw, introduced around 1880, was faster, required less power, and because it was thinner than a circular saw by about ⅛ inch, produced less waste. The amount of wood saved by the smaller "kerf"—the wood cut away by the saw blade—added up fast for saws that made hundreds or thousands of cuts daily. In Alaska, where the cost of timber was low, this saving wasn't initially a major consideration, but as mills upgraded and operators sought efficiency, band saws became more prevalent.[4] The first band saw mill in Southeast Alaska was installed about 1905 at Hadley to support the short-lived copper mining community on Prince of Wales Island.

Frank Heintzleman, after some 35 years of working with timber operators in Southeast Alaska, accurately summarized some aspects of the situation under which the region's sawmills operated:

Small sawmills have rarely been financially successful in the coastal forests of Alaska. Only heavy mill equipment can handle the large logs which produce the high percentage of high-grade lumber manufactured here... All sawmills in this region need drykiln facilities as air drying is not feasible in the wet climate found here. Existing towns offer the best mill locations. Isolated plants have higher operating costs and greater difficulty in holding employees than mills in established communities where better living facilities are available.[5]

Despite their shortcomings, Southeast Alaska's early sawmills were particularly important to the region's economy. As a component of the local manufacturing infrastructure, they provided local employment. The production of lumber and wood products for local use also reduced the need to obtain competing material from Puget Sound, thus keeping money within the region. Finally, lumber and wood products exported by the mills (including cases and boxes used to ship fish) were a source of much-needed capital for the region. Unlike Alaska's canneries and large mines, which were for the most part owned by outside interests, the sawmills of the early era were predominately locally owned, and the profits they earned were to a large extent retained within the region.

As noted above, lumber mills were scattered throughout southeastern Alaska (see Appendix A), but the most important were located in Wrangell, Juneau, and Ketchikan. All three of these mills started out small. But in Wrangell, for example, the mill by 1909 had become sufficiently prominent that the town's *Alaska Sentinel* praised the Willson & Sylvester sawmill as "an institution which for many years has been the pride and backbone of this community and the strong arm upon which, in great measure, the town and its business houses have leaned for support"[6] By 1917, Wrangell's mill was annually cutting some 6 million board feet of lumber and manufacturing about a half-million salmon cases. The plant burned in 1918, and was quickly rebuilt to greater capacity; up to 50 MBF of lumber and 6,000 salmon cases per day. In 1924, the mill cut a little more than 8 MMBF of lumber and its capacity was increased to 65 MBF per day.

The experience at Juneau was much the same. Although the George E. James mill cut just 2 MMBF in 1902, the sawmill's capacity was gradually increased to 6 MMBF per year. It ceased operations, however, in 1918, likely

FIGURE 14. In 1915, the Willson & Sylvester Mill Company was Alaska's premier sawmill and manufactured products that ranged from wooden salmon cases to rough and finished construction lumber to clear Sitka spruce lumber for use in the manufacture of airplanes and musical instruments. (Alaska State Library, Ed Andrews Photograph Collection)

due to competition from the sawmill that was established by Juneau business-man Henry Shattuck and Puget Sound lumberman H. S. Worthen. In 1913, Shattuck and Worthen stated their intention to construct "the most modern and up-to-date [sawmill] in Alaska." The Worthen Lumber Mills Company began operating by 1915, and within a couple of years the Forest Service expected the mill's cut to be in the 9 MMBF range. In 1919, a group of investors headed by Fairbanks lumberman Roy Rutherford acquired the sawmill and renamed it Juneau Lumber Mills. In 1923, the mill employed about 65 men and cut about 50 MBF per day. About 8 MMBF of timber was cut that year. The Juneau Lumber Mills also manufactured salmon cases.

Ketchikan's principal mill had been built as part of a venture to produce and sell electricity as well as lumber. But in 1909, Ketchikan resident James Daly acquired control of the sawmill and began expanding its capabilities. His was the first of three generations of the Daly family that would operate the mill. In 1913, the mill was capable of cutting about 50 MBF per day and manu-factured some 500,000 salmon cases. The Dalys changed the business's name

FIGURE 15. Juneau Lumber Mills in 1929 cut a variety of material that was mostly moved around the yard on two-wheel trucks. Note horse at left center. (Alaska State Library, Postcard Collection)

to Ketchikan Spruce Mills in 1923. Production the following year totaled 10.5 MMBF. In 1925, the *Ketchikan Alaska Chronicle* showed its appreciation for the mill: "The Ketchikan Spruce Mills have always played a big part in the life of the city, furnishing much of the lumber of which the city is built and employment to a large number of men who make their homes in Ketchikan."[7]

The importance of sawmilling varied among other Southeast Alaska communities in the early twentieth century. Petersburg, for example, was a fledgling fishing town in 1902, and sawmilling played an important supporting role to the community's dominant industry. The only local sawmill that year was owned by the Pacific Packing & Navigation Company, the firm that owned the town's cannery. It had a capacity of 6 MBF per day, and had been used to cut the lumber for the cannery's construction. Ironically, the sawmill may not have actually operated in 1902 due to financial difficulties at the cannery. By 1906, however, the mill was replaced by the first of a series of generally medium-sized local mills under various ownerships that kept the community reliably supplied with lumber and fish boxes through World War II. The largest was the Arness Lumber Company's mill, which *The Timberman* listed in 1919 as having a capacity of 50 MBF per day. There were several

FIGURE 16. Ketchikan Spruce Mills was Alaska's most successful and enduring sawmill. The mill was converted from steam power to electric power during the winter of 1924–1925. Image was likely taken between 1943 and 1952. Wigwam burner at right was used to dispose of waste wood. (NARA-Anchorage, RG 95)

attempts during this period to establish large sawmilling operations at Petersburg, but none were successful in the long term.

Sawmilling played a less important role in early twentieth century Sitka. The principal sawmill in 1902, the Sitka Mill Company, was owned by Alaska's governor, John Brady. Brady's mill had a capacity of 10 MBF per day and served mostly the local market. By 1904, it was virtually shut down, likely a victim of its limited market. A series of small, unsuccessful mills irregularly served Sitka's needs for the next three and a half decades.

Outside of the region's major towns, sawmills at one time or another were established in more than thirty locations. These operations, much smaller than those in Wrangell, Juneau, and Ketchikan, were typically established for specialized uses—such as supporting cannery or mining operations— and did not endure long beyond the projects they were tied to.

To some, the industry as a whole promoted sturdy frontier values. A 1917 article by E. T. Allen, head of the Western Forestry and Conservation Association, published in *The Timberman*, recalled William Greeley (who would later

FIGURE 17. Water-powered sawmill at Warm Springs Bay, Baranof Island, 1927. (USDA Forest Service, Alaska Regional Office, Juneau)

become chief of the Forest Service) as having said that, "Lumbering is perhaps the most 'American' of our manufacturing industries. In its individualism, its encouragement of small independent business units, its hearty competition, and the rugged, forceful qualities it has derived, it expresses many national economic and social ideals."[8]

In addition to producing lumber and timbers, Southeast Alaska's sawmills produced waste. "Fishermen in Southeast Alaska complain of the dumping of sawdust and other sawmill refuse into the waters, claiming that this drifts for miles in every direction and drives fish away," observed the federal Bureau of Fisheries in 1908.[9] The problem was sufficient to prompt Alaska's Territorial Legislature to pass the "Territorial Law Regarding Saw Mill Wastes, Etc." during its first-ever legislative session in the spring of 1913. According to the law,

> It shall be unlawful for the proprietor of any mill in the Territory of Alaska, or any employee therein, or any other person, to cast sawdust, planer shavings, or other lumber waste made by any lumbering or manufacturing concern, or

to suffer or to permit such sawdust, planer shavings, or other lumber waste to be thrown or discharged in any manner into any of the streams entering into salt water of this territory, or any bays immediately adjacent to salmon streams, or to deposit the same where high tide water will take the same into any of the waters of this territory.[10]

The Bureau of Fisheries considered the legislation an "excellent measure," and stated that there was "no doubt but that sawdust is very destructive to fish life." Enforcement of the regulation, at least for several years, however, was apparently lax. The bureau's E. Lester Jones complained in 1914 that in some locations sawdust was being dumped into territorial waters "under the very eyes of the local officials." On April 29, 1915, exactly two years after the law had passed, the *Wrangell Sentinel* reported, "We noticed the bay in front of the town was literally covered with sawdust from the local mill."[11]

It wasn't just the territory's law that was not being enforced. In 1899, Congress had passed what became known as the Refuse Act. The legislation made it illegal to "throw, discharge or deposit…any refuse matter of any kind or description whatever other than that flowing from streets and sewers and passing therefrom in a liquid state" into any navigable waters of the United States.[12] The U.S. Army Corps of Engineers was charged with enforcing the legislation but chose—for whatever reasons—to ignore it.

SAWTIMBER SALES

In 1902, loggers, according to John Brady, the lumberman-turned-governor, received $3.50 to $4 per thousand feet at the mills in the southern part of Southeast Alaska and $6 at Juneau, Skagway, and Sitka.[13] F. E. Olmsted, Gifford Pinchot's assistant forester, gave this report of the sawtimber industry in 1906:

Logs are worth from $4.50 to $5.00 per thousand, in raft where cut. No distinction in species is made. The cost of towing to mill is from $2.00 to $3.00, depending on the distance and on conditions of sea and wind. Sawing and handling at mill is figured at about $3.00. Common lumber sells at from $12.00 to $14.00 and finished grades at from $25.00 to $30.00. Common labor is paid $3.50 per day, from which board is deducted.[14]

FIGURE 18. Making up a raft of Sitka spruce logs. (NARA-Anchorage, RG 95)

That year, immediately following the transfer of the forests from the Department of the Interior to the Department of Agriculture, 86 percent of the lumber requirements of Alaska was brought in from elsewhere and 14 percent was cut from the national forests. That reliance on outside sources, however, soon diminished. "By 1919," according to Secretary of Agriculture Henry Wallace in 1923, "these percentages were reversed, and 86 per cent of the local requirements was cut from the national forests and but 14 per cent imported."[15]

The Forest Service's fledgling timber sale program proved generally satisfactory to Southeast Alaska, according to Governor John Strong in 1913: "Fewer complaints are to be noted during the past year as to the administration of the national forest...due to the more liberal spirit that has been manifested in the interpretation of the regulations and the knowledge that the national forest system has become a fixed policy of the Government."[16]

Scaling logs, however, was an issue. The original Forest Service policy was to scale logs at the place of cutting, many of which were distant from towns and off the established routes of travel, causing frustrating and costly delays to operators. The policy was improved in about 1910, when provision was made for the scaling of timber at the mills. Once a log was scaled, "U.S." was stamped on one end, and the log could then be sawed.[17] Though generally more efficient for the Forest Service and appreciated by loggers and millmen, the policy of scaling logs at the mills was to the government's detriment in one respect: logs lost while moored in booms or being towed to the mill were neither recorded nor paid for.

The lost revenue could be considerable. "Last Sunday," reported the *Wrangell Sentinel* in 1911, "while crossing from Zarembo Island with a boom of logs cut by McDonald Brothers, the sawmill tug Alaska, had the misfortune to lose several hundred logs when the boom was broken by heavy seas." And the next year: "Goodwin's log raft which has been ready for the mill nearly two weeks broke up in the Wrangell Narrows recently and destroyed one of the lights marking the channel." Towage fees and payments to those who logged the timber were often also based on the Forest Service scale.[18]

The Tongass National Forest timber sale program changed little in the early years. John D. Guthrie noted, in a 1921 *American Forestry* article defending the Forest Service's timber program in Alaska, that the agency was "selling timber today under practically the same regulations as those of 1905."[19] Frank Heintzleman described the Forest Service's Tongass timber sale procedure in 1921:

> Timber can be purchased from the National forests in any reasonable quantity desired, at any time, and by any American citizen who has the facilities for doing the necessary logging work....
>
> No formal application for timber is necessary. The tract desired is examined by a Forest Officer, who reports on the desirability of making the sale, marks the tract boundaries to make the area a logging unit, cruises the timber and maps the topography. Many of the more desirable timber tracts are examined in advance of application. Timber units exceeding $100 in stumpage value are advertised for sale by sealed bids for a period of at least 30 days, and are awarded to the highest qualified bidder. Sales involving less than $100 stumpage are sold without advertising. The timber is paid for in small installments as cutting proceeds, and on the basis of a scale of the logs made by a Forest officer.[20]

For small timber sales in Alaska, Chief Forester William Greeley thought that the advertising requirement was "simply an irksome delay and an unnecessary piece of red tape" that should be changed. He noted that there was no opportunity for competition, and that timber sales were invariably awarded to the original applicant at the price established by Forest Service officials.[21] Despite Greeley's comments, Forest Service timber sales in Alaska—even small ones—continued to be advertised, just as they were throughout the country.

Bidders on Forest Service timber sales were required to include with their bid a deposit amounting to about 25 percent of the Forest Service's minimum acceptable stumpage price. Additionally, the successful bidder was required to furnish a bond on sales valued at over $1,000 according to the following percentages: $1,000 to $10,000, ten percent of the sale amount; $11,000 to $50,000, eight percent of the sale amount, with a minimum bond of $1,000; over $50,000, five percent of the sale amount, with a minimum bond of $4,000 and a maximum of $50,000.[22]

At first, the quantities produced (as noted above) were too small to meet local demand: "The shipping into Alaska of lumber products from the outside does not prove in the slightest degree that the Alaskan timber is unfit to meet the requirements of local use, either in quality or amount," wrote Forest Service Chief Henry Graves in early 1916. "It indicates that the present economic conditions have not yet justified the development of a manufacturing industry that can compete with the outside material."[23]

Justification soon appeared. Beginning in July 1916, the Forest Service raised the minimum price for stumpage in the Tongass, from $1 per MBF for spruce and red-cedar standing within 2,000 feet of tidewater to $1.50.[24] "No bid of less than $1.50 for spruce within 2,000 feet of tidewater, or $1.00 per M feet B.M. for spruce more than 2,000 feet from tidewater, and 50 cents per M for hemlock will be considered," read a Kupreanof Island timber offering announcement of August 1916. "The reason given for advance of stumpage," according to the *Petersburg Weekly Report*, "was that conditions in the lumber market had improved."[25]

The 1920s were a modest heyday for the lumber industry in Southeast Alaska. The three principal sawmills in the region—Juneau Lumber Mills, Ketchikan Spruce Mills, and the Willson & Sylvester Mill Company at Wrangell—expanded and made major improvements that brought them up to standards of production and efficiency roughly equivalent to comparable sawmills in the Puget Sound area. In addition to serving the local markets for lumber, the region's mills began exporting high-grade Sitka spruce on a regular basis. Much of the material was for airplane construction. Also, the fishing industry was growing and it provided a market for fresh fish and salmon boxes, as well as lumber. A shortage of lumber in Australia during a few of those years provided a temporary but substantial export market for Sitka spruce lumber.

Corresponding to the growth in the sawmilling industry, Forest Service timber sales became larger. One of the agency's biggest sales was in 1923, when it sold some 25 MMBF of timber on the west side of Long Island to Ketchikan Spruce Mills. Approximately 97 percent of the timber was Sitka spruce.[26]

The Forest Service estimated in 1920 that the total cost, exclusive of stumpage, of logging and towing timber to a mill almost 100 miles distant was about $6.50 per MBF. For comparison, figures from a 1914 timber sale broke down the costs of getting timber from the stump into a log boom: felling and lopping tree tops, $1.00 per MBF; hauling and rafting, $1.60 per MBF; wear and tear on equipment, $0.20 per MBF; and miscellaneous, $0.15 per MBF, for a total cost of $2.95 per MBF. Sawmill operations in 1920, including planing and drying, were estimated to total about $6.25 per MBF, with the average value of lumber being about $18 per MBF.[27]

By the end of the 1920s, at least one sawmill was augmenting its workforce by temporarily hiring women, according to a 1929 article in the *Wrangell Sentinel*:

> Women were employed for the first time in the sawmill history in Alaska when a small crew was put to work last week in the Wrangell Lumber & Box Company.
>
> The local mill is getting out an order of butter, cheese and poultry boxes for shipment to middle west points. The boxes are made mostly from salmon box waste, and besides utilizing the waste they provide an additional payroll. Women are particularly adapted for this form of work which requires accuracy and deftness in handling the small pieces with a minimum of muscular effort.[28]

The average annual cut on the Tongass recorded by the Forest Service for the years 1909 through 1930 was about 36 MMBF, ranging from a low of 7 MMBF in 1909 to nearly 54 MMBF in 1925. (see Figure 64 on page 276.) It is likely, however, that a considerable amount of timber was cut without records. In 1928, Frank Heintzleman estimated that 500 men were employed in the forests and mills in Alaska's "Coast Forest," which included Southeast Alaska and the coastal area from Yakutat to Kodiak and Afognak islands.[29] The vast majority of this employment was in Southeast Alaska.

Most of the waters of Southeast Alaska are relatively protected but subjected to storms, usually "Southeasters," particularly in the late fall. Such storms could make towing logs difficult. In September 1923, the McDonald Logging Company tug *Inverness* left Long Island with a tow of 300 MBF of Sitka spruce logs destined for Ketchikan Spruce Mills, 80 miles distant. A severe, prolonged storm slowed the tug's progress to such an extent that it was fully 23 days before it arrived in Ketchikan with its tow of logs.[30] Reporting on the expense of log towing in 1930, Ketchikan District Ranger Wyckoff observed,

> Towage costs vary to a considerable degree depending upon the distance from camp to mill. The charge for this service varies from 50 cents per thousand [board feet] for short tows of 20 miles or less, up to $2 per thousand for the longest tows. The logs are all made up in flat rafts of from 150,000 to 500,000 [board] feet.[31]

In 1937, Frank Heintzleman estimated the cost of towing sawlogs as being about one cent per 1,000 board feet per mile.[32]

SAWMILL DEVELOPMENT

While the Forest Service worked relentlessly to support establishing a pulp and paper industry (see Chapter 8), it did almost nothing to promote Southeast Alaska's lumbering industry other than to note that once a wood pulp industry was established, large-scale pulptimber logging operations would make isolated stands of Sitka spruce timber economically accessible, and this timber could be directed to sawmills. "The saw timber," concurred Frank Heintzleman in *The Timberman*, "comes principally from the patches of large, pure spruce that are scattered through the forests of smaller hemlock, spruce and other trees. These patches frequently have as much as 100,000 board feet of timber per acre."[33]

In 1920, the District 6 forester, George Cecil, who had spent six weeks traveling in Alaska in 1913, discussed the territory's potential demand for lumber and the corresponding need for timber:

> It is probably safe to estimate that the needs of the territory for lumber, piling, railroad ties, box materials, mining timbers, etc., will reach 200 million feet annually within the next 10 or 15 years; and that roughly 20 per cent of the

timber in the National Forests and of growing capacity should be allocated to these purposes. This will represent, broadly speaking, all of the cedar timber and the spruce and hemlock of the largest size and highest commercial quality.[34]

The demand for timber in the territory as a whole did increase, but not to the extent predicted by Cecil.

The Forest Service's view of how sawmills fit into its pulp-and-paper-oriented development plan on the Tongass was described by Frank Heintzleman in 1924:

The development of efficient local sawmills is encouraged. As a whole the Alaskan forests are chiefly valuable for their pulpwood, but the limited good sawtimber areas, if properly handled, can maintain an important local lumber industry. The aim is to have the forest provide all of the saw timber the local markets can absorb, with such additional supplies to the mills for manufacture and shipment outside the Territory as is necessary to keep these mills going throughout the year, and at sufficiently high capacity for economical operation.[35]

A year earlier, he had written that the establishment of pulp and paper mills would foster the expansion of the sawmilling industry:

It is not believed that the spruce lumber industry can be expanded very materially over its present proportion until extensive pulpwood logging operations are started. The many large isolated spruce trees growing with the smaller pulp timber will then be brought to the sawlog market.[36]

By 1927, however, Heintzleman had concluded that lumber from Southeast Alaska had little future except to supply local markets:

It is unlikely that an extensive lumber industry in Alaska could compete successfully with the Pacific Northwest, the greatest lumber-producing region in the world. Our lumber would have to be shipped through or past this producing center under the handicap of a much longer haul. Also, our trees on the average are smaller than those of the forests farther south and will not produce as high-grade lumber.[37]

The future timber needs of sawmills would be met, according to Heintzleman, through the designation of extensive blocks of timber in which no large, long-term pulptimber sales would be made. This never became a regional issue because some pulptimber offerings were never sold. Heintzleman also noted that paper, being a more valuable and less bulky product, could stand a longer freight haul than lumber. He concluded that the development of an extensive sawmill industry in Southeast Alaska was inadvisable.[38] Rather, the conditions in Southeast Alaska dictated that the greatest value of the region's forests was for the manufacture of pulp and paper.

Frank Heintzleman became regional forester in 1937 and continued his effort to develop a newsprint paper industry in Southeast Alaska (see Chapter 8). For Heintzleman, sawmills were desirable—provided they did not hinder in any way the possible development of a pulp and paper industry.

In August 1937, Congress asked President Franklin Roosevelt to prepare a plan for the development of resources and commerce in Alaska.[39] The Alaska Resources Committee—of which Frank Heintzleman was a member—was organized and issued its report the following December. Regarding the lumbering industry, the committee concluded:

> An extensive sawmill development primarily for entering the general markets is considered inadvisable. The pure stands of high grade spruce saw timber are too limited to support a large industry, and it is unlikely that the mill-run of lumber from the predominating hemlock-spruce forests can compete with the material of the same species produced in southern British Columbia, Washington, and Oregon...
>
> The sawmill capacity in Alaska should be gaged to the local demand, and if this is done the supply of high-grade sawtimber is sufficient to maintain a thriving lumber industry. The common lumber can be sold locally, and the clears produced from this select timber will stand the shipping charge to the general markets outside the Territory.[40]

Beginning in the mid-1940s, the Forest Service moved beyond a tacit recognition that sawtimber was a secondary resource and began taking overt steps to constrain the size of the Alaska sawmill industry, as is evident in two major proposals that could have increased the demand for sawtimber. The first involved the Alaska-Asiatic Lumber Mills at Wrangell, which in 1946 was beginning the rehabilitation and enlargement of what had long been known as the "Wrangell Sawmill." The company voiced its intention to more than quadruple the mill's capacity from 35 MBF per day to 150 MBF (about 45 MMBF annually, based on a work year of 300 days), which would have made it by far the largest sawmill in the region. Heintzleman penned a memorandum:

> We have discouraged large sawmills in Alaska as the quality of the timber stands make the material primarily valuable for pulp, and a large sawmill output without any pulp mill capacity would mean an appalling waste in that region; however, that we recognized an obligation to the existing towns in S.E. Alaska, which are partially dependent on the lumber industry, and consequently we are agreeable to having established in the towns, sawmills of such size that they can operate efficiently for most of the year, and arrange for economical transportation of logs to the mill and of lumber to markets; that we considered this size to be around 90 M per day—the capacity of the Ketchikan Spruce Mills.[41]

Alaska-Asiatic Lumber Mills agreed to limit the mill's capacity to 100 MBF per day. However, the increase in capacity did not happen, and the mill was sold in 1949.

The second example comes from 1947, when Milton Daly, owner of the Ketchikan Spruce Mills, the largest sawmill operation in the region, was concerned about losing his supply of sawtimber to pulp operations. He wrote to Heintzleman:

> In view of the increasing interest in pulp timber and the designation of large areas of the Tongass National forest for pulp development, we would appreciate your favorable consideration of an assignment of timber for the future supply of our sawmill operation. While we have not in the past carried timber sales beyond immediate needs, excessive segregation of large areas for pulp sales will restrict the future saw log supply of the existing sawmill industry to the point of acute shortage in relatively few years.
>
> In considering the future operation of our plant we find it essential that we have an assured timber supply to plan and finance the operation... A fifty-year supply of

approximately one and three quarter billion feet would be commensurate with our needs.

In making this request and indicating my belief that a future saw log shortage will exist if excessively large pulp areas are set aside in the Tongass National Forest, I do not wish it to be construed that I am not in accord with a pulp program for the Forest. To the contrary, I believe that a moderate pulp development will be beneficial to our industry and all concerned.[42]

According to records kept by Ketchikan Spruce Mills, Heintzleman never answered the letter. Finally, in March 1954—seven years later—former Chief Greeley's son Arthur, then Alaska's regional forester, wrote to Daly to explain that the Forest Service was "trying to develop an industry whose requirements by log grades are in balance with the supply of logs by grade as they exist within the Tongass National Forest." He added, "We will not, in fact we cannot, set up a long-term sale which consists only of pockets of high-quality material here, there and elsewhere with no requirement for taking out the lower grades, or the intervening timber stands which contain primarily low grades."[43]

Several weeks later, Chief Richard McArdle also wrote to Daly:

It would be contrary to Forest Service policies to make an offering of 1.5 billion board feet of primarily spruce timber on a 50-year contract basis.

The policy of the Forest Service is to make relatively short-term timber offerings to give bidding opportunities for established mills which have gone through their initial depreciation period.[44]

In 1956, Daly hired C. M. Archbold, who had retired from the Forest Service as supervisor of the southern area of the Tongass National Forest. One of Archbold's duties at Ketchikan Spruce Mills was to work toward obtaining a long-term sawtimber allotment, but he never secured one.[45] In the eyes of Frank Heintzleman and the Forest Service, the sawmill industry was clearly a poor cousin to pulp and paper—an industry that would pave the way for the practice of forestry through the broad utilization of the timber resources of the Tongass National Forest.

CHAPTER 6

Sawtimber Finds Its Markets

THROUGHOUT THE EARLY TWENTIETH CENTURY, HIGH TRANSPOR-
TATION COSTS FROM SOUTHEAST ALASKA TO THE MAJOR U.S.
MARKETS, AND THE READY AVAILABILITY OF TIMBER IN WASHINGTON
AND OREGON, PRECLUDED THE DEVELOPMENT OF A SUBSTANTIAL EXPORT TRADE
IN ANYTHING MORE THAN HIGH-GRADE SITKA SPRUCE LUMBER AND SOME
ALASKA-CEDAR LOGS.

The Alaska market for wood products was varied. An article about the
actual and potential uses of Alaska woods in a 1913 issue of the *Forestry
Quarterly* reported that of all the timber sawn at Southeast Alaska's mills in
1912, approximately 52 percent was construction material, 12 percent was
mine timbers, 30 percent salmon boxes, 4 percent halibut and herring boxes,
and 2 percent shingles.[1] The chronic problem facing Southeast Alaska
sawmills was what to do with the substantial amounts of common lumber
produced in their effort to get finer material for export as aero-grade lumber
(suitable for use in manufacturing components for airplanes) and for salmon
cases and fresh fish boxes. Southeast Alaska could absorb only a limited
amount of common lumber, and the mills' inability to sell this material
sometimes limited the amount of timber they were willing to purchase.[2]

At least one Southeast Alaska sawmill began exporting lumber prior to
World War I. In March 1910, the Willson & Sylvester sawmill at Wrangell sent
a trial shipment of high-grade Alaska-cedar to Puget Sound on a scheduled
steamer. Two months later, a second shipment was consigned to the Mare
Island Naval Shipyard in California.[3] The following year, Wrangell's newspaper
wrote that Alaska-cedar "is much sought after and mills pay an exceptionally
high price for logs." The paper noted, however, that "this timber is generally
a little way from salt water and it costs more to log, takes a bigger plant and
so no systematic effort has been made to secure this valuable wood." The
paper also saw opportunity: "Someone will come spying out the land and see
this neglected opportunity and make several fortunes out of the chance."[4]

Henry Graves, then chief of the Forest Service, wrote in 1916 of "a number
of important shipments" of lumber exported from Southeast Alaska in 1913.

Probably the largest of those shipments was 1,200,000 board feet of lumber transported to San Francisco from the sawmill at Hadley (on Prince of Wales Island) aboard the steamer *Melville Dollar*—the first full cargo of lumber exported from the Tongass National Forest. Another shipment consisted of approximately 800 MBF of "first class clear spruce lumber." An early shipment of aero-grade Sitka spruce to the lower 48 states occurred in 1914, when the Willson & Sylvester sawmill exported a load of Sitka spruce boards 36 to 48 inches wide. In 1920, the *Ketchikan Alaska Chronicle* reported that the Juneau Lumber Mills had secured a contract for three million board feet of Sitka spruce lumber to be shipped to the Lower 48. The freight rate for lumber between Southeast Alaska and Puget Sound ca. 1920 was $7 to $10 per MBF.[5]

The Panama Canal opened in 1914, but it was not until 1926 that lumbermen in Southeast Alaska began taking advantage of this comparatively economical route to East Coast markets. That year, a cargo of lumber totaling 2.2 MMBF was loaded at Wrangell and Ketchikan and transported to New York City on the steamship *Commercial Guide*, which was also carrying 2.5 MMBF of Douglas-fir from Washington State.[6]

Like the Panama Canal, the Grand Trunk Pacific Railway was completed in 1914, and its terminus at Prince Rupert, British Columbia, was only 90 miles from Ketchikan. As early as 1917, aero-grade Sitka spruce from Alaska was being shipped to eastern markets via this route, and this mode of transport was periodically an important option for getting the products from Southeast Alaska's sawmills to market. Frank Heintzleman reported that more than 2 MMBF of clear Sitka spruce lumber was exported from Southeast Alaska in 1924.[7]

The growing industry did not escape the notice of Alaska's politicians. The territory's legislature recognized a tax on lumber and shingles as a potential source of revenue for its coffers, and in 1921 it levied a modest tax on lumber mills and shingle mills of $0.10 per MBF for lumber and $0.05 per thousand for shingles. Four years later, the tax on shingles was eliminated. In 1949, Alaska's legislature adopted a new system of taxation that eliminated the taxes such as that on lumber in favor of a general net income tax applicable to all Alaska businesses.[8]

BARRELS AND SALMON CASES

Much of the increase in locally produced lumber in the early 1900s was in response to the demands of Alaska's rapidly growing fishing industry.

Sawmills were built or expanded in large part to meet the canneries' needs for salmon cases and construction lumber. As well, timber operators were often contracted to supply logs and piles for the construction of salmon traps. The well-being of the logging and sawmill industries became very closely tied to that of the fishing industry. "The fishing industry absorbs most of the timber taken from the forest and hence the condition of the above industry has a marked bearing upon the annual cut of timber upon the forest," remarked one Forest Service employee.[9] And although development of the lumber industry had been hampered by the distance from large markets, the *Pacific Fisherman* reported in 1925 that commercial fishing interests "logically look to this source of supply for their requirements, with the result that a thriving business has been built up, supporting a number of first-class mills and logging outfits."[10]

Alaska salmon were processed for the commercial market in three ways: salting, canning, and freezing. Salteries generally preceded canneries and packed their product—mostly salmon and herring—in barrels. It is likely the Russian-American Company's barrel manufacturing machinery was utilized after the transfer of Alaska to the United States. In Ketchikan, Henry Imhoff began manufacturing barrels in 1879, and continued to do so until his death in 1928. By 1907, twelve salteries operated in Southeast Alaska. Stand-alone salteries, however, faded out as salting operations were incorporated into canneries and freezer plants.

Large, red-meated king salmon were normally "split" (filleted) and "mild-cured" (lightly salted) and packed in large "tierces" that held about 800 pounds of fish. The trade in mild-cured king salmon continued in Southeast Alaska until the late 1960s. Coho, chum, and pink salmon for the export trade, on the other hand, were heavily salted in barrels that usually held 200 pounds.[11] The Forest Service's Ketchikan district ranger, J. M. Wyckoff, reported in 1930 that in the early years of Southeast Alaska's salmon business,

> Practically all of the barrels were made locally, no central cooperage plant being established but each saltery or company having its own selected spruce secured adjacent to the saltery. The cooper would select certain trees, cut a fair-sized block out of each above the heavy root swell to determine the suitability for this particular use and if found good in grain and texture they were blocked out in the

woods so that they could be transported to the beach by hand.

By 1930, however, this was no longer the case—all cooperage used in Southeast Alaska was imported.[12]

The first salmon cannery in Alaska was built at Klawock in 1878 by the North Pacific Trading & Packing Company, of San Francisco, which at the same time also built the first post–Russian era sawmill in Southeast Alaska— a 15-MBF-per-day steam-powered facility that was used for making salmon cases and supplying local needs. In 1897, nine canneries operated in Southeast Alaska; by 1913, there were 42. The number peaked at 80 in 1920, followed by decline due to attrition and consolidation.[13] Despite its seasonality and dramatic fluctuations in the catch, salmon canning became the Southeast Alaska's principal industry and eventually gained economic and political control of the region. The industry was largely owned by interests in Seattle and San Francisco.

Prior to the widespread use of cardboard cases, which began in the 1920s, canned salmon were packed in one-time-use wooden cases. Until approximately the turn of the century most of the salmon cases used in Southeast Alaska were manufactured in Puget Sound. As Alaska's salmon canning industry grew and more sawmills became established, the manufacture of salmon cases became a major component of the wood products industry in Southeast Alaska. During 1912–1913, about two-thirds of the salmon boxes used in Southeast Alaska were manufactured in the region. In 1910, the Forest Service's Royal Kellogg estimated that more than one-third of the region's cut of timber was used for salmon cases; he added that "much of the best lumber goes into them." *The Timberman* in 1911 stated that 60 to 70 percent of the lumber cut in Southeast Alaska was used for salmon cases. According to a 1930 *Timberman* article about Southeast Alaska's sawmills, approximately 50 percent of the lumber at Ketchikan, Wrangell, and Juneau was manufactured into salmon cases and fresh fish boxes.[14]

Industry standards in 1919 specified that cases be made of sound, well-seasoned lumber. Knotholes, and loose or rotten knots, could be no greater than one inch in diameter. Alaska-made cases were considered superior to the imported variety because of the high-quality Sitka spruce used in their manufacture. Material for the manufacture of salmon cases was planed, and most was kiln dried. In 1925, all Alaska-made tops were reported to have been made of clear, full-width material.[15]

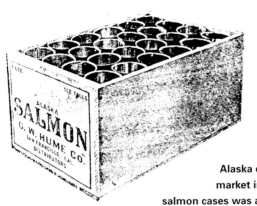

FIGURE 19. Until about 1930, Alaska canned salmon were shipped to market in wooden cases. Manufacturing salmon cases was a primary product for Southeast Alaska's larger sawmills. (left, *Pacific Fisherman*, January 1918; right, *Pacific Fisherman*, January 1921)

Though the cases varied slightly, the following dimensions of material required for a salmon case holding 48 one-pound cans are representative: ends, $7/8$ by $9\frac{1}{2}$ by $12\frac{1}{8}$ inches; sides, $7/16$ by $9\frac{1}{2}$ by $19\frac{7}{8}$ inches; and top and bottom, $7/16$ by 13 by $19\frac{7}{8}$ inches.

Each case required about five or six board feet of rough lumber, and the usual price received at the sawmills, circa 1910, was about ten cents— considered to be equivalent to about $20 per thousand board feet. The annual pack of salmon in Alaska ca. 1910 was about two million cases. Most of the cases were supplied by local mills, and manufacturing them would have required perhaps 10 MMBF of lumber. During the 1920s, about five million salmon cases were used in Alaska each year. The Juneau Lumber Mills in 1927 was a major salmon case manufacturer, and its box factory was reported to be capable of producing 6,000 salmon boxes per eight-hour shift. The mill had storage for 250,000 cases.[16]

Boxes from southeastern Alaska mills were also used for species other than salmon. About 1900, a market developed for large wooden boxes for shipping iced halibut and other fresh fish by steamer to Seattle. The fresh halibut fishery developed at Petersburg, where during the season October 1899 through March 1900, some 5,500 halibut boxes were utilized to ship about 2,750,000 pounds of fresh halibut.[17] The dimensions of the common rough lumber used in construction of a halibut box were: ends, 1 by 16 by 30 inches; sides, 1 by 16 by 54 inches; and top and bottom, 1 by 32 by 54 inches.

Each halibut box held about 500 pounds of fish and required about 45 board feet of lumber. The shook price for halibut boxes in 1910 was $0.65 to $0.70 each, depending upon the quantity ordered.[A] In 1921, a roughly average year for halibut fishing, more than 10,000 halibut boxes were used in shipping,

FIGURE 20. Longshoremen loading cases of salmon aboard freighter. During the 1920s, the average annual Alaska salmon pack was about 4.5 million cases. (San Francisco National Maritime Historical Park)

by steamship to Seattle, some 5.25 million pounds of halibut caught by the Puget Sound halibut fleet fishing in Alaska.[18] The manufacture of 10,000 halibut boxes required nearly a half-million board feet of lumber.

Fresh salmon were packed in ice in wooden boxes holding about 450 pounds. The fresh salmon fishery began at Ketchikan. From January through May of 1905, dealers in Ketchikan shipped some 275,000 pounds of fresh king salmon to Puget Sound ports. The industry quickly expanded, and the following year dealers throughout the region shipped 576,000 pounds.[19]

Frozen fish were also packed in wooden boxes. Petersburg Cold Storage, one of the principal freezer plants in the region, used as many as 20,000

[A] "Shook" or "shooks" is the term used for the unassembled lumber components of wooden boxes (cases) such as were used to pack canned salmon and fresh fish.

wooden boxes a year and continued to pack frozen fish in wooden boxes into the 1960s. For frozen halibut, boxes were sized at 100-pound increments to hold from 100 to 500 pounds of product. Frozen salmon for shipment to Japan and Europe were packed in 100-pound-capacity wooden boxes.[20]

In 1925, the Wrangell Lumber & Power Company reported the manufacture of 691,000 salmon boxes and 14,000 fresh fish boxes, and managers expected to manufacture 800,000 boxes the following year. Supplying the canneries with salmon boxes was so important to mills that in 1916, mill operators in Ketchikan and Wrangell used high-quality Sitka spruce to complete orders for salmon boxes rather than export it to Europe as high-value aero-grade material. In 1925, the salmon canning industry stated that the cost of box shooks represented five percent of the cost of a market-ready case of salmon at its distribution point. Much to the detriment of Southeast Alaska's sawmills, the use of wooden cases for packing cans of salmon gradually came to an end with the introduction of cheaper fiber (cardboard) boxes favored by the brokers, wholesalers, and retailers of canned salmon. Boxes of "solid wood-fibre board" designed specifically for canned salmon became available from outside suppliers by 1918 and presented stiff competition by the late 1920s. The Canning Industry of Alaska complained about availability of wooden boxes and placed a paid statement in the *Ketchikan Alaska Chronicle* in 1925: "There is not a single canneryman who would not be more than willing to buy box shooks and lumber manufactured from Alaskan spruce in Alaska's mills if they were obtainable, but the capacity of Alaska's mills is taxed to produce 10% of the annual consumption." In 1930, Southeast Alaska's sawmills had the capability to manufacture 2.5 million salmon cases annually, but only 1.1 million were produced. Anticipated production for the following year was 500,000. The trend was ominous. To encourage the use of wooden cases manufactured by Alaska sawmills, legislation was introduced in the 1931 Alaska territorial legislature that offered a fish tax rebate of five cents per case to any operator who packed his salmon or clam cans in wooden containers made within the Territory. The bill failed. *The Timberman* reported in 1934 that competition from fiber cases was "being severely felt" at Wrangell's main sawmill, the Wrangell Lumber & Box Company.[21]

Each type of case had its advocates. "All of our California canned foods and salmon from the Pacific Coast come to us in nearly perfect condition because they are packed in fine, strong wood cases, which it is a pleasure to handle and which are sightly and safe," wrote John Lee, secretary of the Western Canners Association, in 1927. Fiberboard interests, not surprisingly,

FIGURE 21. Unloading
salmon from a pile trap.
(NARA-Anchorage, RG 95)

found fault in wooden cases: "The loss of canned goods when boats are being loaded for shipments at Alaskan ports is oftentimes a serious matter," wrote I. Lubersky of Fiberboard Products, Inc., in 1929. "Wooden cases of salmon will be dropped from the hoists, the boxes bursting and the cans rolling on deck where they are sometimes washed away."[22]

By about 1935, cardboard reigned supreme for packing canned salmon in Alaska.

FISH TRAPS

Another major use of timber was in the construction of fish traps used to catch and impound live salmon. Most of Alaska's fish traps were located in Southeast Alaska and were cannery-owned. There were two types of fish traps—pile traps and floating traps—and each used a lot of wood, with the larger traps requiring on the order of 75 to 125 MBF.[23]

Pile traps evolved from hand-driven pole traps and were first used in Alaska in about 1885. A pile trap basically consisted of galvanized iron wire netting secured to piles driven into the sea bottom in a design that channeled salmon migrating along the shore into one or two funnels ("hearts") that led to a holding area.[24] As needed, the impounded salmon were loaded onto scows and vessels for transport to canneries.

A pile trap could be more than 2,000 feet long, with piles driven 10 to 15 feet apart. Piles were held together above the water by a continuous line of wooden stringers, often full-dimension 2 by 12s, which were also used to fasten the netting and to walk on. Western hemlock was the preferred piling material in fish traps (as well as docks and wharves) because it has little taper, is heavier than Sitka spruce, and was thought to be less susceptible to damage from the worm-like marine borers known as teredos (see Glossary) than Sitka spruce. Trap owners needed long piles because they generally preferred to construct their traps as far out from the beach as possible; long piles, however, commanded a premium price because they were harder to obtain and more difficult to transport than shorter piles. The minimum length of piles used for the construction of salmon traps was about 85 feet. Many were between 100 and 130 feet long. Some of the better stands of western hemlock yielded trap piles that averaged 100 linear feet.[25]

To facilitate driving, the big end of a pile was "sniped"—beveled for a length of about four feet—with an axe. Piles cut in the winter were preferred because the bark tended to hold fast and, so it was thought, prevent for a time the ravages of the teredos. Additionally, canneries began driving trap piles in early spring and wanted their supply to be on hand at that time. This fostered employment during the winter, when few other opportunities were present in Southeast Alaska. Piles were sometimes cut by loggers wearing snowshoes or skis.[26] At the end of the fishing season piles were usually pulled, and those that were still sound were put into storage for use the following year. Piles with damaged ends were shortened accordingly and utilized where their length would suffice.

The approved Tongass National Forest stumpage prices for piling in 1912 were ½ cent per linear foot for piles up to 50 feet long; ¾ cent per linear foot for piles between 50 and 80 feet long; and 1 cent per linear foot for piles 80 feet long and over. In the 1920s, minimum stumpage prices for piling were ¾ cent per linear foot for piles up to 75 feet long; 1 cent per linear foot for piles between 75 and 95 feet long; and 1½ cents per linear foot for piles 95 feet long and over. The cost of logging piling (stump to log boom) ca. 1912 was about four or five cents per linear foot. The market price for piling at that time was about seven cents per linear foot. The daily wages received at piling camps

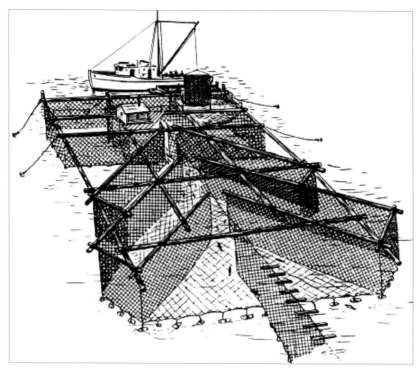

FIGURE 22. Above: In a floating trap, webbing was suspended beneath large Sitka spruce logs. (U.S. Bureau of Commercial Fisheries, *Alaska Fishery and Fur-seal Industries*, 1955) Below: In a pile trap, webbing was suspended between piles driven into the sea floor. (*Pacific Fisherman*, 1903 Annual Number)

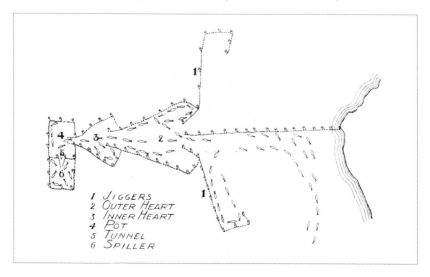

in 1912 were faller, $4; donkey man, $4; and common laborer, $3.50—all less $1 per day for board. The cook was paid $75 per month, plus board.[27]

In about 1930, the base value of cut and sniped piles at camps was 10 to 11 cents per linear foot, with a premium of 10 cents per linear foot for every 10 feet over 100 feet. At a base price of 10 cents per linear foot, a 100-foot pile would have a value of $10; a 110-foot pile, $12; a 120-foot pile, $15; and a 130-foot pile, $19.[28]

The cut of hemlock used for piling in Southeast Alaska for the year ending June 30, 1915, totaled about 2,250,000 linear feet, equivalent to about 13.5 MMBF.[B] District Ranger J. M. Wyckoff wrote in 1923, "The logging of hemlock trap piling to supply the many fish traps driven in southeastern Alaska is an industry almost equal to that of logging sawtimber." Wyckoff gave the approximate annual production of piling from Tongass National Forest at that time as being nearly three million linear feet (about 18 MMBF), while the production of sawtimber was about 23.5 MMBF. Wyckoff did not estimate the amount of timber cut for logs for floating traps. Five years later, in 1928, Frank Heintzleman wrote that about 20 percent of the timber cut on the Tongass was for piling. The apparent reduction in the amount of piling cut reflected a great increase in the amount of timber cut in sawmills for other purposes during the mid-1920s. Some of the piles used in Southeast Alaska's fish traps, especially those of exceptional length, were imported from Canada and the Puget Sound region.[29]

The floating trap was said to have been invented in Ketchikan and was first used in 1907. The principle of operation for a floating trap was the same as for a pile trap. The fundamental design was also similar to that of a pile trap, except that the galvanized iron wire netting was hung from a solid frame of huge logs that were bolted, bound, and braced together. Floating traps were kept in position by cables secured to trees or rocks on the shore side, and by massive anchors on the seaward side. The "floaters" were less costly to build and install than pile traps, could be used where deep water or a rocky bottom precluded driving piles, and could be towed from one location to another.[C] They soon came to dominate the salmon fishery in Southeast Alaska. In 1927, a total of 575 salmon traps were in use in Southeast Alaska. Of these, 474 (82 percent) were floating traps, and 101 (18 percent) were pile traps.[30]

[B] One linear foot of piling was usually considered to represent six board feet, log scale.
[C] In 1925, the average cost of a floating trap was about $6,000. Pile traps were far more expensive.

FIGURE 23. Piling "cold decked" beneath a spar tree. From this location, a float-mounted A-frame would likely yard the piling into saltwater for rafting and transport. (USDA Forest Service, Alaska Regional Office, Juneau)

However, because floating traps were not as rugged as pile traps, their use was limited to relatively protected sites.

Because of their buoyancy, large Sitka spruce logs were the preferred material for the construction of floating traps. Trap logs were usually 80 to 140 feet long with at least a 16-inch top, though the use of logs measuring

170 feet long was not uncommon; such logs were five feet at the butt and tapered to a 16-inch top. Logs sold for trap construction brought a much higher price than if sold to sawmills, and loggers were therefore careful to haul out full tree lengths whenever possible. "The average spruce tree selected by the logger for sale to fish trap builders must be of large diameter and exceptional length," wrote Wyckoff in 1940. "However, characteristics such as rapid growth, pitch pockets, spiral grain and knots do not detract from the value of the piece for a trap log."[31]

The Forest Service scaled trap logs as sawlogs and charged the same for stumpage. In 1929, approximately 4 MMBF of Sitka spruce logs was used for fish trap construction. In 1930, the market price—at logging camps—for selected Sitka spruce suitable for fish trap construction was $17.50 per thousand board feet, while similar material to be used for sawlogs brought about $10. Trap owners in 1936 said that the value of trap logs annually purchased was more than $100,000.[32]

Fish traps, except those located at the Metlakatla Indian Reservation, were outlawed by federal regulators and Alaska's legislature in 1959.[33] The ban was partly a conservation measure and partly to foster the employment of Alaska fishermen.

LUMBER FOR "AERO-PLANES"

A new, important use for Southeast Alaska's Sitka spruce became apparent in World War I. A 1917 issue of *The Timberman* sang its praises:

> Sitka spruce is recognized as the most valuable species of wood in the known world for the manufacture of airplanes. The globe has been searched for a substitute, but up to this time no material has been found to take its place, or even approach it…
>
> Here is a wood that the chiefs of aeronautic science of the world commandeer to make possible the building of their formidable air fleets. To spruce has been accorded the highest privilege for service ever conferred on any wood in the history of the world. It is the autocrat of timbers. Military genius bows before it. Its clean, milky color, fine texture and wonderful strength, combined with extreme lightness, has given it a place in a world's struggle never

enjoyed by any other wood since nations went to war. It bears its responsibilities with grace and ease. Its laurels are easily worn. It is an uncrowned king. Its use value can hardly be measured. Through the creation of great air squadrons the lasting peace of the world is hastened and democracy is made safe. It is a wood which Major R. Perfetti, the head of the Italian special military commission for aeronautics in the United States, says is consecrated by the Creator, to insure the liberty of the world and is the harbinger of peace and good will among mankind.[34]

But the requirements for Sitka spruce lumber to be graded as "aero-grade" or "airplane spruce" were so exacting that the amount of suitable material recovered even from good timber was small. The average airplane in the 1920s required about 800 board feet. Specifications for aero-grade Sitka spruce lumber were listed in the January 17, 1918, *Wrangell Sentinel* and included flat or vertical grain, a width of five inches or more, and a thickness of 15 inches, "though better prices are obtained for that eighteen inches wide and two feet thick."[35] Tongass National Forest Supervisor W. G. Weigle wrote in 1918,

> The very best spruce timber does not produce more than 25 per cent of the whole tree suitable for airplane construction, while the average stand will not yield more than from 5 per cent to 10 per cent and much of the timber will yield less than 5 per cent.

As well as noting that airplane spruce was required to have at least eight annual rings to the inch, he added,

> Owing to the fact that spruce trees do not clean themselves of branches as readily as most species, all spruce logs contain knots in the center, therefore little airplane lumber can be obtained from small logs, the clear lumber all being secured from the outside of large logs. It is also quite common to find large, fine looking spruce trees with spiral growth, which eliminates them for airplane lumber, as lumber wherein the grain varies from a straight line more than one inch in 20 inches cannot be used for airplane

construction. Pitch pockets, curly grain, pocket rot, and wind shake also frequently destroy the value of outwardly fine looking logs for airplane lumber.[36]

Clear Sitka spruce for shipment to European airplane manufactures was priced at Portland, Oregon, in 1914 at $32 per MBF. In June 1915, it fetched $40 per MBF.[37]

Alaska, however, apparently did not participate in the trade until 1917. In June of that year, a shipment of 10 MBF of "picked spruce lumber to be used in aeroplane construction" was shipped east from the Willson & Sylvester sawmill at Wrangell via the Grand Trunk Pacific Railroad at Prince Rupert. A lively industry developed in Wrangell, Ketchikan, and Juneau in the 1920s as buyers from around the world formed a ready market for the material that sawmills in Southeast Alaska had previously been manufacturing into salmon cases. In 1920, the U.S. Government was paying $160 per MBF at Seattle for wing beam stock, and by 1927 the market price for aero-grade Sitka spruce was in the $285 to $350 range, making it probably the highest-priced lumber then available. One trader in aero-grade Sitka spruce noted, "The price of this exceptional spruce is always a secondary consideration."[38]

On March 2, 1918, the Alaska Territorial Council of Defense, in Juneau, notified officials in Washington, D.C., that miners and prospectors acquainted with the country surrounding Lituya Bay had reported a "very large area of most excellent aeroplane spruce."[39] The council suggested the land surrounding Lituya Bay be temporarily withdrawn to ensure government access to the timber located there.

The government was not slow in acting. On May 2, 1918, President Woodrow Wilson ordered "the tract of land in Alaska extending ten miles back from the tide line around Lituya Bay, containing approximately 312 square miles [200,000 acres], be…temporarily withdrawn…for the purpose of supplying timber for use in the construction of aeroplanes for the United States."[40]

At the request of the U.S. Army, in June of that year Forest Supervisor W. G. Weigle and Frank Andrews, a timber cruiser from the General Land Office, spent 11 days examining the timber around Lituya Bay. They found what they described to be "one of the finest stands of spruce in Alaska," and estimated its volume at 50 MMBF. Some individual acres were thought to contain more than 100 MBF.[41] Of this timber, Weigle wrote many years later:

FIGURE 24. Sitka spruce at Lituya Bay was reserved in 1918 as a source of "aero-grade" spruce but was never logged. (USDA Forest Service, Alaska Regional Office, Juneau)

We found the timber about eleven miles up the shore from the bay, where we camped in an old abandoned cabin of a miner. We found it to be a very nice tract of Sitka spruce, covering three or four hundred acres but cut up into several small tracts by glacial streams. A large glacier back of it [probably Fairweather Glacier] had been disturbed, and millions of tons of ice had moved down into the edge of the timber. We took pictures of one of the men with his back against a block of ice 30 feet high and his feet against a five-foot spruce. It was a fine tract of timber, but would have been very expensive to log.[42]

Another problem with that timber was that it was situated outside the withdrawn area. The men recommended an additional withdrawal, and two months later, on August 16, 1918, President Wilson enlarged the previous withdrawal to include the tract.[43] Weigle and Andrews also suggested the possibility of locating a sawmill on the west side of Lituya Bay and constructing a narrow-gauge railroad along the beach to access and transport the timber.

Perhaps owing to its isolation or the ending of World War I, the timber at Lituya Bay was never logged. In 1939, the withdrawn area was included in the President Franklin Roosevelt's proclamation that expanded Glacier Bay National Monument, but the two 1918 executive orders were not revoked until 1949. Frank Heintzleman considered the Sitka spruce forest at Lituya Bay to be "the finest example of coast forest in Alaska."[44]

In the fall of 1923, the barge *Nottingham* loaded 500 MBF of clear Sitka spruce at Juneau and a similar amount at Ketchikan. The material was bound for London, Liverpool, and Glasgow, with the sale handled by F. G. Brynolson's Northwest Spruce Company, located in Seattle. Brynolson became the Seattle agent for the Willson & Sylvester Mill Company and the Ketchikan Spruce Mills in 1922, and the Juneau Lumber Mills the following year. Northwest Spruce Company specialized exclusively in Sitka spruce, and its yard had a fireproof drying shed 350 feet long by 65 feet wide, with a storage capacity of 10 million board feet. The firm sold high-grade Sitka spruce lumber to concerns in the United States, as well as in Great Britain, France, Germany, Italy, Australia, Japan, and China. In 1923, Great Britain alone purchased more than 2 million board feet of clear Sitka spruce. That Great Britain purchased such a large volume of aero-grade lumber was not surprising, given that the country accounted for about 80 percent of global airplane production during this period.[45]

Minor markets for clear Sitka spruce included sound boards for musical instruments; a reference to this use of Sitka spruce from Southeast Alaska was made in 1920 by Governor Thomas Riggs in his annual report to the Secretary of the Interior.[46]

The economics of selling clear (but not aero-grade) Sitka spruce in Seattle was simple. The ca. 1921 price in Seattle for clear Sitka spruce from Oregon and Washington was about $68 per MBF, while similar, Alaska-grown material was selling in Southeast Alaska for $40 per MBF. Freight on scheduled steamships from Southeast Alaska to Seattle was about $10 per MBF, which put the delivered cost of Southeast Alaska's product at about $50 per MBF— some $18 cheaper than that produced in Oregon and Washington. In addition to shipments to Seattle, high-quality Sitka spruce lumber was shipped to Prince Rupert and then by rail to eastern markets. One such shipment, from Wrangell in 1922, totaled 450,000 board feet.[47]

The economics of construction grade Sitka spruce, however, offered a stark contrast to that of the higher grade timber, because selling this product, even locally, posed a problem due to competition from Puget Sound mills. In 1943, the editors of the [Ketchikan] *Alaska Fishing News* wrote that "so far

FIGURE 25. Clear Sitka spruce lumber totalling 400,000 board feet sits on a barge at Juneau Lumber Mills. The wood was destined for Seattle from which it would be sold on world-wide markets, especially in Great Britain. Log rafts in background are raw material for the mill. (USDA Forest Service, Alaska Regional Office, Juneau)

mills shipping from Seattle have been able to deliver lumber cheaper at Annette Island [15 water miles from Ketchikan], for instance, than could be delivered from the Ketchikan mill."[48]

DEMAND FROM DOWN UNDER

One market that in the early 1920s absorbed a considerable amount of Sitka spruce from Ketchikan and Juneau, as well as some from Wrangell, was Australia, where Alaska wood competed against New Zealand white pine (*Dacrycarpus dacrydioides*). The material was cut to Australian specifications for resawing in that country, with common dimensions being 3 inches by 6 inches and wider, 4 inches by 6 inches and wider, 6 inches by 12 inches and wider, and so on. Transportation was a challenge, but one very resourceful buyer of some 10 million board feet employed W. L. Comyn & Company of San Francisco to transport the lumber on wooden sailing ships, each of which could carry about 2 million board feet. W. L. Comyn & Company's Seattle office manager described the challenge:

> It seems easy enough to ship lumber from Alaska to
> Australia, if one does not consider the rather serious

problem of return cargo. We have tried to make both ends meet in the following manner; Our boats take Alaskan spruce to Adelaide, Melbourne or Sydney. In Australia they take on coal for the west coast of South America. In Chile they load nitrates for Honolulu, San Francisco or Tacoma, and in this way a load is assured on each leg of a spherical triangle, at one point of which is the fine spruce of southeastern Alaska. It is a long way around, sometimes requiring a year for completion of the triangular route.[49]

The effort was complicated by the inability of the sailing ships to efficiently navigate the Inside Passage and the amount of time it took the relatively small mills in Southeast Alaska to cut and load lumber. The solution was to man the ships with a skeleton crew and tow them from Seattle to the port of loading. Since the mills maintained little inventory and the maximum cutting capacity of the largest at that time was about 80,000 board feet per day, it took a month or more to load a ship. When loading was completed a full crew was sent north, and the ship was towed to open water where its sails were unfurled for the long voyage to Australia.[50]

One problem with this slow mode of transport is that the lumber being shipped was "weather cured"—stacked unprotected in the yards prior to shipping. The dampness of the wood combined with the heat of the tropics during the three-month (or longer) voyage was conducive to the growth of mold, rendering some of the material stained and unacceptable upon inspection in Australia. Moldy wood was less of a problem for steam-powered ships, which could make the voyage in a month.[51]

At least five vessels all of which were sailing ships operated by W.L. Comyn and Company, participated in this transport:

- *Katherine Makall*, probably 1922 or early 1923, approximately 1.8 MMBF;
- *Anne Comyn*, summer 1922: loaded 1.8 MMBF Sitka spruce lumber sawn at the Ketchikan Power Company;
- *Alicia Haviside*, winter 1922–23: arrived with 800 MBF cargo of Douglas-fir (for Australia), loaded approximately 1 MMBF Sitka spruce lumber sawn at Ketchikan Power Company;
- *Phyllis Comyn*, summer 1923: loaded Sitka spruce lumber sawn at the Juneau Lumber Mills; and,
- *Russell Haviside*, fall 1923: loaded lumber at Juneau.[52]

FIGURE 26. Australia-bound Sitka spruce timbers being loaded aboard sailing ship at Juneau Lumber Mills, circa 1923. (USDA Forest Service, Alaska Regional Office, Juneau)

The demand for lumber of all kinds, in both regional and export markets, ebbed with the onset of the Great Depression.

THE TOLL OF "HIGH GRADING"

Despite the Alaska sawmill industry's modest capacity, decades of cutting substantially reduced the availability of high-grade Sitka spruce timber. The effect of hand logging and A-frame logging along Southeast Alaska's shorelines was noted in the late 1920s by private forester W. J. Frost, who wrote that such "beach combing" had been practiced for years, "and in almost any bay or good booming and rafting grounds we find that most of the handy [high-grade] spruce has been removed."[53]

Just prior to World War II, truck logging was initiated on the Tongass National Forest to access timber situated too far inland to be practical to log directly into the water with high-lead operations. Roads were built to the best timber, and by the middle of the century the region's remaining stands of high-grade timber that could be operated as units had been virtually eliminated. In 1954, Regional Forester Arthur Greeley noted the intensity of the high-grading that had taken place over the years on the Tongass:

FIGURE 27. The original photo caption read: "Good logging in big spruce."
(NARA-Anchorage, RG 95)

All the while sawmills operated largely on Sitka spruce,
supplying part of the Territory's needs and working away at
the relatively small portion of the Tongass National Forest
that bears a high percentage of good quality spruce and not
much hemlock. Thus something like 10 percent of the forest
bore the brunt of the demand for logs.[54]

In a 1956 article titled "Alaska Fosters Green Gold Rush," Greeley observed
that everyone wanted "the cream" of the territory's timber, but few were
interested in "the skim milk alone." A sound forest industry would survive in
Southeast Alaska "only if it can make a go of it on the skim milk. We do not
now need more users of cream."[55]

As late as 1957, Territorial Governor Mike Stepovich echoed a similar
theme:

> The Forest Service continues to receive many inquiries from
> Stateside people who are interested in developing new forest
> industries in the Alaska Panhandle. Most of the inquiries are
> stimulated by the fact that Alaska has an un-earned reputa-
> tion as the home of an inexhaustible supply of high-grade
> timber. The stands actually contain a high proportion of low-
> quality timber, fine for pulp but not so good for lumber or
> plywood. The better pockets of high-grade Sitka spruce have
> been combed over for 50 years and this class of timber is not
> plentiful.[56]

CHAPTER 7

Timber Policy Evolves

THE ORGANIC ADMINISTRATION ACT OF JUNE 4, 1897, STATED THAT "NO NATIONAL FOREST SHALL BE ESTABLISHED EXCEPT TO IMPROVE AND PROTECT THE FOREST WITHIN THE BOUNDARIES, OR FOR THE PURPOSE OF SECURING FAVORABLE CONDITIONS OF WATER FLOWS, AND TO FURNISH A CONTINUOUS SUPPLY OF TIMBER FOR THE USE AND NECESSITIES OF THE CITIZENS OF THE UNITED STATES..."[1] DURING THE NEXT SEVEN DECADES, MORE THAN ONE HUNDRED STATUTES SIGNIFICANTLY INFLUENCED THE PROTECTION AND MANAGE-MENT OF NATIONAL FORESTS, INCLUDING TIMBER POLICY AND PROTECTION. BUT THE ORGANIC ADMINISTRATION ACT REMAINED THE UNDERLYING LEGAL FRAME-WORK FOR U.S. FOREST SERVICE POLICY UNTIL 1976, WHEN CONGRESS PASSED THE NATIONAL FOREST MANAGEMENT ACT.[2]

Forest Supervisor William Langille was responsible for advertising and administering the earliest Tongass National Forest timber sales. Sale notices began appearing in local newspapers by 1909. The following notice in the *Wrangell Sentinel* was typical:

SALE OF TIMBER. Ketchikan, Alaska, March 10, 1911.
Sealed bids marked outside "Bid Timber Sale Application, January 4, 1911, Tongass," and addressed to the Forest Supervisor, Ketchikan, Alaska, will be received up to and including the 15th day of April, 1911, for all or any part of the live timber marked or otherwise designated for cutting by the Forest officer, and available for steam or hand logging on an area of 42 acres situated at the west shore of Zarembo Island and within the Tongass National Forest, Alaska, estimated to be 300,000 feet, B.M., of green spruce saw timber, log scale, more or less. No bid less than one ($1.00) dollar per thousand feet B.M. will be considered, and a deposit of $100.00 must be sent to W. A. Langille, Special Fiscal Agent, Ketchikan, Alaska, for each bid submitted to

the Supervisor. Timber upon valid claims is exempted from sale. The right to reject any and all bids is reserved. For further information and regulations governing sales, address the Forest Supervisor, Ketchikan, Alaska.[3]

Though advertised for public bid, timber sales on the Tongass were hardly competitive: a logger or mill owner typically would scout out a stand of timber to his liking and notify the Forest Service of his interest, after which the agency would prepare a sale that would be duly advertised. The prospective purchaser would then bid the minimum amount, and the timber would be his. One logger's application to the Forest Service—unusual in that it included money—read,

> I wish a permit to cut 1,000,000 ft. B. M. Saw Logs and 17,000 lin. ft. piling. This timber is to be cut along the shore of Tenakee Inlet, Chichagof Island, beginning on the same side of the Inlet that springs is located about 12 miles above the springs and extending along the shore for a distance of about six miles. I enclose $50.00 to cover cost of advertising, and would like a permit to commence cutting at once.[4]

Pursuant to regulations published in 1911, the willful cutting of national forest timber without first obtaining a timber sale was a criminal offense, punishable by a fine of not more than $500, or imprisonment for not more than one year, or both. The measure of damages for willful timber trespass was based on

> the value of the timber when and where found. If, when a willful trespass is discovered, the trees are felled, the assessed damage will be the stumpage plus the cost of felling; if they are cut into logs, the cost of bucking will be added, and if found at the mill the cost of both bucking and hauling will be added. The current value of the lumber will be the basis for assessing damages if the logs have passed through the mill.[5]

In cases of innocent trespass, the measure of damages was simply the value of the timber after it was cut at the place where it was cut. Enforcement appears to have been fairly lenient, however, as indicated in this 1919 letter from Harold Smith of the Forest Service to the Hydaburg Trading Company:

> You state in your letter that the logs have already been cut. This has been done without any formal agreement or permit and therefore constitutes a violation of the Forest Service regulations which must not be repeated in the future. I understand however that some sales have been handled in the past in this manner and for that reason I shall recommend that no action be taken against you in this case.[6]

In 1913, W. G. Weigle, Langille's successor as forest supervisor, reported on loggers' selective cutting practices:

> As is customary in all new countries [large areas of uncut forest], the best timber is always taken first. Spruce is practically the only timber logged at the present time for lumber, hemlock for piling, and cedar for shingles. The timber logged at the present time for lumber is large, running from 2 to 8 feet across the stump and 150 feet tall. It is quite common to scale 15,000 board feet in one tree. This class of timber, however, is quite limited and the time will soon come that smaller timber will be taken.
>
> Every mill man is afraid the other fellow will get some very fine timber that he is not able to use this year, and on account of this he wants to take only the best logs from the nearest and best trees, then run to another stand of timber and do the same thing.[7]

To assist potential buyers, the Forest Service cruised the region's timber and made the information obtained available to anyone interested. In 1924, for example, the Forest Service spent eight months surveying timber on the west coast of Prince of Wales Island.[8]

EXPORT RESTRICTIONS

With the establishment of the National Forest System, owners of timber on privately-held lands were greatly concerned that the Forest Service might flood the market with cheap timber and thereby reduce the value of their

FIGURE 28. Felling a 45,000-board-foot Sitka spruce near Natoma Point, Long Island, in 1941. (USDA Forest Service, Alaska Regional Office, Juneau)

holdings. The agency was forced to walk a fine line—to provide enough timber for development, but not so much as would depress prices.

Early restrictions on the export of logs and lumber from Alaska were born of a concern that material from the territory would end up in the Puget Sound area and further depress prices in a region that suffered from chronic over-production.[9] Under U.S. Forest Service administration, the export of lumber was permitted and encouraged, but the export of unprocessed logs was discouraged.

The federal legislation in 1898 that authorized the sale of timber from public lands in Alaska (see Chapter 2) also prohibited the export of logs and lumber from the territory.[10] A Skagway sawmill owner in 1902 found the restriction on lumber exports onerous:

> When we first put in our plant we expected to be able to ship the bulk of our product over the line into Canadian territory, not knowing the prohibitory law being in effect, and we have been waiting patiently ever since for the law to be changed at Washington, but so far nothing has been done.
>
> So you see we have nothing to depend on for a market except our city here and canneries in the immediate vicinity, and we can cut enough in 90 days to more than supply the present demand.[11]

The 1898 no-export rule was modified by Congress in the 1905 legislation that transferred the administration of the forest reserves from the Department of the Interior to the Department of Agriculture. Regarding Alaska, the legislation specified "that pulp wood or wood pulp manufactured from timber in the district of Alaska may be exported therefrom."[12] The legislation contained no provision to allow the export of sawlogs or lumber.

There seems to have been some confusion about the export law on the part of a federal official who was involved in administering it. In 1906, Assistant Forester F. E. Olmsted inspected the Alexander Archipelago Forest Reserve and reported, "It is not generally known that timber cut from the forest reserve may be exported from the District of Alaska," and "Even the mill men are in doubt about it and have refused to make foreign shipments." Olmsted characterized the local sentiment as "decidedly against" the export of timber, in the belief that local industries would need it all. He himself believed

> never a stick should be exported for which there is a local demand: on the other hand, never a stick should be allowed to go to waste if a market can be found for it anywhere, in case there is not a local demand for it.[13]

Olmsted's statement that implied lumber and timber other than pulptimber from the forest could be exported seems to have been a year premature.

The Appropriations Act of March 4, 1907—the act that changed the name of the forest reserves to national forests—also gave the Secretary of Agriculture discretion to "permit timber and other forest products cut or removed from the national forests of the United States, except the Black Hills National Forest in South Dakota, to be exported from the State, Territory, or the district of Alaska, in which said forests are respectively situated."[14] The Black Hills National Forest provision was rescinded in the Appropriations Act of March 4, 1913, and from 1914 through 1926, a provision affirming the nationwide export policy was included in agricultural appropriations bills.[15]

Through the Secretary of Agriculture, the Forest Service had the authority to limit the export of unprocessed logs from Alaska. In a memorandum to Agriculture Secretary William Jardine, Forest Service Chief William Greeley in early 1928 specified who in his agency would determine whether unprocessed logs from Alaska could be exported. Greeley wrote,

> Timber cut from the National Forests of Alaska shall not be exported from the Territory of Alaska in the form of logs, cordwood, bolts or other similar raw products necessitating primary manufacture elsewhere without the prior consent of: (1) the District forester when the timber sale project involved is within his authorization to sell, or, (2) the Forester when a larger timber sale project is involved.[16]

This regulation became known as the "primary manufacture rule." The circumstances under which the Forest Service might permit the export of unprocessed logs from the Tongass National Forest were outlined later that year by Assistant District Forester Frank Heintzleman:

> Sales of timber will not be made when it is anticipated that the wood will be exported from the Territory of Alaska in the form of logs, cordwood, or other raw products necessitating primary manufacture elsewhere. Export of raw material will, however, be allowed in individual cases where this will permit of a more complete utilization of material on areas being logged primarily for products for local manufacture; prevent serious deterioration of logs unsalable locally because of an unforeseen loss of markets;

permit salvage of timber damaged by wind, fire, insects, or disease; or bring into use a minor species of little importance to local industrial development.[17]

In 1931, Heintzleman defined lumber "manufacture" as, in the least, "breaking the logs down into flitches of dimensional sizes; i.e. cants, 6 or 8 inches thick, with the bark on the two edges."[18] The Forest Service eventually adopted a more restrictive definition that was challenged in 1958 by Alaska Lumber & Pulp Company. The company's sawmill at Wrangell had been cutting cants (flitches) for export to Japan. The *Wall Street Journal* described the episode:

> Under the regulations only 15% of total Japanese lumber shipments from the mill were allowed to be six inches thick or more; the rest had to be sawed up into smaller sizes thereby creating more work at the mill. The Japanese wanted to boost this allowance to 75%. During the dispute the Japanese temporarily shut down the Wrangell mill, threatening the jobs of from 40 to 60 men and a $20,000-a-month payroll. Considerable local pressure was brought on the Forest Service officials to work out a compromise, and a settlement finally was made under which 15% of total shipments sent to Japan can be larger than six inches; 28% just six inches thick, 28% five inches thick, with the rest in smaller sizes.[19]

Though there were periodic efforts to persuade the Forest Service to allow the wholesale export of unprocessed logs—one was by a small but particularly vociferous group of Ketchikan residents during the Depression—none were successful.[20] There are, however, numerous examples of the Forest Service granting specific, even broad, exemptions from the primary manufacture rule.

Alaska-cedar is a point in case. The species has a long history of being exported as unprocessed logs, mostly due to its popularity in Japan, where it is known as "American hiba." It is still used in internal house décor and for religious shrines and temples as a substitute for *hinoki* (*Chamaecyparis obtusa*), Japan's native cypress. The Japanese have an almost unparalleled appreciation for fine wood, especially when they can obtain it in log form and process it to their own standards. The export of Alaska-cedar logs to Asia began during the era of the Russian-American Company, and there are records of Alaska-cedar logs, generally cut into 13-foot (4-meter) lengths, being exported to Japan in the early 1900s.

Forest Service policy regarding the export of Alaska-cedar logs was discussed in a 1923 letter by Frank Heintzleman to Chief Robert Stuart:

> Much interest is now being shown in Alaska cedar by local mills and outside parties. We have been holding to the policy of granting no sales of cedar for foreign export but have construed the provisions of the Timber Sale Policy Statements to mean that sales can be granted in pure cedar stands where the output is primarily for the local mills but the small percentage of firstclass logs available is intended for export.[21]

Heintzleman thought that the growing interest in Alaska-cedar (as well as western redcedar) was due to the "increasing inaccessibility of the remaining cedar stands in Oregon, Washington and British Columbia." The first of several shipments of Alaska-cedar logs to Japan from Petersburg was reported in the spring of 1923, when 114 logs scaling 30,000 board feet were consigned to the Mitsubishi Company and shipped through Seattle.[22]

As of 1960, the Forest Service did not consider there to be a sawlog market in Alaska for Alaska-cedar or western redcedar (neither of which was used in the region's two pulp mills) other than what was needed for the manufacture of what Alaska's Regional Forester P. D. Hanson termed "minor volumes of specialty products." Alaska-cedar and western redcedar trees encountered in early pulpwood logging operations were felled but often left behind. To foster some use of the wood, the Forest Service permitted the broad export of unprocessed Alaska-cedar and western redcedar logs cut in the course of logging pulpwood. In 1959, at least two large rafts that together contained some 3.5 MMBF of Alaska-cedar and western redcedar logs were towed from Ketchikan to Puget Sound. Some of the logs were scheduled to be exported on the foreign market or used as "trading stock," while the remainder were to be cut at the Washington Timber Products mill in Everett. Most exported Alaska-cedar ultimately went to Japan. By about 1970, due almost entirely to Japanese demand, Alaska-cedar had become the most valuable timber in Southeast Alaska.[23]

An example of an emergency exemption for Sitka spruce from the primary manufacture rule occurred in the 1940s. There was at the time a heavy infestation of the spruce beetle on some 6,400 acres of prime Sitka spruce timber on Kosciusko Island. To salvage the dead and dying timber, and to hopefully stem the infestation, the Forest Service proposed "immediate logging in the shortest time possible to avoid as much loss as possible." Since regional mills did not have the capability to deal promptly with the timber, and there was no pulp mill to take the small material, the Forest Service was prepared to export any amount of logs necessary. In 1942, Sawyer, Reynolds & Company had an A-frame operation at Edna Bay (Kosciusko Island), and was exporting Sitka spruce logs to Puget Sound. The Forest Service tried to interest Ketchikan Spruce Mills in purchasing some of the insect-damaged timber, but the company was uncertain of the quality of the wood and declined to do so.[24] This timber was ultimately discounted and sold to the Alaska Plywood Corporation.

"The policy of the Forest Service in Alaska is definitely against the export of round logs," reported *The Timberman* in 1945. Alaska could, therefore, not be counted on as a source of raw material. But exemptions continued to be made, sometimes for experimental purposes. In 1947, something over a million board feet of pulp logs and 278 MBF of "peeler" logs (large-diameter, high-quality logs from which veneer for the manufacture of plywood is peeled) cut on Long Island were shipped to Puget Sound for experimental purposes. Additionally that year, a substantial shipment of western redcedar poles cut by handloggers in the vicinity of Wrangell was exempted from export restrictions and sent to Tacoma, Washington. The following year, the Forest Service granted an emergency exception to its export regulations for an estimated 15–20 MMBF of sawlogs that were surplus to the failing Juneau Spruce Corporation. At least some of the timber had been stored in log booms and was being ravaged by teredos The reduced operation of the Juneau Spruce Corporation also left independent logging operators without a market for timber that they had anticipated selling to the mill. These operators were issued permits to export 23 MMBF of surplus sawlogs to Puget Sound. Apparently alarmed at the volume of timber that was leaving Alaska, the Forest Service announced that it would issue no export permits in 1949.[25]

In the spring of 1951, the Forest Service reported a "concerted effort" by British Columbia and Puget Sound firms to eliminate the agency's restriction of log exports from the Tongass. The restriction, however, had broad public support, and the effort gained little traction. Among the supporters of the export restrictions, the Forest Service was pleased to announce, were all local chambers of commerce as well as the Seattle Chamber of Commerce, Alaska's territorial legislature, and various labor organizations.[26] The Forest Service also received support from the trade publication, *Timberman*, which deemed the general discouragement of raw-log exports a wise decision, declaring,

> In spite of strong pressure from the outside, Alaska succeeded in holding on to her great timber resources in the hope that Alaskan industries would one day have the raw material with which to develop an independent economy for the territory. It would have been easy to permit the export of logs to outside mills, but Alaska wisely clung to the policy of refusing to divert her raw material and reduce her timberlands to unpopulated muskegs.[27]

Despite the intentions of the Forest Service and the popularity of its policy, unanticipated events dictated that exceptions to the export restriction continue. For example, in the late 1950s, logs cut by several small independent operators ("gyppos") with the expectation of purchase by the Alaska Plywood Corporation were permitted to be exported when that business failed. The logs from the different operators were consolidated at Taku Harbor, made into a Davis-type raft, and towed to Seattle, where they were purchased by the Nettleton Lumber Company.[28]

CHAPTER 8

The Forest Service Has a Vision

T HE FOREST SERVICE WAS AND REMAINS BY FAR THE LARGEST
LANDOWNER IN SOUTHEAST ALASKA, AND ANY HISTORY OF LOGGING
IN SOUTHEAST ALASKA BEFORE 1971, WHEN THE ALASKA NATIVE
CLAIMS SETTLEMENT ACT WAS PASSED, IS IN LARGE PART A HISTORY OF THAT
AGENCY IN THE REGION.

With the establishment of the Tongass in 1907, the newly-created Forest
Service faced many immediate challenges. It needed to inventory its remote
domain, formulate an administrative policy, get qualified personnel into the
field, and—to help fend off criticism that it inhibited progress and development
in the region—it required an economic development plan that would
substantially stimulate commerce in Southeast Alaska. The territory's chief
executive was among the political attackers: "The forestry principle involved
is contrary to our traditional system of government," wrote Governor Thomas
Riggs in 1919, "and is setting up a vast feudal domain to be administered for
good or evil according to the will of the lord for the time being."[1] Some agreed
that it was the government's obligation to stimulate development: in 1920,
Secretary of Agriculture Edwin Meredith wrote, "The Government owes it to
Alaska to develop its resources and foster its economic growth."[2]

A great discovery of valuable minerals or a boom in the fisheries would
have been obvious generators of regional economic growth. Lacking that
panacea, however, the utilization of Southeast Alaska's timber resources—for
better or worse—would play a major role in any efforts to develop the region.
The Forest Service's contribution to such development would focus on timber
harvesting. Chief Gifford Pinchot in 1908 spoke to the issue more generally:

> Full utilization of the productive power of the Forests,
> however, does not take place until after the land has been
> cut over in accordance with the rules of scientific forestry.
> The transformation from a wild to a cultivated forest must
> be brought about by the ax. Hence the importance of
> substituting, as fast as is practicable, actual use for the
> mere hoarding of timber.[3]

FIGURE 29. Foresters considered trees such as this large Sitka spruce to be "overmature" in an economic sense. That is, after a certain age such trees were thought to produce less and less wood each year. Eventually, a tree would lose more wood to decay than it would produce in new diameter growth. (USDA Forest Service, Alaska Regional Office, Juneau)

In the early 1900s, there was a commonly held maxim: "Forestry begins with the axe." Old-growth forests, considered decadent and overmature, had to be logged and utilized so that they could be replaced by forests of young, healthy trees that would meet the needs of subsequent generations. "The ax is the chief tool of the silviculturist for increasing forest production," the Forest Service told Congress. On the Tongass National Forest, the Forest Service remained for many decades a consistent adherent to this maxim.

Industry also embraced this approach: "A virgin forest is at a standstill." wrote the *Paper Trade Journal* in 1923. "Decay and windfall balances growth. To produce, it must be cut and properly planted." A 1934 *Pacific Pulp & Paper Industry* article described the challenge:

> The greater part of the timber in Southeastern Alaska is mature and overripe, no longer growing. From an economic standpoint it should be cut to permit young growth to start. In other words, the crop, so to speak, is ready for harvesting, but for the lack of a harvester much is gradually deteriorating, and the yearly increment of young growth is only a small part of what it might otherwise be.[6]

Given the underlying philosophy toward forest management, forest condition surveys and assessments of silvicultural alternatives were among the many ecological and wood utilization research efforts conducted by the Forest Service during the first half of the twentieth century. H. E. Andersen, of the Forest Service's Alaska Forest Research Center in Juneau, delivered a paper on clearcutting as a silvicultural tool to the Alaskan Science Conference in 1954. In that paper, he stated that "approximately 95 percent of the commercial forest land area in Southeast Alaska supports overmature decadent climax forest stands." In these "generally unthrifty" stands, no net growth was taking place because such growth as did occur was largely offset by mortality and decay. Cull (timber with no value due to being misshapen or damaged from injury, insects, or disease) was heavy. According to Andersen in a similar paper presented to the Society of American Foresters a year later, the Forest Service was faced with the challenge of converting these stands into thrifty, managed stands. At the time of Andersen's comments, extensive clearcut logging was already being practiced on the Tongass to support pulp production that had recently begun at Ketchikan. The 1958–1967 Timber Management Plan for the Tongass National Forest stated, "A primary goal of timber management is to convert, as quickly as possible, the old-growth climax stands to a new forest having a desirable age-class distribution of even-aged stands with desirable species composition."[7]

In 1960, A. E. Helmers, of the Alaska Forest Research Center, largely repeated Andersen's comments from 1955, characterizing the vast majority of Alaska's coastal forest as "overmature and defective." *American Forests* took up the refrain: "The Tongass National Forest is very old, rotting and damaged by insects. Foresters say that it should have been cut 300 years ago."[8]

On the Tongass, converting old growth to a desirable species composition meant more Sitka spruce and less western hemlock. Sitka spruce was preferred because of its desirable qualities for lumber and pulpwood. This was reflected in the minimum prices published in an advertisement for a 1922 Forest Service timber sale: "No bid of less than $2.25 per M for Sitka Spruce and $1.00 per M for Western Hemlock will be considered."[9]

Sitka spruce also holds a silvicultural advantage: second-growth stands with a high percentage of the fast-growing Sitka spruce have more large trees and a considerably greater volume of merchantable trees than do similar stands in which the western hemlock is dominant.[10] To increase the proportion of the desired species, the Forest Service advertised premium prices for spruce and "cedar" seeds—but not for hemlock seed—when it advertised in the *Wrangell Sentinel* in 1910:

> The Forest Service will pay the following prices for seed cones, picked between September 15 and October 5: Spruce, 75 cents; Red Cedar, $1.00; Yellow Cedar, $1.00 per bushel, sacked and delivered to any point on any regular mail or steamboat route and marked "Forest Supervisor, Ketchikan, Alaska."[11]

The seed-collecting effort was short lived.

A Forest Service report written circa 1921 stated, "In view of the present large percentage of hemlock on the forest, it is believed that the spruce should be given the advantage whenever the question of cutting methods is concerned, with a view to the regeneration of valuable forest growth." Observations in the field supported it, as it was found that areas containing a greater percentage of spruce produced larger trees. One Forest Service study indicated the proportion of Sitka spruce in a natural second-growth stand between 75 and 100 years of age would be about 50 percent. Pure stands were not desired, however, as the spruce needed some hemlock to keep good form. Research suggested that foresters should seek as high a percentage of Sitka spruce as could be attained without overall degeneration of the stand. This early research informed management decisions through at least the 1950s, when managers aimed to replace the native forest with even-aged stands composed of approximately equal volumes of Sitka spruce and western hemlock.[12]

After its brief effort to collect seeds, the Forest Service determined that the replanting of Southeast Alaska's forest would not be necessary. (The agency changed its mind somewhat after large-scale pulptimber cutting began in the early 1950s and again began collecting seeds.) Experience showed that the region's natural conditions generally fostered rapid regrowth, and that Sitka spruce would regenerate provided that the trees were harvested cleanly to allow sunlight and rain to fall directly upon the soil. "Clean cutting"—a fancy term for clearcutting that was intended to imply efficiency and responsibility—was required on every sale, and where this was impossible or unlikely, the purchase of timber was denied. The Forest Service was sometimes criticized for not selling valuable timber in locations where it was mixed with unmerchantable material. Tongass National Forest Supervisor R. A. Zeller justified the policy: "To leave a canopy of leaves overhead after logging operations," he wrote in 1926, "would be a catastrophe."[13]

The challenge, in order to feasibly convert the original forest to vigorously growing, regulated stands, was to find adequate markets for the Tongass National Forest's timber. Logging in Southeast Alaska in the early 1900s was mostly confined to the small portion of the forest adjacent to saltwater, and mostly supplied relatively small sawmills that primarily cut Sitka spruce and manufactured products for the fishing industry and local needs. Practicing forestry as envisioned by the Forest Service and building a theoretically sustainable industry would require logging on a much greater scale as well as a guarantee that timber would be available over many decades.

FINDING A MARKET

Early efforts to establish a wood pulp industry in Southeast Alaska were largely based on the possible production of newsprint paper. Until a short time after the Civil War, newsprint paper had largely been manufactured from pulp that was made from linen and cotton rags. The paper was a comparatively high-quality product that was both expensive and limited in supply. Increasingly, rags-based newsprint paper was replaced by wood-based paper that was made from a mixture of approximately 80 percent mechanical pulp (basically ground and pressed wood) and 20 percent sulfite pulp (chemically purified wood fibers). The product was of adequate quality, far cheaper, and much more available than its rags-based predecessor. As a result, between 1878 and 1912 the price of newsprint paper fell more than 75 percent.[14]

In 1913, a reciprocal international trade agreement provided for the duty-free import into the United States of Canadian newsprint paper. This agreement retarded the construction of domestic newsprint paper mills (between the years 1909 and 1920, only one newsprint paper mill was constructed in the United States), and by 1919 fully two-thirds of the newsprint paper used by U.S. newspapers was imported or was manufactured from wood or pulp imported from Canada. At that time there were some 90 pulp and paper mills in Canada that produced about 2,100 tons of paper per day. Almost 90 percent of the production was available for export.[15]

An increase in newspaper advertising in the final years of World War I caused the consumption of newsprint paper to surge, and this, coupled with fears of an impending shortage, caused its price per ton (delivered to New York City, the industry-standard price quote) to rise from $42 in 1915 to $113 in 1920. Per capita annual consumption of newsprint paper rose throughout this period. In the U.S. in 1894, it was 9 pounds; by 1919 it had nearly quadrupled to 35 pounds, and by 1924 it had risen again to 50 pounds. In actual tonnage, the U.S. in 1924 consumed some 30 percent more newsprint paper than the rest of the world's nations combined. Some industry observers proclaimed that the nation had entered the "newspaper age."[16]

During the 1920s, the combination of rising demand and the lack of import duties had led to a surge in investments—much of it American capital—in newsprint paper manufacturing capacity in eastern Canada, near the major newsprint paper markets of the U.S. East Coast and Midwest. The price of newsprint paper began a steady decline as the construction of additional mills, mostly in Canada, increased capacity such that it began to exceed even the growing demand engendered in the Roaring Twenties. The 1929 crash of the stock market and the subsequent Great Depression caused demand for newsprint paper to plummet, and by the mid-1930s its price was once again hovering in the $40 per ton range.

The trade in newsprint paper was much like any other commodity business, with market price reflecting a continuous struggle between suppliers and users. Manufacturers were interested in obtaining the maximum price for their output, while newspaper publishers wanted the cheapest possible reliable supply of the material absolutely necessary for their business. This dynamic was set in the greater economic and political arena, where trade restrictions (or lack thereof) and exchange rates played a major part. The price of newsprint paper fluctuated because of a variety of factors, including world wars, the Roaring Twenties, the Great Depression, threats of embargos, or actions by newsprint paper manufacturers or newspaper publishers. While the massive duty-free importation of newsprint paper from Canada tended to keep prices low, by 1947, 40 years after the Tongass National Forest was established, the price of newsprint had rebounded from its Great Depression-era lows to $90 per ton.[17] The general increase in price during the 1940s helped fuel expectations of a successful wood pulp industry in Alaska.

According to John A. Guthrie, an economist who wrote extensively on pulp and paper, the manufacture of pulp and paper in the United States has been characterized by two trends: "The search for better and cheaper raw materials from which to make paper and the tendency of the industry to follow geographically in the wake of the lumber industry." Four fundamental attributes in combination generally define a good location for the manufacture of newsprint paper from standing timber: a nearby supply of suitable, economical timber; inexpensive power; ample processing water; and nearness to market.[18] Eastern Canada possessed all four attributes. Southeast Alaska had timber, hydropower potential, and plenty of water, but its distance from the major markets in the eastern United States was a serious handicap.

Sitka spruce was recognized as a preferred pulpwood. In 1885, A. P. Swineford, Alaska's second governor, wrote that it was the "best-known material for the manufacture of wood pulp." The Forest Service's interest in wood pulp on the Tongass probably had its roots in the 1899 Harriman Alaska Expedition. Former Division of Forestry head Bernhard Fernow was a member of the expedition, and his extensive report on Alaska's forests was included in the series of volumes that documented the expedition and its findings. Under the economic conditions then prevailing, he wrote, Alaska's forest resource did not "offer any inducements, unless it be that the spruce could be turned into paper pulp."[19]

In 1910, Assistant Forester Royal Kellogg, after viewing Southeast Alaska's forest firsthand, confirmed Fernow's judgment and suggested that production of pulp could lead to good forest stewardship: "The annual growth of the coast forests is far in excess of the local needs, and unless methods of utilization are developed which will result in the export of forest products, these forests can not be handled rightly." Kellogg continued: "Utilization for other purposes than for lumber should be encouraged. The most promising of these is for pulp. Both spruce and hemlock are undoubtedly good pulp woods, and, taken together, they comprise almost the entire forest."[20]

During that same year of 1910, William Langille, who had come to Southeast Alaska in 1905 to supervise the government's forest reserve, outlined the region's fundamental assets regarding the manufacture of pulp and made a pitch to potential investors: "There are great possibilities in the development of a pulp-making industry, the abundant suitable forest and unlimited undeveloped water power making its production cheap; and with the cheap transportation to Coast markets this latent resource will no doubt be a considerable factor in the future progress of the region, and deserves the attention of capitalists." Langille recognized that because the financial resources of Southeast Alaska were limited, any major development would require capital from outside the region.[21] A likely consequence of the employment of such capital would be that most of the profits from an operation would leave Alaska.

PRELIMINARY EFFORTS

As early as 1907, the imminent establishment of a paper mill at or near Wrangell was considered by some, including the town's *Alaska Sentinel*, to be an "assured fact." Likewise, the following year the timely establishment of a pulp mill was expected at Tonka (on Wrangell Narrows, south of Petersburg). Both mills were the stuff of rumors and dreams, and neither became anything more. There was also an attempt by several Ketchikan-area men to establish a pulp mill at Carroll Inlet (Revillagigedo Island) in 1905. The men secured 640 acres of timbered land and claimed a waterpower source, but the project proceeded no further.[22]

Nevertheless—and despite their experience with speculative mining and fishing ventures—the communities of Southeast Alaska placed great faith in promoters (including the Forest Service) of the pulp and paper industry. Certainly they welcomed the industry:

> Now, shall we sit with folded hands and let [Wrangell] become a relic of history, or shall we concert our efforts and secure a paper mill that will create a permanent payroll, increase the population, raise the value of property, and insure the future prosperity of the town?[23]

Alaska's territorial governor also promoted the industry. In an issue of *Collier's* magazine (a national weekly) that featured the Alaska-Yukon-Pacific Exposition (1909), Governor Walter Clark speculated somewhat wildly that the time might come when Alaska was able to supply "all of the print paper required by all of the news publishers of the United States."[24]

The first substantial interest in the manufacture of paper in Southeast Alaska came in 1910 when a Norwegian paper manufacturer considered constructing a pulp and paper mill at Thorne Arm (Revillagigedo Island). The Norwegians asked the Forest Service about the possibility of obtaining a 99-year renewable timber concession covering 50 miles of Southeast Alaska's coast. The terms sought involved practically absolute freedom regarding the methods of logging and the condition in which the land might be left, and the time and location at which an unspecified type of mill might be constructed. A royalty of $0.25 per cord, fixed for 25 years, was offered for stumpage. The Forest Service declined to consider the proposition, finding the proposed terms to be, as the chief of the Forest Service, William Greeley, later recalled, "impossible of acceptance."[25]

Industry was content in its certainty that the development of the Tongass National Forest was close at hand. *The Timberman*, which proclaimed itself "Devoted to the Lumber Interests of the Pacific Coast," wrote almost romantically of the situation in 1911: "Secure in the control of the Government, in the isolation of its location and the natural conditions which safeguard it from destruction, this vast forest resource somnolently bides its time, awaiting a call for its utilization in the not distant future."[26]

In a letter to Secretary of Agriculture Henry Wallace in 1921, Chief Forester Greeley recounted the work done by the Forest Service in 1912–13 to facilitate the establishment of a newsprint paper industry in Southeast Alaska. Tests were made of the suitability of Alaska wood for the manufacture of newsprint paper, and pulp and paper specialists made extensive studies to determine the technical and commercial possibilities in the region. They reported that the establishment of a paper industry was not economically possible because of the then-depressed paper market, combined with the region's distance from markets and Southeast Alaska's dearth of skilled labor. The investigators added that the cost of timber was not a factor: even if timber was absolutely free, no profit could be realized making paper in Southeast Alaska. Nevertheless, the Forest Service continued to see opportunities in the future and offered encouragement and assistance to anyone who showed interest. "I cannot conceive of a better long-time investment than one of the many big blocks of pulptimber now in the Tongass National Forest with a five thousand horse-power waterfall in the center of it," wrote Forest Supervisor W. G. Weigle in 1913.[27]

The Forest Service considered the situation in Alaska ripe for pulp and paper in part because demand for lumber was small:

> Ordinarily, the pulp and paper industry has followed the lumbering, and has either had to displace the sawmill through competition or to take the material which the sawmill left. In Alaska, however, cutting operations for lumber and other purposes are small, so that in this respect there would be a greater opportunity for the development of a dominant pulp and paper industry than in the Pacific Coast states.[28]

The process of obtaining pulptimber on the Tongass National Forest was similar to that of obtaining sawtimber. Upon the request of what the Forest Service considered a responsible applicant, the agency would cruise, appraise, and advertise a tract containing sufficient timber for the proposed enterprise.

The Forest Service's first pulptimber offering on the Tongass occurred in 1913, when at the request of local interests who had organized themselves as the Alaska-Pacific Pulp & Paper Company and planned to construct a mill to manufacture paper pulp for the general market, the agency offered a 20-year contract for 300 MMBF of timber along the Stikine River. As a provision of the sale, an additional 300 MMBF of timber was reserved as a future supply for the enterprise. Initial prices for pulpwood were to be $0.50 per cord for spruce, and $0.25 per cord for both hemlock and cottonwood, with price adjustments scheduled at five-year intervals. The price for sawtimber was $1 per MBF for Sitka spruce and western redcedar, and $2.50 for Alaska-cedar.

The sale was heralded in *American Forestry* as "remarkable in that it indicates a beginning of the utilization of Alaskan timber in the general market." The Alaska-Pacific Pulp & Paper Company, however, was unable to secure financing for what was considered a risky venture, and the sale was never completed.[29]

About the same time, a Mr. Lewis Stockley of San Francisco expressed interest in the purchase of a similar amount of timber located at Thorne Arm, not far from Ketchikan on Revillagigedo Island. Stockley, who may have been associated with the *San Francisco Chronicle*, proposed the construction of a $1 million pulp and paper mill, and at his request the Forest Service designed

300,000,000 FEET

National Forest Timber For Sale

AMOUNT AND KINDS: 300,000,000 feet B. M., more or less, of Sitka spruce, western hemlock, western red cedar, cottonwood, and yellow cypress.

LOCATION: Within the Tongass National Forest, Alaska, on the west side of Eastern Passage and on the islands and both sides of the Stikine River.

STUMPAGE PRICES: Lowest rates considered, $2.50 per M. for yellow cypress, $1.00 per M. for Sitka spruce and western red cedar, saw timber, 50 cents per cord for Sitka spruce pulpwood, and 25 cents per cord for western hemlock and cottonwood pulpwood. Rates to be readjusted every five years.

DEPOSIT: With bid $2,500, to apply on purchase price if bid is accepted, or refunded if rejected. Ten per cent may be retained as forfeit if the contract and bond are not executed within the required time.

FINAL DATE FOR BIDS: Sealed bids will be received by the District Forester, Portland, Oregon, up to and including May 10, 1913. This date may be extended 30 days upon the request of any person who desires additional time for the examination of the timber.

The right to reject any and all bids is reserved.

Before bids are submitted full information concerning the character of the timber, conditions of sale, deposits and the submission of bids should be obtained from the District Forester, Portland, Oregon, or the Forest Supervisor, Ketchikan, Alaska.

FIGURE 30. From 1913: The Forest Service's first advertised pulp timber sale on the Tongass National Forest.

a sale for the area, the terms of which were virtually identical to those for the timber offered on the Stikine River. A no-fee water-power permit was authorized by the Forest Service for Stockley's operation, and approval was granted by the Secretary of Agriculture for doubling the size of the timber offering, to 600 MMBF, and a cutting period of 40 years. For lack of financing, this sale never progressed beyond the early planning stage.[30]

Governor John Strong's 1914 report to the Secretary of the Interior maintained optimism:

> Investigations with favorable reports were made during the year throughout the Tongass National Forest relative to the possibilities of the manufacture of wood pulp and paper. The cheap power and vast amount of available timber make it reasonable to assume that southeastern Alaska will in the immediate future be one of the principal pulp and paper centers of the world.[31]

The next year, Henry Graves, second chief of the Forest Service, traveled extensively in the Chugach and Tongass national forests. A report of his findings was published in 1916 in *American Forestry*. According to Graves, "The Forests should be made to serve in the building up of the country, the establishment of industries, and the creation of opportunities for a permanent population." It was "confidently expected," wrote Graves, "that in a short time the sale of timber [in Southeast Alaska] will result in the development of industries manufacturing lumber and wood pulp on an extensive scale." He estimated that "under right handling, that provides for the perpetuation of the forest, not less than five or six hundred million feet could be taken each year, from the Tongass Forest without reducing the total stock, as the new growth would equal the amount cut."[32]

Though there was at that time a temporary shortage of newsprint paper in some eastern and midwestern markets, Graves' confidence was somewhat at odds with an analysis prepared by his own Department of Agriculture. In August 1916, Secretary of Agriculture David Houston wrote to President Wilson regarding Forest Service pulptimber sales in the Pacific Northwest and Alaska:

> Up to the present time it has not been possible to make such sales. The chief difficulty has been that the Western markets have been fully supplied and, in fact, mill capacity probably has been in excess of market demands. Western mills with the advantage of cheap power and cheap timber seemed to be unable to enter the Eastern and Middle Western markets, and the only opportunity for successful enterprise seemed to be the more or less uncertain possibility of being able to develop foreign and chiefly Oriental markets.[33]

While pulp markets were in question, the Forest Service in 1916 inaugurated a national campaign to extensively advertise and promote sales of timber throughout the national forest system. District officers were permitted to use their own ingenuity to carry on the campaign, which included the preparation of pamphlets containing descriptions and maps of timber, as well as instructions to prospective purchasers and sample contracts. In cooperation with the U.S. Geological Survey, the agency also established a number of stream gauging stations in Southeast Alaska to ascertain the power possibilities for the manufacture of pulp and paper.[34]

As noted above, the price of newsprint paper was rising in 1916, and the *Wrangell Sentinel* was expecting great things: "Present Exorbitant Prices of Paper Would Make an Alaskan Paper Mill as Good as a License to Rob Banks," read a headline in September of that year. At the request of an interested party in 1917, the Forest Service offered its largest sale to date—1 billion board feet (BBF) of timber—in the Behm Canal area, near Ketchikan. The prices for the timber were the same as on the previous offerings, and the operating period was 25 years. This sale met the same fate as the others, however: the interested party was unable to secure financing and quickly abandoned its endeavor. Yet Alaskans remained enthusiastic. "Paper mills, pulp mills, pulp and paper, everywhere. That was all I hear on my trip through southeastern and southwestern Alaska," said Charles Flory of the Forest Service in 1919.[35] Headlines from the 1920s expressed certainty that prosperity was just around the corner: "Pulp Mill Sure," read the April 9, 1920, *Petersburg Weekly Report*. The July 29, 1920, *Ketchikan Alaska Chronicle* proclaimed, "Paper Mills for Ketchikan." The *Wrangell Sentinel* announced, "Money for Big Paper Mill at Wrangell Sure" on November 23, 1922. And on March 16, 1923, the *Alaska Weekly* was equally confident: "Pulp Mill at Thane Is Now Certain."

Even after a considerable number of deep disappointments, to some, the development of a newsprint industry remained inevitable: "Alaska needs industry," wrote a forest consultant in 1927. "Alaska was equipped by the Good Lord as a future pulp and paper producing region and no matter what anyone may do or try to do, Alaska will be a big processor of newsprint."[36]

The effort to establish newsprint paper mills in Southeast Alaska had received a big boost in 1918, when Frank Heintzleman arrived in Ketchikan to assume the position of logging engineer and deputy forest supervisor on the Tongass National Forest. Heintzleman was born in 1888 and graduated from the Pennsylvania State Forest Academy. He received a masters degree at Yale Forest School, and in 1910 entered the Forest Service, where he worked in the Pacific Northwest until his transfer to Alaska. Heintzleman had a vision for Southeast Alaska, which, he believed, presented "an excellent opportunity to build up a great, permanent paper making region with modern towns, thoroughly equipped efficient plants and a population of skilled workmen, which will stand for all time as one of the most important sources of pulp and paper supply in the United States." Heintzleman soon personified the Forest Service's determination to establish a wood pulp industry in Southeast Alaska. He traveled far and wide to promote his vision for the region and became known to some as Alaska's "Greatest Salesman." The *Paper Trade Journal* referred to Heintzleman as a "Forest Service official-missionary"; the Associated Press called him "the Territory's 'Mr. Forestry.'" The Forest Service boasted that "no one could listen to 'Heintz' without having the fancied difficulties of logging and paper making in Alaska melt away and be replaced by a real appreciation of the opportunities there." Pulp and paper executives dubbed him "Alaska's Super Salesman."[37] Heintzleman progressed in the agency to become Alaska's regional forester in 1937, and in 1953 became Alaska's territorial governor.

Melvin L. Merritt, assistant forester in Alaska from 1921 until 1934, recalled,

> The main effort of the Alaska Forest Service District during the time I was there was spent in an endeavor to develop a pulp and paper business in Southeast Alaska. B. F. Heintzleman…spent almost his entire time on this work; directing and making timber surveys, stream flow and water storage measurements, and contacting possible timber operators in the States.[38]

In 1919, the Department of Agriculture made a case for providing a domestic supply of newsprint paper. The department released a statement that pointed out the pitfalls of U.S. dependence on foreign supplies of newsprint paper. It read, in part,

> A condition of dependence upon foreign supplies of newsprint paper carries with it serious possibilities not only for consumers of newsprint paper (chiefly our newspapers) but also for other business interests and the public generally. It would afford a dangerous opening for covert interference with the freedom of the press and with untrammeled development of business through advertising.[39]

Among the recommendations to eliminate the nation's dependence on foreign supplies of newsprint paper was the expansion of the newsprint paper industry in the Pacific Northwest and its development in Alaska.

Interest in Southeast Alaska had been stirred by the establishment in 1912 of a groundwood pulp mill at Ocean Falls, British Columbia, only 265 miles from Ketchikan. Though the mill failed financially soon afterward, it was taken over by Pacific Mills, which quickly constructed a state-of-the-art facility to produce newsprint paper. Production began in 1917. If the manufacture of paper could be done at Ocean Falls, why could it not be done in Southeast Alaska, where similar conditions prevailed? Some blamed the Forest Service. Noting that paper mills had gone up in British Columbia, Oregon, and Washington, and that demand for paper was "always in evidence," the *Ketchikan Alaska Chronicle* hinted ominously. "It seems hardly possible that the capitalists would select every place under the sun but Alaska unless there is some reason." The Forest Service responded that capital was presently lacking, but "a substantial development of the paper industry in this wonderful region, combined with an intelligent regard for future timber crops, should settle forever the question of a serious paper shortage in the United States."[40]

The *Wall Street Journal* in 1921 reported that a witness before a House Committee on Territories hearing had produced statistics indicating that the manufacture of paper pulp in British Columbia, which started at Ocean Falls in 1912, had already become the province's second most important industry and sold more than $10 million worth of paper to American newspapers— and that the capital for the development of the industry had come mostly from American sources. "The success of the great pulp mills on the British Columbia coast," wrote Sherman Rogers, industrial correspondent for *Outlook* magazine in 1922, "silences every argument of pessimistic theorists regarding the practicality of paper manufacture in Alaska."[41]

The Federal Power Commission was created in 1920, and to help foster pulp and paper development in Southeast Alaska, the new agency in 1921 undertook with the Forest Service a cooperative project to more fully investigate Southeast Alaska's waterpower resources. The project surveyed many of the known waterpower sites and identified several new and important sources. In 1924, the Federal Power Commission published *Water Powers of Southeastern Alaska*, a report prepared "so as to make available to interested persons all of the known data concerning the waterpower resources of southeastern Alaska and to show the value of these water powers as sources for power for pulp and paper plants." As a standard policy, each application for a power permit was reviewed by the Department of Commerce's Bureau of Fisheries to ensure that the proposed project "would not interfere with the protection or utilization of fishery resources."[42]

A procedure adopted by the Forest Service and the Federal Power Commission linked the sale of timber and the leasing of power sites.[43] Applications for power permits and for pulptimber were advertised simultaneously. Investors applying to the Federal Power Commission for a permit to develop hydropower for timber processing were advised to also apply to the Forest Service for a timber sale, while applicants to the Forest Service for timber were advised to also apply to the Federal Power Commission for a power permit.

As the Alaska District's assistant forester, Frank Heintzleman estimated in 1927 that the production of one million tons of newsprint paper annually in Southeast Alaska would require about 300,000 horsepower continuously throughout the year. It was estimated that the potential developable water-power in Southeast Alaska at the time totaled some 500,000 horsepower. Additional surveying was done by the Forest Service and Geological Survey with the help of the U.S. Navy in 1929. By 1948, the Federal Power Commission and Forest Service's joint estimate of potential average horsepower had roughly doubled to more than one million.[44]

The typical waterpower site in Southeast Alaska had a high, glacier-formed "hanging lake" a short distance from saltwater. The lake provided cheap water storage, especially during the winter months when precipitation was tied up in the form of snow, and only a short conduit was required to connect it with a powerhouse located at sea level. Ideally, transmission lines would be unnecessary because facilities for the manufacture of pulp and paper could be located near the powerhouse.[45]

The Forest Service estimated that development cost per electrical horsepower on the Tongass National Forest in 1927 would be $60 to $75, and total power cost for a pulp and paper mill, including interest, depreciation and operation, would be $0.00125 to $0.0015 per kilowatt-hour. The agency thought it doubtful that lower power costs could be obtained elsewhere in the United States or in Canada.[46]

AN ADVOCATE IN THE NATION'S CAPITAL

In 1920, William Greeley succeeded Henry Graves as chief of the Forest Service. Greeley was an intelligent and practical individual from California who had graduated from the Yale Forest School and quickly rose to become the first regional forester of the timber-rich Northern Region, which included northern Idaho, all of Montana, and parts of the Dakotas. He understood the West Coast lumber, pulp, and paper industries, and he was sympathetic to their needs for raw materials: he once defined forestry as "use and timber cropping; not abstention from use and tree worship."[47]

During his first year as Forest Service chief, Greeley traveled to Southeast Alaska, where he took to the woods with Heintzleman and a cadre of experts from the Forest Service's Forest Products Laboratory to investigate how pulp mills might be brought to the region. He later recalled that "we mapped and cruised the most promising areas of pulp timber. We drafted pay-as-you-go

FIGURE 31. District 6 Forester George Cecil and Chief Forester William Greeley at Mendenhall Glacier, near Juneau, 1920. Greeley was very interested in the federal government's role in fostering the economic development of Alaska. He visited the territory twice while chief forester. (USDA Forest Service, Alaska Regional Office, Juneau)

contracts for 50 years' supply. We drew up prospectuses, advertised in the paper trade journals, and interviewed pulp and newsprint manufacturers on the West Coast."[48]

Greeley was surprised by the amount of high-quality Sitka spruce sawtimber they encountered in the Tongass. He estimated a total of three BBF, and observed, "Very much of it can not be made available until logging is undertaken on a large scale. The answer unquestionably is the establishment of a paper industry in Alaska." Greeley knew a lot about wood and its value, and he lamented that the majority of Southeast Alaska's sawmills were cutting magnificent 60-foot spruce logs into salmon box shook.[49]

Initially, Greeley was very optimistic about the near-term potential of establishing a paper industry in Alaska: "The entire forestry industry of the United States is moving westward, and with it is coming the paper industry." He added, "There is every reason to believe that the present shortage and high cost of all grades of paper will bring the forest industries of Alaska up to the front rank." Whether Greeley was too optimistic or not, wrote the *Paper Trade Journal* in 1923, "there is little question but that Alaska has a pulp producing future which looks promising."[50]

Soon after his visit, though, Greeley tempered his optimism with an analysis that concluded that the establishment of a pulp industry faced two big problems. First, Alaska was, in his words, "a long way up yonder," and the big need was to get Alaska paper into the major markets of New York and Chicago. "Marine transportation to Alaska is not adequate. For the paper industry to succeed there, it must have its own vessels or a plan must be worked out to give the entire territory better service. I regard this as a problem for the United States Shipping Board and do not think it should be left to private concerns." The Shipping Board was organized in 1917 as an emergency agency whose responsibilities included the development of water transportation. Greeley thought that only "an aggressive Federal policy" could resolve the issue. *American Forestry* wrote at this time of Alaska's "enormous transportation difficulties, involving prohibitory freight rates."[51]

The second problem Greeley identified was the lack of development and skilled labor:

> The manufacturer finds in Alaska an enormous country with a population of 35,000 whites. He must develop his enterprise out of the raw wilderness. The nearest real base of supplies with adequate machine shops etc. is Seattle,

670 miles away. The manufacturer must begin by building his own wharves and his own village… Then he must move skilled labor in, build homes and make and keep his employees contented. There is no existing skilled labor to draw upon. It is simply not there.

Greeley did not consider those problems insurmountable to an energetic businessman, however, and he noted that the paper industry had succeeded in British Columbia under substantially similar conditions. Because of high paper prices, he deemed the transportation problem to be "simply one of equipment and organization."[52]

To a certain extent, the challenges Greeley described were mitigated by the potential for inexpensive waterpower and the low prices for which the Forest Service proposed to sell Tongass timber, prices the agency described as being "sufficiently low to make the cost of pulpwood stumpage a relatively negligible factor."[53] Of this, George Cecil, Forester for District 6, which then included Alaska, wrote in 1920,

> Considered as an item of cost of production, the importance of the stumpage is generally largely overrated. During the years 1913 to 1916 inclusive the average cost of producing newsprint paper in Washington and Oregon was about $33 per ton. This cost has increased to a marked extent at the present time. The stumpage price can thus be seen to be a minor factor. With an average stumpage cost of 46 cents per ton of newsprint and an assumed cost of $40 per ton, stumpage is one and one-tenth per cent of the total."[54]

The same year he was made chief forester, Greeley wrote that "the manifest duty of the Department of Agriculture, in view of the shortage of paper in the United States, [is] to take every reasonable step within its power to encourage the development of a paper-making industry in the National Forests of Alaska." The development of a massive wood pulp industry, he reasoned, would bring development and permanent prosperity and provide the Forest Service with an opportunity to prove that its science-based forestry worked. In Greeley's words, "For the first time there is opportunity in a large region and on a big scale to carry out an enlightened, publicly-controlled plan of timber use combined with preservation." "Alaska," he wrote, "needs the application to her forest problems of the experience, technical knowledge, and organization provided by the Forest Service." Reiterating Secretary of Agriculture Edwin Meredith's 1920 comment about the government's obligation to foster development in Alaska, Greeley later added, "The Forest Service feels a special responsibility to aid in the development of Alaska."[55]

The development of Southeast Alaska's forest, Frank Heintzleman wrote in 1927, "can best be depended upon to constitute a rich heritage passing along from generation to generation, and to insure the continued existence and wellbeing of the community."[56] The newsprint industry would bring stability, he predicted: it would have "none of the stampede and wild-cat characteristics of many earlier Alaskan ventures. It will be systematic, orderly development, its only romance being that of constructive and the solid growth of thriving communities."

Overdevelopment would not be allowed to happen. According to Heintzleman, the Forest Service was committed to the policy of managing the Tongass National Forest according to the principles of forestry to provide a sustained output of timber. The proposed industry would conservatively match the region's expected ability to produce timber, and at least one mistake of the past—overcutting—would not be permitted:

> Many localities in the Lake States and other parts of the country, which enjoyed a short period of high prosperity while the virgin timber lasted but which thereafter immediately lapsed into poverty and have continued in that condition, are a striking object lesson of the folly of overcutting and the failure to apply forestry in the handling of the timber resources of a region. This mistake will not be repeated in Alaska.[57]

Greeley, too, stressed the perpetual returns: "What is Alaska worth?" he asked. "1,000,000 tons of newsprint delivered to the presses of the United States every year for all time to come." Furthermore, cutting the "decadent" timber would further his agency's conservation goal: "In many ways, the prospective conversion of Alaska's virgin forests into newsprint paper will stand out as a significant chapter in the story of forest conservation."[58]

The *Ketchikan Alaska Chronicle* termed Greeley a "good friend of Alaska," writing that the fact that he "is doing everything in his power to interest capital in the pulp and paper possibilities of Alaska is shown by the many periodicals with articles by him, and the numerous interviews that he gives in which he gives material aid to the Territory in development of its resources."[59]

Greeley's Alaska district forester, Charles Flory, was likewise a strong proponent of the establishment of a pulp and paper industry in Southeast Alaska. In 1922, Flory was glowingly described by the industrial correspondent for *Outlook* as

one of the most efficient timber men I have met in many a day. He knows his business, is intensely practical, cuts red tape to the bone, and his whole heart and soul are wrapped up in the securing of substantial pulp and paper mills near the great Alaskan water-power projects. There is nothing hazy about his ideas—nothing impractical about the timber development policies he advocates. He realizes as every one familiar with Alaskan timber resources does, that the southeastern section of our northern empire is the coming Mecca of cheap paper manufacture. He realizes, at the same time, that every year of delay in securing these mills means a heavy loss, not only to the progress and prosperity of Alaska, but also to the taxpayers of the United States, who would greatly benefit though stumpage sales.[60]

And he was ready to see the industry begin cutting timber: "Conservation and reservation are two different things," Flory said. "To conserve doesn't mean to tie up the resources, it means to use them properly without waste, in this generation, and not in years to come. We must conserve here, but not reserve."[61]

James Anderson, writing in *Scientific American* in 1921, estimated the cost of pulpwood delivered to mills in Southeast Alaska—even considering the region's primitive logging methods—would normally average $4 to $6 per cord. He gave the 1918 average per cord cost of pulpwood at mills in Washington, Oregon, and California as being $8.95. In an article in *American Forestry* in 1920, the Forest Service's A. F. Hawes stated that the cost of wood for the domestic newsprint paper industry ranged from 30 to 40 percent of the cost of newsprint paper.[62]

The future stumpage price adjustments written into contracts for Tongass timber, however, caused uncertainty and were cited as a deterrent by some prospective investors. Their preference was for contracts for 30 years at a fixed price—"No sane American business man would invest the large sum of money necessary for the establishment of a paper mill in Alaska with the forestry service reserving the right to change the charges for the raw material every three, five or seven years," said Representative Charles F. Curry (R-California), chairman of the Committee on the Territories—but the Forest Service considered that option "out of the question."[63]

The sense of urgency in the Forest Service to establish a wood pulp industry in Southeast Alaska continued to grow. George Cecil wrote in a 1920 issue of *The Timberman*:

> The Forest Service is not content to await applications for timber from possible manufacturers of pulp and paper who may learn of the opportunities in Alaska. It has been and is pursuing an aggressive policy of bringing these opportunities to the attention of paper makers and users. This is done not only by articles, but also by personal interviews and individual letters. It has actively assisted interested men in investigating, on the ground, the pulp chances and water power resources in Alaska. The Service has been and is actively pushing, with all the varied means at its disposal, the use of the forests of Alaska to meet the urgent needs of the nation for an increased paper supply and to establish in the territory a large and permanent industry.[64]

The Forest Service wanted to be an accommodating partner for potential pulp and paper manufacturers. The agency advertised itself as being "in a position to guarantee permanent supplies, on reasonable terms as to price, and made available as needed."[65]

The *Ketchikan Alaska Chronicle* reported in 1920 that 20 or more organizations were investigating pulp and paper manufacturing opportunities in Southeast Alaska, including at least one from London. Though Chief Greeley noted that some of the inquiries his agency received were from "promoters who hoped to sell somebody something," he was optimistic: "You are going to have some big pulp and paper mills here in the near future, or I miss my guess a long way."[66] Local newspaper editors remained upbeat:

> Optimism and More Optimism. During the last few days conditions have been looking up materially. The good news of the coming immediately of the pulp industry has made the future look more rosy than ever. It will not, it appears now, be necessary for Ketchikan to wait for years and years for the new needed industry to increase its greatness.[67]

But it would be more than three decades before the stars aligned and the dream of committing the Tongass's timber to pulp production was even partially realized.

CHAPTER 9

Optimism Fuels Expectations

THE MANUFACTURE OF WOOD PULP WAS NOT ENVISIONED BY THE INVESTORS WHO INCORPORATED THE SPEEL RIVER PROJECT IN CALIFORNIA IN 1915 UNDER THE LEADERSHIP OF EUGENE KENNEDY, FORMERLY THE ASSISTANT GENERAL MANAGER OF JUNEAU'S TREADWELL MINE. THE MAIN OBJECTIVE OF THE SPEEL RIVER PROJECT INVOLVED THE DEVELOPMENT OF WHAT THE BUSINESS ESTIMATED TO BE 100,000 HORSE-POWER OF YEAR-ROUND HYDROELECTRIC POTENTIAL FROM SEVERAL SOURCES NEAR THE SPEEL RIVER, AT PORT SNETTISHAM, WHICH IS SOUTH OF JUNEAU ON ALASKA'S MAINLAND. THE INVESTORS CONTEMPLATED THE FORMATION OF THE SPEEL RIVER ELECTRO CHEMICAL COMPANY TO DEVELOP THIS HYDROELECTRIC POTENTIAL. THE ELECTRICITY WOULD BE USED TO REFINE IRON ORE OR BAUXITE, WHICH THE INVESTORS WERE CONFIDENT WOULD BE FOUND IN REASONABLE PROXIMITY. IN ADDITION TO ENGINEERS WHO HAD BEEN INVOLVED IN THE CONSTRUCTION NEAR JUNEAU OF THE HYDROELECTRIC FACILITIES AT SHEEP CREEK (1910) AND NUGGET CREEK (1912), FAMED FINANCIER AND INDUSTRIAL ORGANIZER J. P. MORGAN WAS REPORTEDLY A SHAREHOLDER.[1]

Neither the iron ore nor the bauxite ever materialized, and hydroelectric facilities were never developed, but with the price of paper increasing rapidly, the group decided to construct a small pulp mill, the returns from which would be used to expand operations to include the manufacture of paper.[2] For this purpose, the Alaska Pulp & Paper Company was organized in San Francisco in 1920. That same year, the business expressed to the Forest Service an interest in 100 MMBF of timber situated on two tracts totaling about 10,000 acres.[3] The first tract was at Port Snettisham, and the second on the southern portion of Admiralty Island's Glass Peninsula.[4] The Forest Service subsequently prepared and advertised a sale based on the company's interest. Alaska Pulp & Paper, the sole bidder, secured the timber at the Forest

Service's minimum acceptable bid of $1 per MBF for Sitka spruce and Alaska-cedar, and $0.50 per MBF for western hemlock. The Forest Service reserved approximately one billion board feet of additional timber for the expected expansion of the operation.[5]

The sale was lauded as a harbinger of a prosperous future for Southeast Alaska, but Alaska Pulp & Paper from the beginning was greatly handicapped by its initial decision to produce undried, mechanically-ground pulp, known as groundwood pulp—a heavy, fairly low-value product that was very costly to transport. Newsprint paper, a lighter and more valuable product, could have been transported considerably more economically.

Near the Speel River at Port Snettisham's north arm, Alaska Pulp & Paper built a small mill of lumber powered by water routed from nearby Tease Lake (elevation 1,000 feet) to an approximately 1,000-horsepower water wheel at the mill site. The mill was equipped with two pulpwood grinders.[6] Grinders of the sort used at Alaska Pulp & Paper consisted of a grindstone ("pulpstone") 54 inches in diameter by 27 inches wide mounted vertically on a horizontal water-driven shaft. The grindstone was encased in a strong, circular cast-iron housing, turned at 220 to 230 revolutions per minute, and was constantly bathed in water. Around the upper circumference of the housing were four openings or pockets some 26 inches long and 12 inches wide. Two-foot lengths of debarked and cleaned wood were placed into the pockets sideways against the revolving wheel. Hydraulic pressure was then used to force the wood against the stone. The wood's fibers were torn away by abrasion and transported in a slurry to a series of screens for dewatering.[7]

The Alaska Pulp & Paper Company mill employed about 60 men and was capable of producing about 40 tons of wet groundwood pulp—the equivalent of about 15 tons of industry-standard dry groundwood pulp—per day. The mill turned out the very first pulpwood to ever be manufactured in Alaska on January 24, 1921. Made completely from Sitka spruce, the pulp was con-sidered to be of excellent quality. Ninety-nine tons of it was shipped to Puget Sound in May.[8]

Pulpwood for the mill was not cut on the tracts under contract with the Forest Service, but was instead purchased from Sawyer & Reynolds, an estab-lished contract logging firm with an operation near Angoon, on the west side of Admiralty Island.[9] The production of 15 tons of groundwood pulp per day required only a modest amount of timber. Calculated in board feet (log scale),

FIGURE 32. Alaska Pulp & Paper Company pulp mill at Port Snettisham, circa 1921. Logs were drawn up sloped "loghaul" on right into the wood room, where they were cut to size before entering the pulp mill. (Alaska State Library, Eugene Patrick Kennedy Collection)

the theoretical operation of the mill for a year of 300 working days would have required about 2.4 MMBF.[A]

Transportation was from the beginning a problem for Alaska Pulp & Paper, whose market was the Washington Pulp & Paper Company of Port Angeles, Washington (owned by the Zellerbach Paper Company).[10] Both the Alaska Steamship Company and the Pacific Steamship Company initially showed little interest in transporting the pulp, but ultimately Pacific Steamship agreed to do so at the prohibitively high price of $9 per ton. Pulp from Alaska Pulp & Paper Company's mill had a moisture content of about 55 to 60 percent, and thus the transportation cost per dry ton of pulp would

[A] Calculation based on the utilization of Sitka spruce pulpwood, with a conversion factor of 1 CCF = 2,100 lbs. of groundwood pulp. Calculation: 30,000 lbs. daily production @ 2,100 lbs. per CCF =14.3 CCF per day; 1 CCF = 550 bf (log scale); 14.3 x 550 bf = 7,865 bf; 7,865 bf x 300 days/year = approx. 2.4 MMBF/year (log scale).

have been on the order of $16 to $20. The cost of transportation was ultimately negotiated down to $5 per wet ton—equal to about $11 to $12.50 per dry ton, but with the pulp valued at about $35 per ton delivered to Puget Sound, it was still impossible to operate profitably, and the company soon ceased operations.

Despite the transportation challenge, production was resumed in late 1922. In early November 1922, the plant was reported as working three shifts daily. Some 200 tons of pulp was at that time stored on the dock and ready for shipment on a freighter that was expected to call in mid November. The company was also increasing its pulp storage area and making improvements to speed up cargo loading. Alaska Pulp & Paper, however, could not overcome the transportation obstacle it faced, and the plant shut down permanently in November 1923. A freight rate of $3 per ton for wet pulp from northern Southeast Alaska to Puget Sound was later established by the steamship companies, but it was never used.[11]

Total exports of wood pulp from Alaska for 1923 were reported to be 6,374 tons valued at $113,655 ($17.83 per ton). The Alaska Pulp & Paper Company's contract with the Forest Service was cancelled by mutual consent in 1925.[12]

Many years later, Frank Heintzleman, angling for the establishment of a pulp mill near Juneau, commented that the site at Speel River was not desirable for a large plant, but that electricity generated there could easily be brought by transmission line to an operation on Gastineau Channel.[13]

TIMBER ON FAVORABLE TERMS

By 1920, the Forest Service had constructed a comprehensive and optimistic development plan for Southeast Alaska. Agency-authored articles outlining the region's timber resources and potential began appearing in trade publications. Under the *Ketchikan Alaska Chronicle*'s headline "Hundred Million Board Feet Pulp Timber Will Be Sold," the agency's Fred Ames was quoted as saying, "The Forest Service is now, as it has always been, extremely anxious to see the resources of the Alaska national forests developed in every legitimate way, and especially that the national forests may play their part in helping to relieve the present acute shortage of newsprint paper."[14]

In the spring of 1921, the agency published the bulletin *Regional Development of Pulpwood Resources of the Tongass National Forest, Alaska*, by Forest Inspector Clinton Smith. The bulletin was prepared to "aid those who wish information on the timber and other resources of the Tongass

National Forest in Alaska, to indicate the capital and organization necessary for the development of Alaskan pulp and paper mills, to show what data on the timber resources of that region have been and are being collected by the Forest Service, and to outline the conditions of purchase of timber on the National Forests." It was distributed to some 850 firms nationwide.[15]

The Forest Service publication included a map in which most of Southeast Alaska was divided into 14 pulptimber sale units, each with a mill site and waterpower potential and located within a "working circle" of timber theoretically able to supply the perpetual operation of a mill capable of producing 100 tons of newsprint paper daily.[16] For the Forest Service, this map was a starting point. The agency was willing to reconfigure units at the request of potential purchasers.

The timber in each unit, as described by President Warren Harding after his 1923 visit to Alaska in the company of his Secretary of Agriculture and the chief of the Forest Service,

> was enough to keep the plant operating for forty-five years. That period is selected because it is the one in which the lands first cut over will, with proper care, reproduce their timber. Thus, after the whole tract has been cut over in the forty-fifth year, the new crop of trees on the land cut over in the first year will be ready for another cutting.... The allowance of timber, although made on the basis of forty-five years, is really sufficient for sixty to sixty-five years.[17]

The rotation period stated by Harding was a coarse estimate: the Department of Agriculture had simply used an estimated growth rate for coniferous forests in New England as a best guess for growth rates in Southeast Alaska.[18]

Nevertheless, at least one trade publication echoed the Forest Service's enthusiasm: "Think of it!" wrote the *Paper Trade Journal* in 1926. "A magic circle encloses within its bounds your timber for fifty years. You log it all off and then go back over it again (you, or your successors) and find a second merchantable stand ready for its conversion into paper."[19]

Sherman Rogers, an industry correspondent for *Outlook*, also relayed a positive message during this period: "I spent ten days tramping over part of the timbered areas of Admiralty Island, which doesn't contain the best Alaskan pulp timber, by any means. I was amazed at the size and quality of the spruce; trees sound as a bullet and six feet in diameter were not

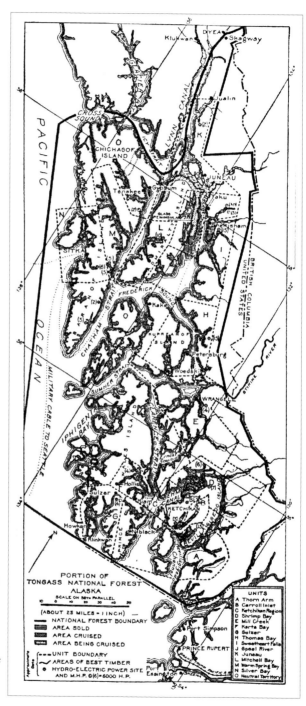

FIGURE 33. Forest Service Tongass pulp timber units, 1920. (USDA Forest Service, *Regional Development of Pulpwood Resources of the Tongass National Forest*)

uncommon, and one of the main tracts I looked over contained a large area that would cruise better than one hundred thousand feet to the acre."[20]

The Forest Service had from the beginning considered itself an authority on what the territory needed, but now its approach became even more cavalier. In his report, Clinton Smith had written that "Alaska's first need is capital," and added that "large-scale operations are essential." Smith felt that "scarcely any other part of the country offers a field for the upbuilding of a permanent pulp and paper industry equal to that afforded by Alaska," and, he said, "The value to Alaska of a pulp and paper industry on the National Forests can scarcely be overstated." Smith also touted the easy disposal of mill effluents into tidewater as a regional advantage held by Southeast Alaska over the great majority of pulp mills in the U.S., which tended to be located along rivers.[21] The *American Lumberman* went further, stating that Alaska's timber could "solve the world's acute pulp and paper situation"[22]

In consideration of the domestic shortage of newsprint paper, the Forest Service justified a more flexible pulptimber sale policy: 30-year contracts plus an additional 15-year reserve supply where sufficient timber was available. The minimum acceptable prices on pulptimber stumpage were about $0.50 per cord for Sitka spruce and $0.25 for western hemlock. Stumpage reappraisals were to be at five-year intervals and based on the then-current value of similar timber in Southeast Alaska, with full consideration given to the quality and accessibility of timber included in a particular contract, as well as other physical factors in a particular operation. To guarantee a certain amount of price stability to the purchaser during what was felt to be the critical first 10-year period of mill operations, a provision was made that the amount charged for the timber in the second reappraisal period would under no circumstances exceed double the charges for the first period.[23]

Since timber could be purchased and paid for as it was used, major up-front investments in timber were unnecessary. This substantial incentive would enable a company to concentrate its investments in productive capacity, such as a pulp mill, rather than timber. Because of interest charges, taxes, and protective costs (mostly from fire, which was not a factor in Southeast Alaska), owners of timberlands on the Pacific Coast at that time generally considered their investment cost in standing timber to double every eight or ten years, depending upon the rate of interest employed in calculations.[24] Added to this was the continual demand by stockholders for dividends on their investments in private timberland.

In a 1915 paid advertisement, the Forest Service had highlighted the favorable terms: "Lumbermen have bought National Forest stumpage in preference to acquiring privately owned timber for an operation because of the desirability of working under a Government contract. They have not found the requirements of the Forest Service impractical or burdensome. An important factor is that National Forest stumpage is paid for only as it is cut— no heavy initial investment in timber is required."[25]

There was yet another advantage to purchasing timber from the Forest Service. Logged-off lands in the West were generally considered a liability, and it had been the custom of lumbermen to sell them cheaply or, if no buyer could be found, abandon them. Since timber purchased from the Forest Service was for stumpage only, the problems associated with logged-off lands would be the agency's.

"It is now possible," wrote J. M. Wyckoff, Ketchikan district ranger on the Tongass National Forest, in 1920,

> to secure large areas of public timber and public water power in Alaska, for long terms at very reasonable rates upon otherwise liberal terms in the forest reservations for pulp making or other manufacturing uses. These laws enable the pulp maker to obtain timber and power without any investment in land, without payment of taxes or interest on investments in lands, and without danger of ruin by loss in forest fires. They permit him to limit his investment to the minimum amount needed to build his power plant and pulp mill, and machinery and the cost of installation, while the government carries his necessary initial investment in timber, lands and power site. He is not obliged to pay for his timber until it is cut for pulp making and his expense is thus reduced to the minimum.[26]

In saying that the investor could "limit his investment," the Forest Service was understating the problem. However favorable the terms may have been for acquiring timber, contracts for its purchase included a stiff requirement: the timely construction of an appropriately sized pulp mill—certain to be an expensive proposition, given Southeast Alaska's remoteness and the severity of its topography and weather.

In 1921, Chief William Greeley offered another rosy vision of Southeast Alaska's possible future: "Wisely handled, a paper industry can be developed in Alaska as permanent as the paper industries in Scandinavia, and capable of supplying a third of the present paper consumption of the United States… Alaska 30 or 40 years hence should be a second Norway, with permanent mills supported by an assured supply of timber with stable industrial communities."[27] Secretary of Agriculture Edwin Meredith was even more confident: "With her enormous forests and rapidly growing species suitable for pulp, her water power, and her tide-water shipment of manufactured products, Alaska will undoubtedly become one of the principal paper sources of the United States."[28]

That vision for Southeast Alaska was quickly challenged, however, by the American Pulp & Paper Association, which in a letter to its members stated that "considerable doubt is expressed by paper manufacturers that paper from Alaskan wood can be used profitably for paper sold in eastern United States, in view of the long freight haul. It would necessitate coming into competition with paper manufactured from wood cut in the eastern Canadian provinces and Scandinavian paper."[29]

FOREIGN COMPETITION

The American Pulp & Paper Association represented new and established paper-manufacturing concerns who were at that time making heavy investments in the production of newsprint paper, particularly in eastern Canada. They viewed the potential production of major amounts of newsprint paper in Alaska as a threat to their investments, at least in the short term. In the long term the industry expected the development of a pulp and paper industry in Southeast Alaska, but not until the market had expanded sufficiently to justify the additional production.

The growing production of newsprint paper in Canada was an issue for the Forest Service, and the evolution in the agency's timber sale policies was in large part an attempt to be competitive with opportunities in Canada. In 1921, William Greeley wrote, "The terms offered by the Forest Service to paper manufacturers in Alaska are, indeed, more flexible and more favorable to the operator than in the case of any public timberlands in Canada, with which comparisons have frequently been drawn."[30]

Among the unfavorable terms for obtaining timber in Canada referred to by Greeley was the program introduced in 1905 that provided for transferable 21-year timber-cutting licenses, known as "timber limits," on square-mile tracts of Crown lands. Costs to operators for these options involved ground rent, a fire protection tax, and a royalty on timber cut.[31]

Despite those terms, during the years 1905 through 1907, timber-cutting licenses were issued on some 14,000 square miles of Crown lands in British Columbia.

In about 1920, the province of British Columbia offered 30-year leases for pulptimber. In exchange, the operator was required to produce a $350,000 bond and construct a mill within two years. The operator also paid a royalty of $0.85 per MBF for the better grades of timber and $0.25 for the lower grades. This royalty was subject to adjustment. An annual fee of $0.10 per acre for ground rent and $0.015 per acre for fire protection was also paid.[32]

In 1921, Alaska's territorial legislature offered a solution of its own to the problem of establishing a pulp and paper industry in Southeast Alaska. In a joint memorial to Congress, the legislature wrote, "Whereas, Private capital has not been enlisted to meet the needs of a larger paper supply available at fair prices, it is desirable that the Federal Government should enter the industry as a direct method of controlling the situation."

The legislature requested that Congress "enact necessary legislation for the construction and operation of a pulp and paper factory in the Territory of Alaska."[33] The legislature's petition, however, was futile, as was a similar request that year by the City of Ketchikan, which wrote the following to Senator Miles Poindexter (R-Washington):

> Inasmuch as the Government uses tons and tons of paper for their own printing which they have to purchase it seems a great folly to permit private enterprises to take up all the power sites which the Government now owns and controls in this great Territory when the Government could, having the power and needing and using all kinds of paper and great amounts of it, build a paper mill of its own [at or near Ketchikan].[34]

The sale terms developed by the Forest Service were embodied in two 1921 pulptimber sales. The first was for 335 million cubic feet (almost 2 BBF) of pulptimber on the west side of Admiralty Island. The sale area encompassed a strip of land 2 to 4½ miles wide running from the head of Hawk Inlet to Mitchell Bay, a distance of about 50 miles. It was prepared in response to an application by Bartlett Thane of the Alaska Gastineau Paper Company. The second was for 100 million cubic feet (approximately 550 MMBF) on 45,000 acres that extended some 55 miles along the shore of Behm Canal on Revillagigedo Island.[35]

Bartlett Thane was general manager of the Alaska Gastineau Mining Company, a gold mining and milling interest that had come on hard times. Thane hoped to salvage his investment by establishing a 200-tons-per-day newsprint paper mill at the company's mill site at Thane, about five miles south of Juneau. At the site were warehouses, shops, a wharf, and a large amount of company-owned housing. As well, the company owned a substantial hydroelectric plant on Salmon Creek.[36]

The Forest Service prospectus divided the timber on Admiralty Island into three general types:

- **Overmature trees with understory of young timber.** This type constituted about 50 percent of the timber area in the sale. The large hemlock component in this classification was considered to be highly defective, but the small hemlock in the understory was considered excellent. Western hemlock made up about 90 percent of this forest type, with the remainder being Sitka spruce, which contained little defect.
- **Mature timber.** This forest type, which encompassed about 25 percent of the sale area, was described in the prospectus as "dense stands of large, thrifty spruce and hemlock" with little defect and no understory of younger timber.
- **Young hemlock and spruce.** This forest type accounted for the last 25 percent of the sale area. It comprised young hemlock and spruce trees ranging from 8 to 24 inches in diameter. The trees possessed little defect and were considered by the Forest Service to be "a most excellent class of pulp timber."[37]

The terms of the sale required the successful bidder to establish a pulp manufacturing plant of not less than 100 tons daily capacity. The minimum bid acceptable to the Forest Service for the timber was $0.60 per hundred cubic feet (CCF) for Sitka spruce and Alaska-cedar, and $0.30 per CCF for western hemlock. The sale period was for 30 years, with stumpage prices adjusted at five-year intervals.[38] No consideration was made to manufacture the timber into anything other than pulp.

Though the prospectus did not specify clearcutting, it was understood that all 90,000 acres of "merchantable timberlands" in the sale area, except for designated seed trees, would be clearcut. As Frank Heintzleman said later, "You cut it as clean as possible, but leave patches of timber here and there throughout the cutting areas for reseeding purposes."[39] Such regeneration

efforts remained for many years the Forest Service's general prescription for the entire Tongass.

FIGURE 34. "Thrifty" timber: a relatively young stand of mixed Sitka spruce and western hemlock that contained little decay and was highly suitable for the manufacture of pulp. (USDA Forest Service, Alaska Regional Office, Juneau)

Bartlett Thane promoted his idea of a pulp mill at Thane for seven years, and though he managed to attract a group of Japanese to inspect the area, he found no serious interest in his proposal. Admiralty Island's timber was not purchased, the Alaska Gastineau Mining Company was foreclosed upon and sold in 1924, and Bartlett Thane was eventually blacklisted as a promoter.[40]

The Alaskan-American Paper Corporation (New York) was the successful bidder on the Revillagigedo Island pulptimber and had received a permit from the Federal Power Commission for a waterpower development at Shrimp Bay. The company's contract required the construction of a pulp mill with a daily capacity of 25 tons, and the company initially anticipated a cut of 2.5 million to 3 million cubic feet of timber annually. Alaskan-American, however, soon proclaimed its intention to start work immediately on a pulp and paper mill with nearly five times the capacity of that originally envisioned—a $4 million facility capable of producing 120 tons of newsprint paper per day. Expectations and optimism ran high in Ketchikan, with a *Ketchikan Alaska Chronicle* headline proclaiming, "Shrimp Bay Paper Pulp Plant Assured."[41] But, as had happened before and would happen again, the project never progressed beyond the planning stage.

CHAPTER 10

Warren G. Harding Visits Alaska

IN THE YEARS IMMEDIATELY FOLLOWING THE ESTABLISHMENT OF THE TONGASS NATIONAL FOREST, PERCEIVED INEFFICIENCIES IN THE VARIOUS DEPARTMENTS AND BUREAUS THAT MANAGED ALASKA'S RESOURCES LED TO FRUSTRATION AMONG DEVELOPMENT-MINDED INDIVIDUALS. MINERS WERE PARTICULARLY AGGRAVATED. IN A 1911 SPEECH TO THE AMERICAN MINING CONGRESS THAT WAS REPRINTED BY THE U.S. SENATE, A MINING REPRESENTATIVE FROM VALDEZ, ALASKA, SPOKE OF GOVERNMENT OFFICIALS IN WASHINGTON. AS "LOCUSTS OF OFFICIALDOM" WHO ENDEAVORED TO "GATHER ALL OF THE PUBLIC LANDS OF OUR TERRITORY AS AN ESTATE—ITS FORESTS, ITS MINES, ITS WATER POWER, ITS WHARF SITES—TO BE MANAGED BY A VAST HORDE, A DEVASTATING HOST OF GOVERNMENT UNDERLINGS AND WORKED BY OUR CITIZENS AS TENANTS, LICENSEES, AND LESSEES."[1]

Some in government were ready to rectify the perceived wrongs. In 1914, a movement was afoot in the Department of the Interior and Congress to put control of all governmental affairs connected with natural resources and development in the territory into the hands of a "Development Board" under the control of the Secretary of the Interior. The effort failed, and it is worth noting that despite a claim by the *Washington Post* that Alaska was being kept as a "breeding ground for fads in forestry," few complaints with the Forest Service were heard in Alaska, where administration was considered satisfactory and efficient. Practically all questions relative to the forests were settled by local officials, and when it was necessary to seek outside approval, a prompt response was the norm. Forest Service Chief William Greeley stated in 1921 that "ninety-five per cent of the business of these National Forests does not pass beyond Alaska."[2]

President Warren G. Harding visited Alaska in 1923 and endorsed the Forest Service's efforts to establish a pulp and paper industry in Southeast Alaska. He added his perspective to the Forest Service's pulptimber contracts: "I venture, with some knowledge of conditions in various paper-making countries, to state that no better contract, indeed none so good, can be

FIGURE 35. President Warren Harding at Metlakatla, Alaska, in 1923.
(Alaska State Library, President Harding's Trip to Alaska, 1923 collection)

secured in any of them." Of the pulp and paper industry's potential in Southeast Alaska, Harding predicted, "We are, in short, on the eve of an expansion which, if not rapid, will be sound and permanent."[3]

Harding, a Republican and the first U.S. president to visit Alaska, had good reason to be in the territory. The seemingly perpetual argument over for what ends and by which agency Alaska should be administered had intensified and was causing deep conflict within his own administration.

Historically, one of the federal government's primary functions with public lands was to transfer them into private ownership, which the Department of the Interior, considered by some to be the government's real estate agent, did generously through numerous mechanisms, such as the homestead program, mining claims, the timber culture program, and grants to railways. The Forest Service, however, was charged with a very different function— the stewardship of public lands such that they retained their productivity for the benefit of future generations. One of its goals was to ensure a perpetual supply of timber through science-based forestry. This was in marked contrast to the general management of private forestlands at a time when little consideration was given to purposeful forest management.

The idea of any restrictions or limits on access to public timber and other resources in Alaska did not sit well with the proponents of development, both within and outside Alaska. Many had become comfortable with the liberal land use and disposal policies of the Department of the Interior and considered the administration of the two national forests in Alaska by the conservation-minded Forest Service an unconscionable hindrance to the region's development. References were made to the "dead hand of administrative mismanagement," Alaska as a "treasure-house whose door had been sealed against legitimate enterprise by a lot of short-sighted bureaucrats," and the "intolerable muddle of Federal mismanagement." For a fact, the population of Southeast Alaska had been declining throughout the twentieth century, but this was mostly because of the loss of population at Skagway after the height of the Klondike gold rush.

One individual who shared those no-hindrance-to-development views was Albert Fall, Harding's Secretary of the Interior. Fall advocated the consolidation of the administration of Alaska under Interior, and he wanted to allow private business to immediately exploit the territory's natural resources. For Fall, the Alaska issue was part of a broader effort, which was the return of control of the national forests to Interior. He considered the Alaska situation to be, in the words of William Greeley, "the prize exhibit of a shackled empire."[4]

As one might expect, Fall was vehemently opposed by Greeley, who charged that Fall's proposed scheme invited "unrestrained and destructive exploitation."[5] Henry Wallace, Harding's Secretary of Agriculture, shared Greeley's sentiments.

Wallace described the developers' Alaska complaints in a September 1923 article in *The Timberman*: "It was reiterated incessantly that the natural resources of the territory were locked up by impractical theories of conservation; that bureaucratic red tape had the young empire of the north lashed hand and foot; that its resources could not and would not be developed," and that "the very life blood of the territory was being sucked out by federal vampires."[6]

There were, of course, others who saw no problem with the Forest Service in Alaska. One of those was Dan Sullivan, Alaska's non-voting delegate to Congress, who in 1922 said,

> I don't think the service could be improved... I have talked with representatives of the largest paper companies in the United States relative to the leasing provisions allowed by the Forest Service, and they tell me the provisions are liberal enough to accommodate them in every respect. As far as I have heard, there has not been any criticism of the Forest Service in Alaska.[7]

Sullivan's comments were reinforced by the *Ketchikan Alaska Chronicle*, which in 1923 wrote,

> It may be said that the forestry bureau in Alaska which comes in direct contact with a substantial portion of the public has given such eminent satisfaction that it is one of the few bureaus of the national government of which one never hears complaint—a rare accomplishment in this age. Its treatment of the public is always fair, prompt and courteous.[8]

P. S. Ridsdale, secretary of the American Forestry Association, held a similar opinion. The Forest Service, he wrote, "has established a competent and efficient decentralized organization, has developed sound and workable policies, and has the confidence of the country at large." Ridsdale warned that if Secretary Fall was successful, Alaska might become the "promised land of bilk and money for the interests."[9]

The issue became very heated, and in 1923 Harding scheduled a three-week excursion to Alaska aboard the Navy troop transport ship *Henderson*, in part so he could personally examine the situation. Accompanying him on the July trip were Secretary of Agriculture Henry Wallace and Secretary of the Interior Hubert Work (Albert Fall had resigned in March because of the Teapot Dome scandal, which involved his leasing of federally owned oil fields in exchange for "loans"), as well as Herbert Hoover, then Secretary of Commerce; Frederick Gillette, Speaker of the U.S. House of Representatives; and Forest Service Chief William Greeley. While in Southeast Alaska, the group visited the communities of Metlakatla, Wrangell, Juneau, Skagway, Sitka, and Ketchikan.

"There is no sea trip in the world to equal it," Harding said of his visit. "There is no lure of mountain, stream, valley, and plain to surpass it anywhere."[10] Upon returning from Alaska, Harding made in Seattle what was to be his last formal appearance before a group of his fellow citizens. To an audience estimated at 35,000 to 40,000 gathered at the University of Washington stadium, he reported to the nation on the situation in Alaska. "I must confess," Harding said,

> I journeyed to Alaska with the impression that our forest conservation was too drastic, and that Alaska protests would be heard on every side. Frankly, I had a wrong impression. Alaska favors no miserly hoarding, but her people, Alaskan people, find little to grieve about in the restrictive policies of the federal government. There is not unanimity of opinion, but the vast majority is of one mind. The Alaskan people do not wish their natural wealth sacrificed in a vain attempt to defeat the laws of economics, which are everlasting and unchanging. I fear the chief opponents of the forest policies have never seen Alaska, and their concern for speedy Alaskan development is not inspired by Alaskan interests.[11]

In his speech the president had clearly rebuked Albert Fall and endorsed the conservation-minded viewpoint of his Secretary of Agriculture.

Several days after giving his report on Alaska, President Harding died of natural causes and was succeeded by Calvin Coolidge. Former Secretary of the Interior Albert Fall was later convicted, fined, and imprisoned for his part in the Teapot Dome scandal.

Among those who agreed with Harding's handling of Alaska was George Cornwall, publisher of *The Timberman*, who wrote a front-page editorial in his July 1923 issue titled "Forest Service Should Retain Control of Alaska's Forest Resources." Cornwall echoed the rationale for considering forestry a form of agriculture: "The growing of timber is similar to that of growing any other crop, and therefore more properly should be administered by the Department of Agriculture" with its staff trained in the technicalities of forestry. He believed that the Forest Service had an effective organization in Alaska and that "the lumbermen of Alaska...by reason of years of association, have come to understand the viewpoint of the Forest Service, and as a consequence, a mutual respect has been established." Cornwall praised the Forest Service as having been "freer from political control than possibly any other branch of the government."[12]

CHAPTER 11

The Forest Service Perseveres

I N MARCH 1923, THE FOREST SERVICE OFFERED A PULPTIMBER SALE OF 334 MILLION CUBIC FEET, COMPARABLE IN QUANTITY TO ITS OFFERING ON THE WEST SIDE OF ADMIRALTY ISLAND IN 1921 THAT NEVER SOLD. THE CASCADE CREEK UNIT COMPRISED TWO TRACTS NEAR PETERSBURG, IN CENTRAL SOUTHEAST ALASKA. THE FIRST TRACT, 43,000 ACRES, WAS ON THE MAINLAND SIDE OF FREDERICK SOUND AND INCLUDED PART OF THOMAS BAY, THE LOCATION OF CASCADE CREEK. THE SECOND TRACT, 300,000 ACRES, WAS ON THE NORTHERN PORTIONS OF KUIU AND KUPREANOF ISLANDS.[1] TIMBER ON THE SECOND TRACT WAS DESCRIBED BY FRANK HEINTZLEMAN:

> A large portion of the unit, especially in the northeastern part, carries a stand of 75 per cent of which is clean, smooth hemlock averaging between 20 and 30 inches in diameter breast high, and 90 to 120 feet in merchantable length to a six-inch top, together with Sitka spruce from 20 to 40 inches in diameter breast high and 95 to 150 feet in merchantable length.[2]

A potential buyer had been expressing interest: "We have been investigating Cascade Creek for the past two years," Frank C. Dougherty, a representative of a San Francisco brokerage firm, was quoted as saying in summer 1922, "and all figures secured from the preliminary surveys look favorable and now after looking it over it is even better than we thought. In fact, it is almost too good to be true."[3]

The Forest Service set minimum acceptable stumpage prices the same as those on the West Admiralty Island offering: $0.60 per CCF for Sitka spruce, and $0.30 per CCF for western hemlock. During the 30-year operational life of the contract, timber prices were to be adjusted at five-year intervals, with a special provision that pulpwood prices for the second period could not exceed twice those of the first period. To ensure that practically no investment in timber was necessary, advances for stumpage were limited to $20,000. On August 24, 1923, the sale was conditionally awarded to the San Francisco brokerage firm of Hutton, McNear and Dougherty, which had bid the minimum acceptable rates and spoke of spending $10 million on the construction of the largest pulp and paper mill ever built in the United States. The company also expressed its intention to build a 100-MBF-per-day "cargo mill" to cut lumber specifically for the maritime export trade.[4]

The *Paper Trade Journal* reported in early 1923 that at the Department of Agriculture, negotiations were under way "which are expected to initiate the establishment of an important industry in southern Alaska capable of furnishing perpetually a large proportion of the print paper demands of the United States," and that the Secretary of Agriculture expected "soon to sign contracts with a number of responsible concerns for the purchase of pulpwood from Alaska."[5] Secretary Henry Wallace himself was quoted in the *Ketchikan Alaska Chronicle* as saying, "The physical conditions in southeastern Alaska—presence of cheaply developed power, an enormous supply of inexpensive wood, and the availability of water transportation—are the very factors which make inevitable the expansion of pulp and paper manufacturing in the Territory."[6]

Though there was much fanfare over the awarding of this contract—which would have been the largest sale of pulptimber ever made by the Forest Service—the sale failed to progress beyond being "conditionally awarded." Hutton, McNear and Dougherty, which the *Paper Trade Journal* later derided as "a firm of promoters," failed to carry out the development program required by the contract, not even making a financial showing for the purchase of timber.[7] The concern's Federal Power Commission power site license was revoked in the spring of 1926.[8]

Despite increases in the consumption of newsprint paper, serious obstacles remained to the development of large-scale pulp and paper operations in Southeast Alaska. The September 27, 1923, lead editorial of *Paper Trade Journal* was titled "Alaska's Problem." "Nature seems to have brought a number of unfortunate obstacles to bear," the editorial observed, "and it will take a good deal of ingenuity to overcome them." The obstacles cited were "little waterpower, a limited area of richly timbered land and bad transportation." The development of an Alaska pulp and paper industry, in the eyes of the *Paper Trade Journal*, "would hardly pay until paper and timber become much more valuable than they are now."[9]

Although waterpower overall was adequate, according to the *Paper Trade Journal*, it consisted of many relatively small sources, the tying together of which would be an "impossibly expensive feat." It was certainly true, as the journal asserted, that all of Alaska "clear up to the Arctic Circle" was anything but heavily timbered, but the publication misrepresented the extent of Southeast Alaska's resource by stating that the only timber of value laid in a narrow strip around Juneau and Skagway. Certainly, the transportation infrastructure of Alaska was pretty much as described—extremely primitive, though to serve the distant markets that some envisioned, transport by water held some advantages over rail.[10]

The September 1923 trade-journal article caused great consternation in the Forest Service. Secretary of Agriculture Wallace wrote to the publication in protest, asserting that its editorial contained "certain misstatements" regarding waterpower, timberlands, and transportation.[11] Actually, the development of a pulp and paper industry in Alaska faced even more problems than were noted by the *Paper Trade Journal*.

First, shipping costs were high. The Forest Service eventually reduced its estimated potential annual production for Southeast Alaska to a million tons of newsprint paper, which amounted to over 2,700 tons (5.4 million pounds) per day.[12] The marine transport of this amount of material would require a dedicated fleet of vessels. As Charles Flory, forest supervisor on the Tongass, wrote in 1920,

> Eastern capitalists who are figuring on the establishing of paper plants in Alaska wish to make the shipments by boat through the Panama Canal to the Atlantic coast, and that is a problem which has not been settled. If they run their own boats they will be compelled to go into the common carrying business, in order to have return cargoes.[13]

Southeast Alaska was at a transportation disadvantage even with Swedish manufactures of pulp and paper. The distance from Juneau to New York City via the Panama Canal is some 6,900 miles. The distance from Stockholm to New York City is less than 4,000 miles.[14]

The freight cost per ton for full cargo shipments of newsprint paper from Southeast Alaska was later estimated at $4 to Seattle, $6 to San Francisco, and $12 to $13 to Gulf of Mexico ports via the Panama Canal. The economics of trans-Pacific transport, however, were more favorable, since the sailing time between Ketchikan and Tokyo was about 15 percent less than between Seattle and Tokyo. This gave Alaska a slight advantage over the Puget Sound area in transportation to the Orient, provided there was no backhaul involved.[15]

Another option was to utilize the Canadian Pacific Steamship Company's ships, which regularly plied the waters from the rail terminus at Prince Rupert north to Ketchikan (just 90 miles away) and on to other Southeast Alaska communities. The Grand Trunk Pacific Railway, which connected Prince Rupert to points in eastern North America, was completed in 1914 and theoretically could provide economical transport to the major markets in the U.S.'s Midwest and East Coast.

That option, however, was stymied with the passage of the Merchant Marine Act of 1920. Section 27, known as the Jones Act, was designed to protect U.S. maritime interests by requiring that all passengers and cargo transported between U.S. ports be aboard U.S.-built and U.S.-operated vessels. The way the law was written, Southeast Alaska operators could not ship products destined to U.S. markets to the railhead at Prince Rupert aboard Canadian vessels. Efforts to secure an exemption for Alaska wood products proved futile.

The possibility of operating a dedicated railroad car ferry or barge service between Alaska ports and Prince Rupert was also explored. Lumber was occasionally shipped via the Canadian rail system after being transported to Prince Rupert by U.S. carriers.

A second problem facing the industry in Alaska was, paradoxically, the high prices for newsprint paper in the early 1920s. The price run-up prompted huge, speculative investments of largely U.S. capital in eastern Canada, which possessed vast forests on fairly level ground, abundant waterpower, and good transportation by rail and water to the main U.S. markets. Though the consumption of newsprint paper in the United States rose during the Roaring Twenties, the production of newsprint paper rose even faster. Canadian production increased from 800,000 tons in 1920 to four million tons in 1930, causing the supply of newsprint paper to exceed demand. Prices fell, and producers of newsprint paper were faced with the challenges presented by overcapacity. There was little incentive to consider, therefore, massive investments in new production facilities in Southeast Alaska, despite the Forest Service's enticements.

In 1929, Alaska Governor Scott Bone blamed the Forest Service for the absence of a pulp manufacturing industry in Southeast Alaska. He charged that "manufacturers were investing huge capital in Canada, but were looking askance at Alaska, because the regulations of the Forest Service were regarded as too stringent." In fact, at the time of Governor Bone's remark, the overbuilt Canadian industry was seriously distressed, and authorities in Ontario and Quebec had let it be known that new mill construction would be discouraged.[16]

A third problem was consolidation within the industry. A few large players, such as the International Paper Company with its interests in Canada, and later the Crown Zellerbach Corporation (the result of the 1928 merger of the Zellerbach Corporation and the Crown Willamette Paper Company) controlled much of the industry.[17]

Fourth, pressure was beginning to develop to utilize the immense quantities of pulp-quality timber going to waste in logging and sawmill operations in Washington and Oregon.

Fifth, though it was of no immediate consequence, the increased use of waste paper as a raw material for paper plants was being seriously discussed and researched during this period in response to dire yet unfounded warnings of a future "paper famine."[18]

Finally, the lack of a skilled labor force and the high cost of doing business in Southeast Alaska remained a problem.

Considered together, the obstacles to the development of a pulp industry in Southeast Alaska were pretty much insurmountable. Even a Forest Service forester admitted in 1922 that "economic conditions during the past year have not been favorable to the launching of a pulp and paper industry in Alaska. Business in the Territory has encountered the same difficulties as in the States."[19] It appeared that the only factors that might improve the territory's timber fortunes were the possibility of an unexpected actual or potential shortage of newsprint paper, or a scientific discovery or technological development that would stimulate the need for and value of wood pulp. Nevertheless, while the small Alaska Pulp & Paper Company enterprise along the Speel River operated for only a short time, and mills in Canada struggled with the problems associated with overproduction, the Forest Service continued its efforts to develop a massive pulp and paper industry in Southeast Alaska.

Despite the lack of success in its early efforts, the establishment of a wood pulp industry in Southeast Alaska remained the Forest Service's priority. Frank Heintzleman reiterated earlier Forest Service statements in 1928: "Aggressive action will be taken to interest prospective investors in the pulp-timber and water-power resources." He believed that the management of the timber resources of the Tongass National Forest had two objectives: "The development and maintenance of a permanent pulp and paper manufacturing industry commensurate with the available waterpower and timber resources," and "The furnishing of a permanent and convenient supply of timber for local consumption, with such an additional supply to the local sawmills for the general lumber markets as may be needed to justify efficient milling facilities and provide yearlong operations."[20]

CHAPTER 12

The Agency Promotes Efficient Utilization

L ATE IN 1925, THE FOREST SERVICE ANNOUNCED A LIBERALIZATION OF THE TERMS OF PROPOSED ALASKA PULPTIMBER CONTRACTS. UPON THE RECOMMENDATION OF THE FOREST SERVICE AND WITH THE APPROVAL OF THE SECRETARY OF AGRICULTURE, THE AMOUNT OF TIMBER THAT COULD BE OFFERED IN ALASKA PULPTIMBER CONTRACTS WAS INCREASED FROM TWO BILLION BOARD FEET TO A MAXIMUM OF THREE BILLION. ADDITIONALLY, THE PERIOD OF TIME BEFORE STUMPAGE PRICES COULD BE ADJUSTED WAS EXTENDED. EARLIER CONTRACTS STIPULATED THAT OPERATORS HAD A SEVEN-YEAR PERIOD IN WHICH TO CONSTRUCT A PLANT AND CARRY ON OPERATIONS BEFORE STUMPAGE PRICES COULD BE ADJUSTED. UNDER THE LIBERALIZED TERMS, THIS PERIOD WAS EXTENDED TO THE TIME REQUIRED FOR THE CONSTRUCTION OF A PLANT PLUS TEN YEARS.[1] "COLONEL GREELEY," SAID THE *KETCHIKAN ALASKA CHRONICLE* OF THE FOREST SERVICE CHIEF, "IS MORE DETERMINED THAN EVER TO OPEN UP THE ALASKAN [PULP AND PAPER] INDUSTRY."[2]

One of the ways the Forest Service supported possible development of a pulp and paper industry was to make increasingly detailed timber inventory information available to prospective investors. By 1927, the agency had cruised and mapped some 600,000 acres of the Tongass; this effort had been materially helped by the U.S. Navy which—at the request of the Forest Service and the U.S. Geological Survey—had made an aerial photographic survey of about two-thirds of Southeast Alaska during the previous summer. District Forester Charles Flory referred to the aerial survey as "the greatest single step ever taken for the industrial development of Southeastern Alaska."[3]

The resulting photographs served to help construct a more comprehensive map of Southeast Alaska's timber and show in a general way its character. Additionally, the survey provided much information regarding developable waterpower sites that might be of use to potential pulp and paper operators. The Forest Service estimated that the aerial surveys yielded more

FIGURE 36. By about 1926, the Forest Service had reduced the number of pulp timber units to seven that were each roughly twice the size of the 1920 units. (Library of Congress)

information in a few seasons than could have been gained in 20 years of ground surveys.[4]

HARVEST ROTATIONS AND EMPLOYMENT ESTIMATES

Obtaining an accurate inventory of the Tongass's timber resource was essential, but another important piece of information that would be critical for developing any timber-based industry in Alaska was how fast trees would grow back following harvesting. Investigating regrowth rates was one of the first research endeavors of the fledgling Forest Service early in the twentieth century but was slow to come to Alaska despite the mounting pressure for industrial development. By the 1920s, there had been no studies that had progressed far enough to definitely establish the rate of regrowth of pulptimber in Southeast Alaska. The 1924 estimate of a 50-year rotation period, which had been based on coniferous forest regrowth in New England, was abandoned. In its place by 1927 was an 85-year rotation period that Frank Heintzleman characterized as "very conservative," implying that the actual rotation period would likely be shorter. By the following year, however, Heintzleman had increased his estimate to 85 to 100 years. During this period, he wrote, "virgin timber will be entirely cut," followed by a new growth of timber that he predicted would "have a [pulptimber] volume per acre at least twice the average volume now found in the virgin commercial forest." This assumption still held in 1960, when Regional Forester P. D. Hanson said, "We know we can produce about double the forest crop in the second rotation. Heintzleman's "conservative" plan (in 1927) had estimated a sustained annual yield on the Tongass of some 900 million board feet of pulpwood. An attractive facet of the short rotations that Forest Service officials proposed was that little decay occurs in Sitka spruce and western hemlock trees until they are about 100 years old.[5]

W. J. Frost, a logging engineer employed by private interests to cruise and assess Southeast Alaska timber in the late 1920s, was far more conservative than Heintzleman. In an unpublished report, Frost estimated that "it will take at least 150 years for this forest to reforest to a size suitable for [pulpwood] operation. Of course the reforestation will be more rapid in some localities than others, but the entire forest will not reproduce in less than 150 years."[6]

The first substantial study to scientifically estimate regrowth on the Tongass began modestly in 1924 as a part-time effort by rangers who were mostly busy with other duties. They had little to work with, because logging at that time was still dominated by single-tree removal and there were very few second-growth stands that had resulted from logging. The rangers had to work largely in patches of second growth that originally had been cleared by windstorms or landslides, and in abandoned Native villages and former mining sites. Over the following years, however, the research was gradually intensified, and the results were published in 1934. The Department of Agriculture's *Yield of Second-Growth Western Hemlock-Sitka Spruce Stands in Southeastern Alaska* suggested that second-growth stands in Southeast Alaska could be cut for pulpwood at an age of approximately 75 years, but it also stated that rotation periods had not yet been determined for the region. The 1937 *Report of the Alaska Resources Committee*, of which Frank Heintzleman was a member, cited studies that indicated a rotation period "which should be about 75 to 80 years." *Pacific Pulp & Paper Industry*, in a 1944 article that Frank Heintzleman may have helped write, spoke of a rotation period on the Tongass that averaged 78 years. By 1949, Heintzleman had modified his estimate, writing that a Tongass timber program based on "timber cropping," with a rotation period of 80 to 85 years and an annual cut of approximately one billion board feet of timber per year, was a volume sufficient to make "at least a million tons of chemical pulp, plus considerable quantities of high-grade lumber and plywood."[7]

In 1946, Forest Service Chief Lyle Watts commented that in Alaska, "Pulpwood possibilities are especially promising," and because most of the timber was located on national forests that "[i]ts sound development and orderly use is therefore a Forest Service responsibility." On the other hand, Watts noted the absence of organized research in Alaska and wrote, "Extremely limited information is available as to the best methods of cutting and management to assure perpetuation of desirable growth."[8]

Heintzleman's initial projections of potential economic activity in Southeast Alaska were as optimistic as his predictions of regrowth in the forest. In 1927, he wrote that the annual production in Southeast Alaska of one million tons of newsprint paper would require 13 plants, each with a daily capacity of 250 tons and a capital value of $10 million. Each plant (including logging operations) would employ about 750 men, for a total direct employment of almost 10,000. Indirect employment by the mills would add considerably more, which Heintzleman thought "should result in the eventual addition of at least fifty thousand to the present population."[9]

By 1949, Heintzleman had revised his estimate downward: "If fully developed, the [newsprint paper] industry could support, directly or indirectly, a total of 30,000 persons in southeastern Alaska."[10]

The lack of skilled labor in the region was no impediment to the development of the industry, according to Heintzleman:

> The recruiting and holding of a large force of woods and mill workers on a paper project here will present no serious difficulties either as regards technical and skilled men or common laborers.... As regards common labor, it is the normal condition for more men to be coming into this country than the local industries can absorb. Seattle is a large center for woods and sawmill workers and one that is easily drawn upon at any time by the southeastern Alaskan industries. A majority of the workers in this region are Scandinavians, an excellent class of laborers. In addition, we have a native Indian population of over 5,000, and the mines, sawmills and canneries have found that they constitute an inexpensive and good supply of common labor.[11]

BOOSTERS AND NAYSAYERS

Within Southeast Alaska, the solid support for the establishment of a wood pulp industry was tempered with a provincial concern born of experience with salmon canneries and mining interests: that large "outside" interests could gain political and economic control of the territory. Legislation introduced in the Territorial legislature accordingly sought to discourage outside investors. Though such bills were introduced in the session that ended in the spring of 1927, none passed. Frank Heintzleman, however, thought that outside capital was required: "Unlike many parts of the West, where family and individual efforts could be depended upon in the pioneer days to build up the country, the development of the southern coastal section of Alaska appears to require a corporate industry and large capital investments."[12]

Outside Alaska, however, there were some serious reservations and opposition to the establishment of a large pulp industry in Southeast Alaska. One objection involved the qualities of Sitka spruce, particularly as a wood for use in the construction of airplanes. There was a concern expressed in

The Timberman that it was "truly an economic waste to work this fine grade of wood up into paper fiber." Forest consultant B. T. McBain, however, countered with a distortion that persisted for decades: "Alaskan timber is not saw timber, but good for pulping purposes only."[13]

Another objection had to do with the effect a large pulptimber industry in Southeast Alaska would have on the Pacific Northwest. In a May 1, 1927, front-page editorial, the Seattle, Washington-based *West Coast Lumberman*, the self-proclaimed "Pioneer Authority of Logging and Lumber Manufacturing Interests in the West," stated,

> It is in a way unfortunate that the government at this time
> has seen fit to proceed with plans for the development of
> the pulp and paper business in the North. There should be
> no occasion for crowding the splendid pulp timber of Alaska
> onto the market. That wonderful timber should rather be
> held in reserve until such a time, at least, as the pulp and
> paper industry in Washington and Oregon, now in its
> infancy, gets thoroughly established. While everybody
> believes and hopes that this infant industry here in the
> Northwest will thrive, nevertheless, it would have been
> infinitely better for all concerned, including the
> government, had the latter refrained from stepping into the
> picture just at this time.[14]

The "infancy" of the industry in the Pacific Northwest was actually quite robust. In 1921, Washington had produced a negligible amount of pulp from seven mills, but by 1929 the state had at least 17 mills, with more under construction and even more planned. The state of Washington, by this time, had quickly risen to rank fourth in pulp production, behind the states of Maine, Wisconsin, and New York, and contributed to the overproduction that plagued the industry.[15]

The official government stance on timber sale promotion, nationwide, was stated by William Jardine, Secretary of Agriculture: "The policy of the Department of Agriculture in the sale of national forest timber is not to crowd Government timber on the market in any locality, not to sell at bargain prices, but to make sales at not less than fair, carefully appraised prices and after public advertisement."[16] Nevertheless, sawmill owners and owners of timberland in the Pacific Northwest were concerned that new pulptimber was being offered by the government while vast amounts of pulpable timber and sawmill waste were going unutilized in the Pacific Northwest.

"Overproduction is not development; it is willful waste," according to one observer, writing in *Pacific Pulp & Paper Industry* in 1927.[17] In fact, the objections to Alaska production had less to do with to an aversion to waste than with a desire to see the Pacific Northwest's sawmill industry integrated with a pulp industry. Such an integration would generate revenues through the utilization of sawmill waste and the pulping of logs not considered suitable for sawing.

And there was a lot of waste. A Forest Service survey in the Douglas-fir region of western Washington and Oregon that was completed in 1928 showed waste in logging operations to average almost 21,500 board feet per acre—"probably more waste per acre left after logging than in any other timbered region that has ever been logged," according to an article in *The Timberman*. Annually, this totaled to more than three billion board feet of cordwood-size and larger material. A writer in the *Paper Trade Journal* in 1930 noted that wood, the "universal raw material" for newsprint, was "so relatively cheap and abundant that few serious efforts have been made to use it intelligently and economically."[18]

Part of the problem in Washington and Oregon was that the predominant commercial tree was the very large Douglas-fir. Logging individual trees that held 10,000 to 20,000 board feet of lumber required massive equipment that could not efficiently be used to harvest the generally smaller western hemlock that grew in the same stands.

In 1928, it was obvious to pulp operators that the "high-powered logger with his big machines...skinning out the prime logs and laying a wake of ruined pulpwood" was jeopardizing the future of the pulp industry. Twenty years later, in its own eyes, the pulp industry considered itself to be more responsible and efficient, since its operations clean-cut the forest and utilized timber considered useless by lumbermen. Pulpwood logging operations on the Tongass, where the forest was to be cut "as clean as possible," would therefore be a paradigm of responsible efficiency.[19]

The issue, as it related to pulp development in Southeast Alaska, was considered in a 1927 *Timberman* article by forest consultant B. T. McBain:

> The cut of spruce and hemlock [in Oregon and Washington] is 13.2 per cent of the total of 12 billion feet or over one and one-half billion feet of these woods. With the usual waste of

one-third cord per 1,000 feet in sawmill and one-half cord per 1,000 feet in logging camp, this was: Sawmill, 500,000 cords; logging camp, 750,000 cords; total, 1,250,000 cords.

In 1926 Oregon and Washington pulp mills used less than 600,000 cords, so there were wasted in Oregon and Washington twice as much pulp wood as was used. This is true every year.

We do not need more government pulp wood or logs, but we do need to stop waste in our present operations, logging, sawmills, and pulp mills… So let's get together—not find fault with Uncle Sam or feel sorry for ourselves because Alaska is getting some help—but help ourselves by helping those now in the lumber industry to stop waste and get out of the red ink.[20]

American Forestry in 1920 reported that only about three percent of the sawmill waste generated nationally was used to manufacture pulp.[21]

In Southeast Alaska, the agency saw a waste problem of a different nature. William Greeley spoke in 1927 of "the immense quantities of mature timber now going to waste in Alaskan forests."[22] A 1931 article by Emmanuel Fritz, editor of the *Journal of Forestry*, noted the contradictory impulses:

The paper industry…has been urged to develop the pulpwood resources of Alaska, while at the same time in the United States proper raw materials suitable for pulp are being wasted. One government report decries the wastefulness of the lumber industry, another encourages the development of virgin resources elsewhere.[23]

CHAPTER 13

Major Newsprint Interests
Make a Commitment

I N LATE 1925, FRANK HEINTZLEMAN TOOK LEAVE OF ALASKA WITH THE
INTENTION OF SPENDING SIX MONTHS IN THE LOWER 48 STATES AND
CANADA TO PROMOTE THE DEVELOPMENT OF A PULP AND PAPER
INDUSTRY IN SOUTHEAST ALASKA. THE FIRST ITEM ON HIS SCHEDULE WAS AN
ADDRESS IN NEW YORK CITY TO THE WOODLAND SECTION OF THE AMERICAN
PULP AND PAPER ASSOCIATION, AFTER WHICH HE WOULD MAKE OFFICIAL
CONTACT WITH PULP AND PAPER MANUFACTURERS AS WELL AS NEWSPRINT
PAPER CONSUMERS THROUGHOUT THE UNITED STATES. IN EARLY APRIL 1926,
HEINTZLEMAN WAS IN SEATTLE AND ON HIS WAY BACK TO ALASKA BUT WAS
DIRECTED TO RETURN TO NEW YORK, WASHINGTON, AND CHICAGO TO PERSUADE
PAPER MANUFACTURERS TO SEND REPRESENTATIVES TO MAKE INSPECTION
TOURS OF SOUTHEAST ALASKA THAT SUMMER. AFTER VISITING THOSE CITIES,
HE SENT WORD BACK TO ALASKA IN LATE MAY THAT HE WOULD BE RETURNING
SOON WITH A PARTY OF FOUR POTENTIAL INVESTORS, AND HE REQUESTED THAT
THE FOREST SERVICE'S VESSEL *TAHN* BE PLACED AT HIS DISPOSAL. THIS EFFORT,
HOWEVER, BORE NO FRUIT, AND IN THE FALL HEINTZLEMAN WAS ON THE ROAD
AGAIN, THIS TIME TO SAN FRANCISCO AND ON TO NEW YORK AND OTHER
EASTERN CITIES.[1]

In November newspaper magnate William Randolph Hearst filed an
application with the Federal Power Commission for a power site at Port
Snettisham, and interests of the Zellerbach Corporation, a large manufacturer
of paper, filed an application for a number of power sites near Ketchikan.
Hearst never proceeded beyond the permit application, but the *San Francisco
Chronicle* reported the following month that the Zellerbach interests were
seriously contemplating the construction near Ketchikan of a paper mill with
a capacity of 500 tons per day. The Zellerbachs were confident of the project's

feasibility but estimated it would take two years to survey and analyze the situation before a firm commitment could be made.[2]

The Zellerbach Corporation had some history in Alaska. It owned Washington Pulp & Paper Company, the business that had purchased ground-wood pulp from the Alaska Pulp & Paper Company in the early 1920s. In addition, the president of the corporation, Isadore Zellerbach, had looked in 1923 into the purchase of timber in Southeast Alaska. That year, at a meeting with Representative Charles Curry (R-California), then chairman of the Committee on the Territories, he spoke of building what the *Ketchikan Alaska Chronicle* headlined as the "Greatest Paper Mill Plant on U.S. Territory." Zellerbach also met at that time with Secretary of Agriculture Henry Wallace and Forest Service Chief William Greeley.[3]

In January 1927, the Forest Service offered for sale two units of what it termed pulptimber under exceedingly liberal terms. These were by far the largest timber sales the Forest Service had ever attempted. Secretary of Agriculture William Jardine announced the sales and emphasized the government's commitment to them when he stated, "We invite and will protect the investment of capital necessary to establish large units of paper manufacture."[4] A prospectus outlining the sales was widely distributed to businesses and individuals nationwide. The *Washington Post* reported,

> **The Department of Agriculture is on the lookout for two or three individuals, firms or corporations with a knowledge of paper-making and each with a capital of at least $8,000,000. To such the department offers the opportunity of a lifetime: enough timber to keep a couple of 500-ton mills busy turning out newsprint for 50 years.[5]**

The timber being offered consisted of 835 million cubic feet of timber (5 BBF) in northern Southeast Alaska in what was referred to as the "Juneau pulp timber tract," and a similar amount in the southern part of the region, the "Ketchikan pulp timber tract." The Forest Service was very flexible on timber selection, and prospective purchasers were given the option to select whatever timber within the tracts pleased them, subject to the approval of the district forester. The timber on each unit was considered adequate to keep a mill of 500-tons-per-day capacity operating 300 days a year for 50 years, and sufficient to provide for the amortization of an operator's initial investments in power development and plant construction. The Forest Service provided "Assurance of a permanently available supply of timber under the

Government's policy of limiting sales for the regional wood-using industries to a total cut no larger than can be maintained in perpetuity by the growth of timber." In the words of an unnamed Forest Service official, the supply of timber on the Tongass would last "from the time the mill starts running till the land sinks beneath the sea."[6]

Though there was no precedent in the national forest system, under a revised policy the duration of the contracts was extended to a firm 50 years, twenty years longer than the Cascade Creek offering in 1923. Greeley explained the reasoning behind extending the contracts in a letter to the Secretary of Agriculture:

> The Forest Service fully appreciates the need of liberal provisions in its timber sale agreements and selling policies to meet the conditions peculiar to the establishment of a pulp and paper industry in Alaska. I can see no objection to making agreements for a supply of timber sufficient for periods as long as fifty years. Fifty years in everyday business is a common allowance for amortization; also, many bond issues run for this period of time.[7]

Another consideration was that the 1920 legislation that created the Federal Power Commission also authorized the agency to lease waterpower sites in Alaska for periods of up to 50 years.[8]

Advances for stumpage were limited to $30,000, an arrangement similar to those made in Canada. The minimum acceptable bids established by the Forest Service for these units were $0.60 per CCF for Sitka spruce and $0.30 per CCF for western hemlock. During the term of the sale, stumpage rates were to be redetermined at five-year intervals, but limits were placed such that the maximums that could be charged 20 years into the sale were $1.07 per CCF for Sitka spruce and $0.54 per CCF for western hemlock. The purchaser was further protected with a clause in the timber contracts that stated that for paper delivered to a common market, "the prices fixed for any five-year period shall not make the cost of paper from Alaska mills higher than the average cost of paper of similar kinds and quality from mills in other paper-producing regions of the Pacific Coast." The Forest Service estimated the cost of timber delivered to a pulp mill, including stumpage, logging, rafting and towing, would not exceed $6 per CCF.[9]

The timber was offered at a time that a hard reality was beginning to be acknowledged: there was serious excess capacity in the North American paper industry. Yet, at the same time these concerns about excess capacity were expressed, the newsprint paper industry shared a common expectation that newsprint paper prices would eventually rebound. Major players from as far away as Great Britain came to Southeast Alaska to examine what the Forest Service was offering, and in 1927 the Service accepted conditional bids for the two offerings. The bids were notably from interests on the West Coast, where Alaska newsprint paper would be, at least theoretically, more competitive than on the more-distant East Coast.

The publisher of the *San Francisco Chronicle*, George T. Cameron (representing himself as well as Harry Chandler of the *Los Angeles Times*, and C. A. Morden, of the *Portland Oregonian*) was the sole bidder on the Juneau pulptimber unit. His bid for pulpwood was the minimum acceptable to the Forest Service: $0.60 per CCF for Sitka spruce; $0.30 per CCF for western hemlock. The *Portland Oregonian* withdrew its interest in the endeavor shortly thereafter.[10]

Isadore Zellerbach and his son, J. D. Zellerbach of the Zellerbach Corporation of San Francisco, successfully bid on the Ketchikan unit. The unit had received one other bid, from the Alaska International Paper Company, a subsidiary of the International Paper Company. The Zellerbach bid for pulpwood was $0.80 per CCF for Sitka spruce, and $0.40 per CCF for western hemlock.[11]

Each contract required the completion of a 200-ton paper mill by April 1932, which was to be expanded to a minimum of 400 tons within the succeeding ten years. The Forest Service reasoned that for Southeast Alaska to compete with operations in Canada, plants would have to be built big to benefit from the same economies of scale as the huge Canadian plants. By the Forest Service's estimate, a 500-ton paper mill would require about 50,000 horsepower.[12]

For their part in securing the contracts, though conditional, and their efforts in encouraging the development of a pulp and paper industry in Southeast Alaska, District Forester Charles Flory and his assistant district foresters, Frank Heintzleman and Melvin Merritt, were praised in a May 1927 resolution passed by an appreciative Juneau Chamber of Commerce. The resolution praised Heintzleman's efforts in particular as being "crowned with success."[13]

Cameron (from the *San Francisco Chronicle*) and his associates were interested in following the example of the *New York Times* and the *Chicago Tribune*, which like a number of other papers obtained their newsprint paper

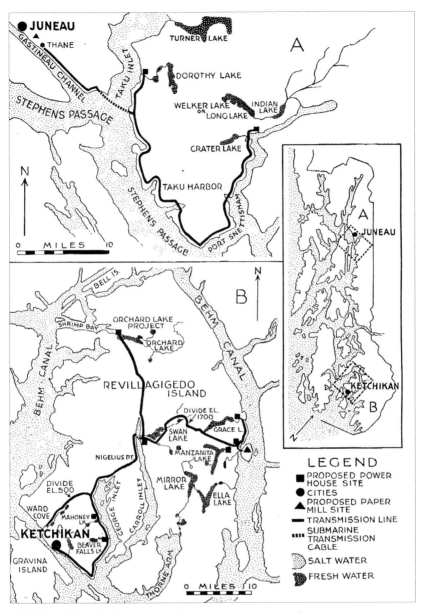

FIGURE 37. Map showing expected newsprint paper manufacturing developments in the Juneau and Ketchikan areas, 1930. (*Pacific Pulp and Paper Industry*, December 1930)

through affiliated paper mills in Ontario. Producing their own newsprint paper—the *San Francisco Chronicle* and the *Los Angeles Times* together consumed about 60,000 tons of newsprint paper annually[14]—under a long-term Forest Service timber contract seemed a way to guarantee a reliable, long-term supply at a competitive price. An advantage operators in Alaska would have over Canada was that Alaska paper would not be subject to future import or export duties, or embargoes, such as were occasionally threatened by the Canadian government.

The delivered price for newsprint paper in New York City had peaked in late 1920 at $112.60 per ton. At the time the Cameron-Chandler and Zellerbach interests entered into conditional contracts with the Forest Service, the average contract price per ton for newsprint paper delivered to New York City was $71.80. By 1928, the recognized base price of newsprint paper had fallen to $65 per ton, though most was being sold for considerably less.[15]

In the summer of 1927, Royal S. Kellogg toured Southeast Alaska. Kellogg had left the Forest Service and was head of the Newsprint Service Bureau, which represented American and Canadian newsprint paper interests. He counseled that the capacity of the newsprint paper industry of North America already exceeded consumption and was still growing. He told Alaskans, "I find everyone out here quite warmed up over the pulp and paper. There are many things to be considered before embarking in the business. The most thorough investigation into matters of production and marketing are necessary, and your enterprises should proceed with caution."[16]

So far as the forest resources available for the manufacture of pulp and paper were concerned, Kellogg considered Southeast Alaska to be in a preeminent position but thought the region's waterpower resources "not quite on par with eastern Canada." Over the long term, however, Kellogg was optimistic, stating, "When the newsprint manufacturing industry is brought to the Territory, and it must come eventually, it will be the greatest thing that ever happened to Alaska."[17]

NEW PROMOTIONS AND CONTRACTS

Many in Southeast Alaska, even sawmill interests, were eager for the pulp and paper industry to become established in the region. George Weisel, assistant manager of the Juneau Lumber Mills, wrote,

> The need of Alaska today is for an industry, or industries which will furnish year-round employment. No country can prosper on seasonal employment. And Alaska has been

suffering greatly from the latter class of industries. First it was fur, under the Russian regime, and it was trapped out. Next came the bonanza gold fields which played out. Next came fish, and I want to say if it wasn't for the U.S. Bureau of Fisheries, you would be dead again.

Now ahead lies the development of the forests. No extended utilization of them is possible from the sawmills, because there isn't a sufficient quantity of the higher grades of lumber capable of being produced from the average tree. The local forests are primarily valuable for pulp wood and the policy of the Forest Service in making them available for that purpose, in promoting that kind of utilization, is far-sighted.[18]

But in the summer of 1928, *Pacific Pulp & Paper Industry* quoted an anonymous newsprint paper executive to have said, "Except for some of the mills most favorably situated, several newsprint concerns are barely covering cost of production, and that can't go on indefinitely." The executive added, "We're all in the dark."[19]

Others saw a silver lining in the situation. That same year, the *Paper Trade Journal* quoted the comforting opinion of the *Financial Times*: "The time is now arriving when the test of the ability of the manufacturing companies to work out the problem of overproduction will be faced. That the outcome will be more satisfactory than was at one time hoped for is reasonably assured." The publication then added the opinion of a well-known Montreal stock brokerage firm: "News print stocks are favorites and recent figures have tended to confirm the bulls that it won't be so very long before a tremendous upward swing commences in these stocks."[20]

Despite the uncertainty of the situation, all concerned understood that decisions to establish pulp mill operations would be made not simply with the next few years in mind, but in consideration of the fact that pulp mills were investments that were expected to span generations.

In 1928, the Zellerbach Corporation and the Crown Willamette Paper Company merged into the Crown Zellerbach Company to become the nation's second-largest paper manufacturer. Actually a holding company, Crown Zellerbach owned paper manufacturing operations that extended from Los Angeles, California, to Ocean Falls, British Columbia.[21] The company's interest in Southeast Alaska was part of a pattern of rapid expansion and an opportunity to establish itself as a ground-floor player in Alaska.

But there was a potential problem. A 1930 Federal Trade Commission report cited "the possible monopoly or tendency toward monopoly" of Crown Zellerbach in the Pacific coast states and touched upon the effect of Crown Zellerbach possibly producing newsprint paper in Alaska. The report noted that "if Alaska newsprint can be delivered to New Orleans thru the Panama Canal at $59 a ton, it can be sold in Cincinnati and even in Chicago at $64.60 a ton. The current delivered price in these two cities is $62.00. The increase proposed by the governments of the Provinces of Quebec and Ontario would establish a price of $67 a ton, giving Alaskan paper an advantage."[22]

The Forest Service, of course, had spent years promoting the advantages of producing newsprint paper in Southeast Alaska and had been courting firms such as Crown Zellerbach. The concerns of the Federal Trade Commission, however, were not without substance: as early as 1923, the commission had filed a complaint against the Zellerbach Paper Company for suppression of competition, and later, in 1934, when it formally charged Crown Zellerbach as being in violation of the Clayton Antitrust Act over its acquisition of the Crown Willamette Paper Company, the corporation controlled the manufacture and sale of about 80 percent of the newsprint paper in the Pacific coast states. The company considered the charge to be one of "various suits and proceedings arising in the ordinary and normal course of business."[23]

In the summer of 1927, a reconnaissance party representing both the Zellerbach and Cameron-Chandler interests made a preliminary investigation of Southeast Alaska's timber and potential hydropower resources. Over the following two years, the interests cooperated and completed thorough field investigations of the two pulptimber units. Don Meldrum, a logging engineer based in Seattle, was in charge of surveying timber, and R. A. Kinzie, a San Francisco associate of Cameron and previous general manager of the Treadwell Mine at Juneau, had charge of waterpower studies. The waterpower studies were done year-round, and occasionally float planes were employed for transportation to remote lakes. Meldrum, with a crew of about ten experienced timber cruisers, worked out of a floating camp consisting of a house scow attended by two gasoline-powered boats.[24]

During the summer of 1928, Secretary of Agriculture William Jardine visited Southeast Alaska to "study the pulp problem." In the late fall of that year, the Newsprint Service Bureau's Royal Kellogg questioned in the *Paper Trade Journal* the feasibility of the development of a newsprint paper industry in Southeast Alaska of the scale and at the cost envisioned by the Forest Service. The Forest Service's estimated perpetual production of a million tons of newsprint paper annually, and the existence of 400,000 horsepower of potential waterpower were, according to Kellogg, "very large assumptions." Likewise, he thought the Service had underestimated the costs of establishing a paper manufacturing operation by about half.[25]

Kellogg observed that Forest Service pulptimber offerings on the Tongass—none of which had been successful—had been "very widely press-agented as indicating immediate large-scale production of newsprint paper," such that "the general public is very largely of the belief that paper is being made in Alaska today." He considered it "unfortunate that public expectations of developments in that region should continuously be led so far in advance of actualities." The *Paper Trade Journal* noted that during the past ten years many "Alaska chickens" had been counted as hatched "when, as a matter of fact, the period of incubation has scarcely begun."

Kellogg charged that early Forest Service pulptimber sales had been "made to promoters instead of practical paper manufacturers." He nevertheless praised the Zellerbach and Cameron-Chandler interests for prompting "the first thorough study of the paper-making possibilities of the territory made by practical paper manufacturers of sufficient resources to tackle the job if they find it worth while."[26]

The work of Meldrum and Kinzie was enhanced in 1929 by the joint Forest Service and Geological Survey effort that continued the work on the aerial survey project begun in 1926. The 1929 work resulted in the discovery a power site on the east side of Taku Inlet that was said to exceed 20,000 horsepower. There was apparently a great degree of cooperation with the contractors by the Navy and the two government agencies, because Meldrum was able to survey extensively from Navy airplanes. The surveys were marred by an October 1929 incident in which John Thayer, a Forest Service employee cruising timber at the south end of Admiralty Island, died after being mauled by a grizzly ("brown") bear. Thayer Lake, located on Admiralty Island, was named in his memory.[27]

Though Cameron in particular made no commitments, when the surveys were completed in October 1929, it was commonly accepted that the successful bidders were satisfied with the conditions and would soon complete the contracts. The Forest Service had estimated that each Tongass contract would "require an initial investment of at least $8,000,000 or $10,000,000 in waterpower development, manufacturing plant, and logging equipment," though the chairman of the board of the Zellerbach Corporation, M. R. Higgins, had stated in 1927 that he expected the company to spend close to $30 million on the Alaska project, adding, "This has to be done in a big way or not at all." Per worker employed, the cost of constructing a pulp mill exceeded all other industries, save those making chemicals. Consideration was given by the Cameron-Chandler and Zellerbach interests to the possibility of jointly building and operating a newspaper plant, probably located near Juneau. *The Timberman*, however, reported in 1930 that the government "did not look with favor on the plan, believing it would be to the best interest of Alaska that the two companies should operate independently."[28]

William Greeley, who would soon resign from his position as chief of the Forest Service to become secretary-manager of the West Coast Lumbermen's Association, stated in his 1927 annual report, "These large pulp-wood sales in Alaska represent the culmination of 15 years of work by the Forest Service to get the newsprint industry extended to Alaska." In early 1930, Frank Heintzleman, assistant district forester in Alaska, characterized the undertaking of the Cameron-Chandler and Zellerbach interests as "nothing short of stupendous."[29]

In reality, the forest engineers who had done the surveys urged Alaska investors to be cautious. Meldrum calculated that because of the high costs of operation in Southeast Alaska, "unless a high percentage of spruce is considered very essential, an operator could afford to pay a high stumpage rate in Oregon and Washington and still be in a stronger position than an operator with the apparently cheap stumpage in Alaska." He suggested consideration be given to the Olympic National Forest in Washington, where there was reported to be 150 BBF of mostly pulptimber available—nearly twice as much as was purportedly available on the Tongass.

There were other considerations as well. Though Meldrum liked the idea of a perpetual timber supply purchased from the Forest Service on a stumpage basis, he considered disadvantageous the Forest Service's supervision of logging operations and its right to examine the financial records of the paper company for the purpose of adjusting stumpage prices.[30]

In late 1930, the Federal Power Commission granted 50-year licenses for the development of waterpower for the two prospective mills. The Zellerbach power (about 48,000 horsepower) was to be generated at five separate sites on Revillagigedo Island, while the Cameron-Chandler power (about 60,000 horsepower) was to be generated at two sites on the mainland southeast of Juneau.[31]

The Cameron-Chandler interests favored locating a mill along Gastineau Channel, near Juneau, but studiously avoided the announcement of a specific site. Though sites at Lemon Creek (just north of Juneau) and at Treadwell on Douglas Island were under consideration, the most likely location looked to be the site of the defunct Alaska Gastineau Mining Company at Thane, about five miles south of Juneau. This was the same site Bartlett Thane, who died in 1927, had unsuccessfully promoted for years. Speculation that this was the chosen location was bolstered in 1931 when the Forest Service allocated $200,000 from its Forest Highway Fund to pay for the reconstruction of the Glacier Highway between Juneau and Thane. Frank Heintzleman was confident that the timber from Admiralty Island alone could perpetually sustain a mill that manufactured 440 tons of newsprint daily. The mill, he wrote, would provide year-round employment for 1,100 men.[32]

The Zellerbach interests likewise avoided committing to a specific site for their mill, though Thorne Arm, located east of Ketchikan on Revillagigedo Island, seemed a good prospect.

THE GREAT DEPRESSION

On the eve of the stock market crash of 1929, reports on the health of the industry exuded enthusiasm: "It is safe to prophesy that progress and prosperity will continue to march steadfastly forward," wrote the *West Coast Lumberman* in 1928. "The future of American business is bright."[33] And according to the *American Lumberman,*

> It is believed that at no former time in the industry's history has a year begun more auspiciously than that just opening. The industry is in possession of the knowledge and the machinery for making prosperity during 1929, and with the proper application of that knowledge in combination with the machinery already set up a profitable twelve months is assured."[34]

But the pulp mills just over the horizon were not to be. In the Great Depression, newsprint paper prices plummeted, and the lure of manufacturing paper in Southeast Alaska lost its luster. By mutual consent, the Forest Service contracts with the Cameron-Chandler and Zellerbach interests, which had been extended for a year, were quietly dissolved in 1933. It was estimated the Cameron-Chandler and Zellerbach interests had spent approximately $250,000 in cruising timber and making power surveys.[35]

That year the average contract price per ton for newsprint paper delivered to New York City had fallen to $41.25, forty-three percent less than the $71.80 price of 1927, when the Cameron-Chandler and Zellerbach interests had entered into conditional contracts for Tongass timber. During the Great Depression, the production of newsprint paper in the United States fell to half its 1926 peak. The situation was similar in Canada, where some mills operated at just 40 percent of capacity, and almost 60 percent of the industry was either bankrupt or in some form of voluntary financial reorganization. "The mortality of the pulp mills has again been quite severe," reported *Lockwood's Directory* in 1930. "No less than fourteen mills during the past year have gone idle or ceased to exist altogether."[36]

Though the Department of Agriculture assured the public in 1933 that "had it not been for the depression substantial pulp and paper developments would have already been under way [in Southeast Alaska]," it may have been fortuitous that the Cameron-Chandler and Zellerbach interests were not successful in their endeavors, for the Forest Service years later concluded that "the plants would likely have proved to be sub-marginal and unable to withstand the competition of those lean years."[37] In the wretched business climate of the Great Depression, the possible development of a pulp and newsprint paper industry in Southeast Alaska was now considered by many to be a threat to hard-pressed established concerns. In the summer of 1931, the American Paper & Pulp Association, which had questioned the viability of a newsprint paper industry in Southeast Alaska ten years earlier, sent the following letter to President Herbert Hoover:

> Your attention is respectfully invited to a condition which constitutes a menace to the newsprint industry…
>
> We refer to the pending negotiations for the sale of national forest timber in Alaska for conversion into newsprint, together with the necessary water-power rights. We are in hearty sympathy with the high motives of the government in undertaking to promote the commercial and industrial development of Alaska, and the paper industry looks forward to the time when the rich timber resources of Alaska will provide needed raw materials on a sound economic basis.…
>
> [But to] thrust unwanted production upon an overexpanded industry struggling with a diminished

demand and vanishing profits would make a bad unemployment situation worse and further depreciate a capital investment of over $800,000,000 in the United States, Canada and Newfoundland, of which more than $600,000,000 is estimated to have been invested by citizens and companies of the United States....

We appeal to you, therefore, for suitable action to remove the disturbing threat of government timber in Alaska and establish it as a great reserve for the future.[38]

Though it was probably a moot gesture, the Forest Service ceased its promotional efforts, still arguing that

this over-production, to the extent it exists, is due largely to importation from other countries, which presents the question of whether it is best to favor the industrial operations which have been founded upon such importations, or to insist that the Territory of Alaska is entitled to full consideration with these other sources of wood pulp and newsprint and, therefore, will not be placed under any restrictions other than those of economic law and principle.

The Forest Service added that Alaska "should be given an equal opportunity with Canada to liquidate her vast holdings of high-grade pulpwood."[39]

A letter to the editor of *Pacific Pulp & Paper Industry* from George W. Houk of the Oregon-based Hawley Pulp & Paper Company reiterated the Alaska threat:

You may be sure that the Alaska paper mills will be the first cause for closing newsprint mills in Washington and Oregon, as this will be Alaska's logical market, and more important payrolls in Washington and Oregon will be jeopardized. I assure you that Alaskan paper mills will in fact sign the death warrant of the present generation of the United States Pacific Coast plants.[40]

In late 1934, *Pacific Pulp & Paper Industry* reported that the Forest Service was "marking time on the development of the pulp and paper industry in Alaska because economic conditions were still uncertain," and early the following year *The Timberman* reported the status of the development of a pulp industry in Alaska to be "dormant."[41]

Unemployment in Alaska was a problem, however. The 1,677-page "National Plan for American Forestry," published by the Department of Agriculture in 1933, considered imported pulp and paper to be the "equivalent to the 'exportation' of full-time jobs for 70,000 American citizens." Legislation introduced in the U.S. Senate in 1934 by Clarence Dill (D-Washington) would have placed an embargo on the import of foreign pulpwood and newsprint paper.[42] The territorial legislature took up the issue of unemployment in 1935, and passed a joint memorial calling for the establishment of the pulp industry so that Alaskans could

> be furnished gainful employment and that the Territory may be developed through the utilization of this great natural resource, which for more than twenty years has been so bottled up as to result in great economic loss and expense and which might otherwise…be the source of employment to more than thirty thousand persons and bringing the further result of development and prosperity to this great and undeveloped Territory of the United States.[43]

In the summer of 1937, Congress asked President Franklin Roosevelt to prepare a comprehensive plan for the development of Alaska's resources. Roosevelt was very interested in Alaska and delegated the preparation of the plan to his National Resources Committee. The committee, subsequently known as the National Resources Planning Board, operated out of the president's office and was chaired by Harold Ickes, Roosevelt's Secretary of the Interior. The committee was to work with government agencies and public and private institutions to "collect, prepare and make available to the President, with recommendations, such plans, data and information as may be helpful to a planned development and use of land, water, and other national resources."[44]

The National Resources Committee organized the Alaska Resources Committee to prepare the plan. Among its six members was Frank Heintzleman, of the Forest Service. After several months of organizing and analyzing data, *Alaska—Its Resources and Development* was delivered to Congress in January 1938. The 213-page plan's discussion of Alaska's forest resources consisted in large part of the Department of Agriculture's 1928 publication, *Pulp-Timber Resources of Southeast Alaska*, written by Heintzleman. The committee believed the pulp and paper industry offered the best near-term prospect for development in the territory, and among its recommendations was that a study be made of the technical and economic problems associated with the establishment of a pulp and paper industry in Alaska.[45] The study was never done. The committee also recommended the completion of the aerial survey work done in Southeast Alaska in 1926 and 1929. Perhaps in response to this recommendation, high-quality aerial photographs were taken of all of Southeast Alaska in 1948.

Frank Heintzleman advanced to the position of regional forester in 1937 and, ever-determined and focused, he immediately resumed promoting Southeast Alaska, citing increases in the price of paper and paper products that were due to improvements in the U.S economy. Heintzleman reasoned that the expansion of the industry into eastern Canada had been logical, but the projects there of "outstanding merit" had been taken up, and eastern Canadian timber was becoming more costly to produce. Southeast Alaska's turn to supply the nation with newsprint paper was at hand. His optimism faced a stern economic reality: the Great Depression was not yet over, and the increases in the price of paper and paper products, though impressive, proved to be temporary. Between 1937 and 1938, newsprint paper production in the United States declined fully 25 percent, to the lowest level in 35 years.[46] Until the economy strengthened very substantially, an Alaska newsprint paper industry would stay on the drawing boards.

During the summer of 1939, Forest Service Chief F. A. Silcox made a five-week survey of Alaska's timber. He told Alaskans that he ultimately expected the territory's timber to be "an important factor in the economics of the world."[47] But upon returning to the States, Silcox added,

> Because of the ample stands of pulpwood in the Pacific Northwest, the huge investments recently made for the production of pulp in the south and international trade conditions, I doubt if Alaskan pulpwood will show itself in the market for twenty years or more. Attempts to place Alaska's pulp products on the market at this early date might injure rather than help the situation.[48]

That fall, Adolph Hitler's army invaded Poland, triggering World War II. Ventures such as the development of a pulp and paper industry in Southeast Alaska were put on hold.

CHAPTER 14

Alaska's Contribution
to the Battle for the Skies

T HE VALUE OF SOUTHEAST ALASKA'S SITKA SPRUCE FOR AIRPLANE CONSTRUCTION HAD BEEN PROVED IN WORLD WAR I AND ALL THE YEARS SINCE. NOW WORLD WAR II HAD AGAIN CREATED A CRITICAL NEED FOR SITKA SPRUCE. "AVAILABLE SUPPLIES OF SPRUCE OF AIRPLANE QUALITY IN THE PACIFIC NORTHWEST ARE INSUFFICIENT TO MAINTAIN THE PRODUCTION NEEDED," READ A 1942 FOREST SERVICE REPORT. LOGGING AERO-GRADE SITKA SPRUCE IN ALASKA WAS PROBLEMATIC, HOWEVER: "IT WAS RECOGNIZED THAT UNUSUAL FINANCIAL RISKS, AND EVEN DIRECT WAR RISKS OF JAPANESE RAIDS OR ATTACK, WERE INVOLVED; BUT IT WAS FELT THAT THE WAR NEEDS FOR AIRCRAFT LUMBER WERE SUCH AS TO JUSTIFY TAKING THOSE RISKS," ACCORDING TO THE HEAD OF THE DIVISION OF TIMBER MANAGEMENT AT THE FOREST SERVICE.[1]

The World War II Alaska Spruce Log Program (ASLP) had its roots in the demands of World War I efforts in the states of Washington and Oregon to obtain aero-grade Sitka spruce for use in military aviation.

During World War I, the supply of aero-grade Sitka spruce had been considered critical to the Allied aircraft program, in which the United States alone anticipated the construction of 30,000 aircraft. The need to rapidly secure large quantities of the material led to the formation in 1917 of the U.S. Army's Spruce Production Division. The year-long, $46 million program enlisted some 29,000 Army personnel to help build infrastructure (including railroads) and to log and mill nearly 150 MMBF of aero-grade, "super-selected" spruce, which came almost exclusively from Washington and Oregon. An article in *American Forestry* in 1923 stated that it was probable that "never in the history of nations has a tree been more eagerly sought after than was the Sitka spruce" during the war.[2]

Mills in Southeast Alaska sold some 589,000 board feet of aero-grade Sitka spruce to the Spruce Production Division, which in early 1918 offered $160 per MBF delivered at Seattle for stock in lengths of 20 feet or more that was suitable for wing beams. Later that year, the offered price for wing beam stock was reported by the Forest Service to be $180 per MBF.[3]

The Spruce Production Division was disbanded at the close of World War I. The emergency effort had come at a steep price: a huge amount of valuable Sitka spruce was wasted. The majority of timber felled did not meet the government's standards, and much was left to decay on the ground, where it attracted insects and disease. A 1940 article in *American Forests* stated that "Probably not in the history of American logging—and it has many black pages—has such reckless, useless waste of a valuable and limited timber resource been recorded."[4]

After making an inspection trip to Alaska in 1921, Chief William Greeley stated that the territory "probably contains the largest quantity of clear high grade spruce to be found in the United States." The following year, however, when the War Department asked him about the advisability of reserving all "airplane spruce" stands in the Tongass National Forest to supply the department's future needs for airplane lumber, Greeley demurred; such timber, he responded, was to be found "in larger quantity, in better quality, and in more nearly solid stands" in Washington and Oregon.[5]

By the onset of World War II, aluminum and steel had largely replaced wood in the construction of aircraft. But there were important exceptions in which Sitka spruce was used, and military demand for aero-grade Sitka spruce rapidly developed. Sitka spruce would once again do service in preserving democracy.

The most notable application involved the de Havilland Mosquito (DH-98), which came into service in 1942. The Mosquito was constructed of wood because the British feared that strategic metals might become scarce during wartime. Also known as the "Wooden Wonder" and the "Termite's Dream," this highly-maneuverable, twin-engine aircraft was perhaps the most versatile combat aircraft of the war, serving day and night as a fighter, a bomber, and a reconnaissance craft. Achieving speeds in excess of 400 mph, it was at the time the fastest operational machine in the world. Though the cost of such construction was extremely high, some 8,000 Mosquitos were built in factories in Great Britain, Canada, and Australia, where luthiers, cabinetmakers, and other artisans skilled in working with wood were often employed. The plane was primarily constructed of birch plywood, balsa, and Sitka spruce.[6]

Another use for Sitka spruce was the Boeing/Stearman Kaydet (PT-17), a two-seat biplane trainer used by the U.S. Army and Navy and the Royal Canadian Air Force. The fuselage of this aircraft was constructed of a welded

Gliders, transport planes, invasion barges, carried the Sicilian invasion!

"FLYING LOGS"
LAND TROOPS IN EUROPE!

5000 MORE LOGGERS NEEDED!

The War Manpower Commission is calling loggers out of the shipyards, airplane factories, and other war industries— for no work is more important than log production right now!

FANTASTIC? Not at all! America is building the greatest fleets of troop-carrying transports and gliders the world has ever known. And they are made of WOOD! England is mass-producing Mosquito Bombers that out-speed the Nazi's fastest fighter planes. They're made of WOOD! Literally — logs are flying — from the woods right into enemy strongholds. THAT is why America so desperately needs more saw logs,

more peeler logs, more pulp logs NOW!

In addition to planes and gliders, there MUST be wood for thousands of invasion barges, pontoon bridges, sub chasers, shell cases and aircraft carrier decking. Logs are making munitions from nitro-cellulose to shell cases. Every logger can serve America best by sticking to his ax and saw...and by urging his logger friends who've quit the woods, to return to the forest fighting front!

KETCHIKAN SPRUCE MILLS

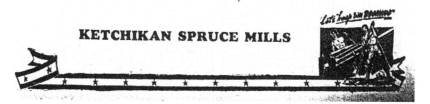

FIGURE 38. Ketchikan Spruce Mills advertisement, 1943.

steel framework covered with fabric, but the framework of the wings was wood. More than 10,000 Kaydets were built before the end of 1945. After the war they were favored as stunt planes and crop dusters. As well, the U.S. developed the XP-77, an experimental fighter aircraft constructed of Sitka spruce plywood. The government hoped the XP-77, like the Mosquito, would be capable of speeds in excess of 400 mph. It was not, and, for this reason and others, the project was abandoned in late 1944.

British Columbia's Queen Charlotte Islands had been logged to help supply Great Britain's need for aircraft material during World War I and for pulpwood in the following years. As World War II loomed, substantial stands of Sitka spruce remained. A major selective logging operation commenced on the islands in 1939 under the Canadian government's Aero Timber Products Program, which estimated a need to build up to 60,000 aircraft. Supply failed to match demand, even though the program took most of the best Sitka spruce timber on the islands.[7]

In Washington and Oregon, the demands of World War I had largely exhausted the easily accessible supply of Sitka spruce. There were two exceptions. The first was the so-called Blodgett Tract in Oregon. This 13,000-acre timber stand in Lincoln County (along the Pacific coast, west of Corvallis) contained a considerable quantity of high-quality Sitka spruce timber and was practically intact. Efforts were made to have the tract transferred to the Siuslaw National Forest to be reserved for future military needs. The efforts failed, however, and the timber was sold to a private company. The company logged most of the area, which then suffered the effects of several fires.[8]

The second exception was Olympic National Park, which Congress in 1938 had upgraded from national monument status. The park encompassed 835,000 acres on Washington's Olympic Peninsula and was home to a stand of about 750 MMBF of prime-quality virgin Sitka spruce, all of which was off-limits to logging.

Lumbermen decried the creation of Olympic National Park as a "colossal national folly" and in the spring of 1943 they called for the removal of some 500,000 acres from the park—acres that could be opened to logging to supply the war effort.[9] Their endeavor was unsuccessful in the face of unequivocal opposition by President Franklin Roosevelt's administration, which maintained an initiative promoting conservation even during wartime. The only domestic recourse yet available was to obtain the needed material from the Tongass National Forest.

The amount of aero-grade material, even in good timber, is small. On the Tongass it amounted to an estimated two percent of the nearly 11 BBF of standing Sitka spruce timber over 24 inches in diameter—about 220 MMBF. The actual proportion was even smaller, since about two-thirds of the aero-grade material sent to airplane factories was typically rejected because of imperfections.[10]

The cut of aero-grade Sitka spruce on the Tongass in 1939 was only about 360 MBF. After World War II began in Europe that September, U.S. officials recognized that a major effort targeting the best of the Sitka spruce would have to be quickly established. In December 1940, a conference was scheduled in Washington, D.C., during which U.S. and British representatives would work to coordinate the distribution of Sitka spruce between the two countries. The increasing demand for aero-grade Sitka spruce caused the price to rise dramatically, and in April 1942 the federal Office of Price Administration (OPA) established price ceilings for "aircraft spruce"—defined by the government as "spruce lumber, plank, board, flitches and cants produced within the United States, exclusive of Alaska, suitable for use in the construction of aircraft"—at prices that corresponded to the levels prevailing in October 1941. For some of the select grades and sizes, the price ranged to $690 per MBF.[11]

The need for aero-grade Sitka spruce became so critical that in July 1942 all U.S. supplies of grade 1 and grade 2 Sitka spruce logs, cants, and flitches were by law frozen in the hands of their owners, their processing prohibited without official authorization. In Canada, the government considered workers in the aero spruce industry "essential" and issued an order in October 1942 that forbade them from seeking other work.[12]

Lumber of all kinds was considered critical to the war effort. The estimated U.S. requirement of lumber for the war effort in 1942 alone was 42 BBF, a number that exceeded production that year by 6 BBF. To make up for this shortage, programs to increase production were instituted, as was an effort to decrease consumption through conservation. The Lumber Branch of the War Production Board allocated the material that was available.[13]

THE ALASKA SPRUCE LOG PROGRAM

Looking to satisfy the need for aero-grade Sitka spruce, the War Production Board requested the Forest Service to study the possibilities of utilizing Sitka spruce from Alaska. The Forest Service dispatched its own

FIGURE 39. Sitka spruce logs destined for Seattle for evaluation for use in airplane construction. (USDA Forest Service, Alaska Regional Office, Juneau)

man, James Girard, an internationally known authority on Sitka spruce, to Southeast Alaska, where he and a timber survey crew spent January through March 1942 cruising some 200 MMBF of Sitka spruce. Girard found the Sitka spruce of Southeast Alaska to be of "unusually good quality," with the northwestern coast of Prince of Wales Island and the adjacent Kosciusko and Tuxekan islands offering, in his estimation, the best possibilities.[14]

The Alaska Spruce Log Program was officially created by the Secretary of Agriculture on June 5, 1942, with the goal of providing 100 MMBF of aero-grade Sitka spruce per year. The program was administered by Forest Service and provided with a $2.5-million revolving fund to cover the operating costs of the enterprise by the Department of Agriculture's Commodity Credit Corporation. Alaska Regional Forester B. Frank Heintzleman was named director of the program, and Assistant Regional Forester Charles G. Burick was made general manager. J. M. Wyckoff supervised field operations as assistant manager until he was transferred to Seattle to handle lumber sales.

He was replaced by C. M. Archbold, who had been in charge of cruising timber on the Tongass.[15]

Because the mills in Southeast Alaska were not capable of cutting the quantity of Sitka spruce needed for this program, it was decided to tow logs to Puget Sound, where they would be sold at public auction to mills certified by the government for the cutting of Sitka spruce airplane stock. Fred Brundage, West Coast administrator for the War Production Board, was responsible for the administration of this facet of the program.[16]

The organization of the Alaska Spruce Log Program was based on advice given to Frank Heintzleman by personnel at the Crown Zellerbach Corporation who had prepared a report, "Plans and Feasibility of Logging and Transporting High Grade Sitka Spruce From Alaska to Puget Sound for Airplane Stock." On the advice of Crown Zellerbach's logging manager, Heintzleman chose to contract with small independent operators ("gyppos") to do the logging. Contractors provided their own equipment, but the program provided fully-equipped camps, docks, and communications equipment. A 1942 Forest Service announcement for the program read, "High grade spruce logs are urgently needed in airplane construction and it will be necessary to do selective logging to get the big volume that is needed… No hemlock, cedar or low grade spruce need be taken."[17]

Though Army tugboats (some "gunned-up like a battleship") were sometimes used, the program chartered independent tugboats for towing the logs to the main camp at Edna Bay (on Kosciusko Island) and to Puget Sound. As well, it chartered a large cannery tender and power scow for local transportation of men, supplies, and mail. Several Forest Service vessels were used for administrative purposes, including the 52-foot *Beaver*, which was built at Petersburg specifically for the project.[18]

The War Manpower Commission assisted the independent operators in recruiting labor, and help was also provided in the procurement of transportation, food, equipment, repair parts, and other goods and services necessary to their operations. The Commission rated loggers as "essential" workers in the war effort, and, as such, they were granted deferments from the military draft.[19] To save time for the important logging being done, coal was used instead of firewood to heat the buildings in land-based camps.

The Nettleton Logging Company of Seattle was contracted to build the headquarters camp at Edna Bay. The camp, with two miles of truck roads bearing names such as Spruce Street, Hemlock Street, and Alaska Way, had facilities to house and feed 250 men, a machine shop, a fueling station, a small

FIGURE 40. Alaska Spruce Log Program, Camp No. 3, Edna Bay, Kosciusko Island. (USDA Forest Service, Alaska Regional Office, Juneau)

FIGURE 41. Alaska Spruce Log Program, Camp No. 5, Calder Bay, Prince of Wales Island. (Rasmuson Library, University of Alaska, Fairbanks, Angela Janzen Burke Papers)

FIGURE 42. Experienced loggers felling Sitka spruce, Alaska Spruce Log Program.
(USDA Forest Service, Alaska Regional Office, Juneau)

commissary, a shipping wharf, and a special booming area to construct the huge Davis rafts in which logs would be towed the 700-mile distance to Puget Sound. Buildings at all the camps were based on 12-by-24-foot housing units prefabricated in Seattle. Eight additional outlying camps (four floating and four land-based) were established at logging sites. (Camp No. 1 was located

FIGURE 43. Alaska Spruce Log Program loggers bucking log with electric chain saw. (USDA Forest Service, Alaska Regional Office, Juneau)

at Tuxekan, an abandoned Indian village that was still graced by totem poles.)[20]

Production was based on eight-hour workdays and a six-day workweek, but because the camps offered few diversions, some loggers preferred seven-day workweeks. The men were usually paid by the hour. Logging ceased during the coldest part of the 1942–1943 winter, and many loggers took the opportunity to travel south to spend the holidays with their families. Since most healthy young men were off fighting the war, loggers in the Alaska Spruce Log Program tended to be older men. Though chainsaws were available for felling timber, these men usually preferred the two-man crosscut saws to which they were so accustomed.[21]

The Forest Service initially intended to have Sitka spruce logged selectively, taking only grade No. 1 and grade No. 2 logs from the woods, but eventually it employed the standard practice of clearcutting. Non-aero-grade material (western hemlock and grade No. 3 Sitka spruce) was provided to mills in Southeast Alaska, which were busy cutting lumber mostly for war-related construction. Mills at Juneau, Ketchikan, Metlakatla, and Sitka and

FIGURE 44. Ten MBF truckloads comprised of one to three logs were not uncommon. (USDA Forest Service, Alaska Regional Office, Juneau)

possibly Hidden Falls were scheduled to receive logs. Logs were also provided to U.S. Army engineers at Prince Rupert. An individual who worked for the ASLP was impressed by the number of one and three-log truckloads he saw hauled out of the woods at Edna Bay.[22]

Logs were purchased by the Alaska Spruce Log Program in the water at the contractors' respective camps. A flat rate was paid, but the price varied with the location of the timber being logged and the grade of logs produced. For the Bowen & Dillinger Logging Company, which was operating at Clam Cove on Prince of Wales Island, the August 1943 contract prices per MBF were $28 for grade 1 spruce, $24 for grade 2 spruce, and $14 for grade 3 spruce, hemlock, and cedar.[23]

Scaled flat rafts of logs were towed to Edna Bay, where the logs were constructed into rafts for the tow to Puget Sound. The rafting system for this operation had been invented and patented by a Canadian, G. G. Davis, in about 1911. In charge of a logging operation on the west coast of Vancouver Island, Davis had needed a secure means of rafting logs to be transported through the commonly rough waters of the area. His design involved a rectangle constructed with "sidesticks" 120 or more feet in length, chained to end boom sticks of 70 feet. Three tiers of logs 40 feet long were arranged inside the rectangle and were interlaced with cables

FIGURE 45. A layer of logs interwoven with steel cables constituted the base mat of a Davis raft. (USDA Forest Service, Alaska Regional Office, Juneau)

fastened to the side sticks, forming a large, flexible woven mat of logs, upon which additional logs were layered and secured. A degree of longitudinal stiffness was obtained by staggering the butts of adjacent logs. It has been said the construction of a Davis raft was an art in itself. A Davis raft regularly contained 1 to 1.5 MMBF of timber, drew 20 to 30 feet of water, and could be towed by a large tug at a speed of approximately three knots.[24]

The first Davis raft, towed by the tug *Sandra Foss*, departed Edna Bay on December 20, 1942, and after a trip interrupted by gales finally arrived at the Morrison Mill Company at Anacortes, Washington, on January 15, 1943. The raft contained 912,810 board feet of logs, of which 49,300 board feet were western hemlock for experimental purposes. The average Sitka spruce log scaled just shy of 2,000 board feet, and fully 38 percent of the shipment's Sitka spruce logs was grade No. 1. In addition to the Morrison Mill Company, the following business purchased Alaska Spruce Log Program rafts: American Mills, Aberdeen, Washington; Bloedel-Donovan Lumber Mill, Bellingham, Washington; Hart Mill Company, Olympia, Washington; Pacific Mills Company, Ocean Falls,

FIGURE 46. Alaska Spruce Log
Program Edna Bay Davis raft
construction area. (USDA Forest
Service, Alaska Regional Office,
Juneau)

British Columbia; Robinson Manufacturing, Everett, Washington; and Walton Brothers Timber Company, Anacortes, Washington.[25]

Though the Davis raft was successfully used throughout the life of the Alaska Spruce Log Program, it was not infallible. On January 19, 1944, the *Ketchikan Alaska Chronicle* reported, "Something less than a half-million feet of spruce logs that should be headed for Germany as parts of Mosquito bombers instead today were washing up on the beaches north of Ketchikan after a raft broke away from the chartered tug Andrew Foss yesterday."[26]

Though most of the logging concerns contracted by the Alaska Spruce Log Program were from the lower 48 states, two—the Alaska Co-operative Loggers and Bowen & Dillinger—were Alaska-based.[27] The program also purchased logs and lumber from established Alaska loggers and mills.

Information regarding the nine camp locations, contractors, employees, and planned production as of May 1, 1943, was published by *The Timberman*[28]:

Camp	Location	Operator	Type	Workers	Planned (MMBF)
1	Tuxekan	Gross Logging Co. (after Spiering)	tractor	25	7
2	Tuxekan Island	Walker Spruce Co.	A-frame	45	13
3	Edna Bay (West)	Buol Logging Co.	truck	60	41
4	Edna Bay (East)	Nelson-Deierlein Logging	truck	20	12.5
5	Calder Bay	Hamilton Bros.	tractor	25	6.5
6	Clover Pass	Redding, Hooper & Ashley (after Saddler)	tractor	10	10.5
7	Tuxekan Pass	Alaska Co-operative Loggers ("Alcop")	A-frame	16	6.5
8	El Capitan Pass	Mougin Spruce Log Co.	A-frame	12	4.5
9	Clam Cove	Bowen & Dillinger Logging	A-frame	18	9
Total				**231**	**110.5**

As of August 1943, the West Coast Lumberman reported a total of 187 employees at the camps. The number included 84 Alaskans, of whom 23 were Indians. Hourly wages at the Buol Logging Company were $1.25 for a choker setter, $1.65 for a donkey engineer, $1.50 for fallers and buckers, and $1.50 for a logging truck driver.[29]

In early 1944, the war in Europe was thought to be nearing its end, and the production of Allied aircraft ceased almost completely. Prices for aero-grade Sitka spruce fell as existing contracts were completed but not renewed, and in February 1944 the War Production Board recommended the discontinuance of the Alaska Spruce Log Program. Cutting ceased in the following month, and by October the operation was over.[30] A large part of the aero-grade Sitka spruce lumber inventory was suddenly considered surplus.

During its year and a half of existence, the Alaska Spruce Log Program shipped to Puget Sound some 38.5 MMBF of high-quality Sitka spruce. Additionally, a total of 46 MMBF of western hemlock and grade No. 3 Sitka spruce was delivered to mills in Southeast Alaska.[31]

FINANCIAL FALLOUT

When the Alaska Spruce Log Program terminated, the facilities and equipment owned by the program were sold at auction by the Forest Service.

FIGURE 47. Chief Forester Lyle Watts at Edna Bay with Sitka spruce log. During the summer of 1944, Watts traveled to Southeast Alaska for a first-hand look at the Alaska Spruce Log Program. The visit also furthered his understanding of the Tongass National Forest's industrial potential. (USDA Forest Service, Alaska Regional Office, Juneau)

Three of the logging companies contracted to the Alaska Spruce Log Program—Walker Spruce Company, Gross Logging Company, and Alaska Cooperative Loggers—found themselves, in the words of the E. E. Carter, the Forest Service's Chief of Timber Management, as "having not only lost their own meager financial resources in the undertaking, but also of finding themselves in debt to the government."[32] In a letter pointing out the beleaguered companies' need for financial relief, Carter explained,

> The financial risks involved in going into this relatively inaccessible, heretofore undeveloped region made such a project unattractive to private commercial loggers; hence it was necessary for the Government to undertake the job and produce the logs regardless of the risks involved… The risks and uncertainties were greatly increased by the necessary departures from normal methods of logging; [for example] the [initial] logging of only selected trees from the stands in order to speed up production of aircraft quality logs…
>
> The Alaska Spruce Log Program, in entering into these contracts, knew that these abnormal risks were involved. It also knew of the limited financial resources of the contractors, and that if the logging costs proved to be unusually high, there was little possibility of recovering the full amounts advanced. In other words, the need for producing the logs, the necessity of entering into contracts to produce them, the possibility of financial losses and the inability of the contractors to make such losses good if they occurred, all were envisioned by the Alaska Spruce Log Program in negotiating the contracts.
>
> Likewise, the contractors knew that unusual risks were involved in engaging in logging operations in this remote region, and they were actuated by patriotic motives as well as the possibility of reasonable profits in entering into the contracts. They also knew of the Government's authority under the First War Powers Act to reappraise the value of the work to be done and to amend the contracts if logging costs unavoidably were substantially higher than anticipated, and assumed that the Government would be fair and equitable.[33]

Each of the three financially troubled contractors subsequently found work producing logs and lumber for the war effort. The Forest Service determined that the cancellation of the contractors' debt was the best course of action, since the contractors had few assets and were actively contributing to the war effort, and it would be a waste of government time and money to prosecute cases in which there was no apparent possibility of obtaining tangible results.[34]

An appreciation of the contractors' efforts appeared in June 1944. Writing in *The Timberman*, Fred Brundage, the administrator for the program, said, "Certainly aircraft spruce producers have made a very vital contribution to our war effort. If conditions have now changed and the product is no longer needed, this in no way detracts from the real service the spruce industry has rendered." In early 1945, however, the government temporarily came back into the market for very large quantities of Sitka spruce, likely because of uncertainties over how much longer the war might last. This situation greatly upset Frank Heintzleman, since the Alaska Spruce Log Program, had it not been terminated, could still have been producing.[35]

In addition to the Alaska Spruce Log Program, considerable quantities of piling and other rough wood products were cut for the use of military authorities without charge for stumpage. A considerable amount of piling and some sawlogs were also tractor-logged on Glacier Bay National Monument land near the military shipping base that had been built at Excursion Inlet. All this wartime activity revived interest in Southeast Alaska's timber, and some hoped the logging infrastructure that had developed would remain to log for pulp mills.[36]

STATEHOOD PROPOSALS

World War II, to a large extent, brought Alaska into the modern world. During the war, military and civilian construction battalions built an infrastructure of roads (the Alaska Highway being the most notable), port facilities, airfields, and more. All told, something like a million men served in the territory, and after the war ended many decided to make Alaska their home. The time for Alaska to become the 49th state had arrived.

A long-held complaint and rallying point was that restrictive federal policies had for too long inhibited the development of Alaska's natural resources. Statehood would rectify this situation by transferring much of the control over those resources to a more development-friendly state government.

National security was also a factor. The territory had figured very prominently in World War II, and the Japanese invasion of several remote, lightly populated Aleutian Islands was evidence of its vulnerability. Additionally, with the onset of the Cold War, Alaska gained importance as a strategic and military bulwark of defense against potential aggressors in the North Pacific and Arctic areas. This alone ensured a continued military presence.[37]

A populated, developed area would be easier to defend. "It is a military axiom that a populous area with a developed economy and a stable citizenry and political institutions is a bulwark against invasion, militarily and psychologically," reported the Senate Committee on Interior and Insular Affairs in 1950. This had been considered by National Resources Planning Board, which wrote in 1942, "Economic development in Alaska cannot be divorced from strategic considerations," and that there was "an urgent need for backing up the military defense with a strong transportation system and a better balanced civilian economy."[38]

If resource development was considered the key to populating and developing the territory, it was also a key to statehood, for in its relatively meager economy largely based on a seasonal fishing industry, Alaska lacked sufficient resources to finance the services a state government would require. Of the territory's resources, timber seemed to offer the best opportunity for near-term development. The establishment of a pulp and paper industry would offer year-round, full-time jobs for Alaskans and create a substantial financial basis for statehood.

"A great deal of [timber-related] development has gone into Canada that we feel could properly have gone to Alaska, which is under the American flag…," Governor Ernest Gruening told a U.S. House committee in 1947. "We would like to see some in Alaska where it can pay taxes to the Federal Treasury and employ American citizens."[39]

Frank Heintzleman, the Forest Service's pulp industry architect and later a territorial governor, was not in a hurry for statehood, however. As governor-elect in 1953, Heintzleman told the U.S. Senate Interior Committee that the territory should build up its industry first. At that time a pulp mill was being constructed at Ketchikan, and the Japanese and others were negotiating for additional mills in Alaska. Heintzleman thought Alaska was "fast reaching" financial wherewithal to provide the services state government would require, and he hoped statehood would come in the "next few years."[40]

Alaska would become a state on January 3, 1959.

CHAPTER 15

A Window of Opportunity
Opens…Briefly

EVEN AS WORLD WAR II WAS BEING FOUGHT, TIMBER INTERESTS WERE
ANTICIPATING OPPORTUNITIES IN THE POST-WAR ECONOMY AND
TRYING TO POSITION THEMSELVES ADVANTAGEOUSLY. ONE OF THE
RESULTS OF THE ALASKA SPRUCE LOG PROGRAM WAS THAT IT BROUGHT
SOUTHEAST ALASKA'S TIMBER RESOURCES TO THE ATTENTION OF WOOD-USING
INDUSTRIES.

In 1944, the American Paper and Pulp Association issued a report
warning of an impending "catastrophic shortage of paper" in certain regions
due to pulpwood shortages, and early in the following year a lead editorial in
Pulp & Paper Industry stated, "The most urgent and most important problem
facing the management of a pulp and paper industry today is its future supply
of pulpwood."[1]

For the Forest Service and others promoting a pulp and paper industry
in Southeast Alaska, a window of opportunity, however fleeting and narrow,
had just presented itself, and the moment to begin the utilization of the
region's timber resources might finally be at hand. But it was still going to
be a hard sell, for the newsprint paper shortage was temporary, Southeast
Alaska was known to be an expensive place to operate, and the industry had
not forgotten the painful consequences of overexpansion during the 1920s.
The region, from one point of view expressed in *The Timberman*, presented
a "tragic economic picture of an area well able to support a thriving industrial
empire based upon forest resources and yet unable to accomplish such a goal
on account of economic difficulties which apparently cannot be remedied
for a long time." Some industry leaders at the time contended that those
"economic difficulties" could be overcome by towing logs from Southeast
Alaska to mills in Puget Sound, where they could be "converted into pulp
more cheaply than by the construction of mills in Alaska."[2]

In the spring of 1944—more than a year before the end of the war—the
Forest Service revived its efforts to establish major pulp and paper operations
in Southeast Alaska. In its "Proposal for the Development of the Pulp and Paper-

Making Industry in Alaska," the agency invited "industry representatives and any other interested parties to investigate the timber resources, operating conditions, and Forest Service timber sale policies and proposals in their relation to the post-war establishment of pulp and paper-manufacturing enterprises on the Tongass National Forest, southeastern Alaska."[3]

The agency realized that because of the wartime difficulty in obtaining equipment and supplies, the construction and operation of mills would necessarily have to wait until after the war, but its leaders believed that if preliminary arrangements could be completed, work could begin promptly at war's end. "The great future forest industry of Alaska is the manufacture of pulp and paper," read a 1944 agency promotion, "The extensive forests are well suited to this use, abundant water power is available for the conversion of raw material, and the region has other advantages that will contribute to the success of the industry."[4]

That same year the Forest Service offered a 50-year contract for approximately 1.35 billion cubic feet (7.5 BBF) of pulptimber in the vicinity of Ketchikan. The volume was considered sufficient to supply a sulphate or sulfite pulp mill having a daily capacity of 150 tons for 10 years and 525 tons for the remaining 40 years. A provision in the proposed contract required the purchaser to build a 150-ton daily capacity mill within three years after the signing of the contract, or within three years after the cessation of hostilities, whichever proved to be longer. In any event, however, the mill was required to be completed before April 1, 1949. Another provision stated, "To the extent that such local laborers have the necessary skills and are practically available, the woods and other crews for the purchaser's operations will be required to be recruited from among all residents of southeastern Alaska."[5]

The Crown Zellerbach Corporation, which had first expressed an interest in Southeast Alaska in 1923 and had conducted extensive timber and waterpower surveys in 1927–1929, was one of the firms invited by the Forest Service to consider the agency's most recent offering. The company, however, seems in the intervening years to have developed misgivings about operating a pulp mill in Southeast Alaska. According to Irving Brant, of the *St. Louis Star-Times*, the pulp operations manager at Crown Zellerbach had told him in 1938 that the idea of building a pulp mill in the region had been "abandoned as unfeasible" and that if any use was to be made of Alaska timber in the future, it would be rafted to Puget Sound for processing. Despite this assessment, a group of top officials from Crown Zellerbach, as well as several of the company's logging and forestry experts, journeyed to Southeast Alaska in August and

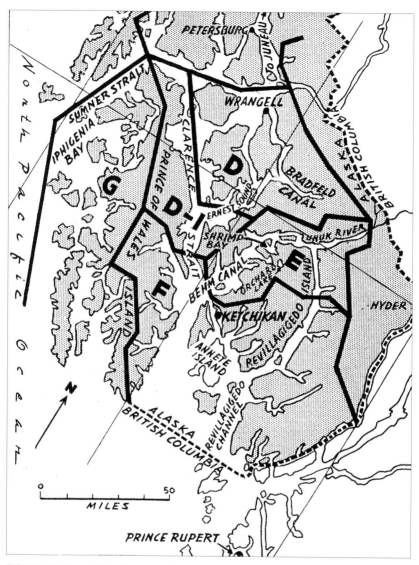

FIGURE 48. Forest Service map of proposed Ketchikan area pulp timber sale area.
(*Pacific Pulp & Paper Industry*, May 1944)

September of 1944 for a firsthand examination of the timber being offered. Officials from the St. Helens Pulp & Paper Company also visited Southeast Alaska about that time. Time, Inc. was also reported to be interested in the region's paper prospects. The interest expressed by Crown Zellerbach was at first encouraging, but the company announced later that year that it had decided against "pioneering the pulp industry in Alaska," according to C. M. Archbold, the Tongass National Forest's southern area supervisor at Ketchikan.[6] At this point, no one else was willing to do so either.

In early 1945, confusion over exactly how the war in Europe was progressing led the Forest Service to postpone—but not withdraw—its pulptimber offering. That same year, to address interest in the post-war commercial development of Alaska's forests, Forest Service Chief Lyle Watts called for a "program of research to provide a sound foundation for the utilization and management of Alaskan forest resources." In July 1948, confident that the development of a wood pulp industry in Southeast Alaska was finally at hand, the Forest Service established the Alaska Forest Research Center in Juneau. An early focus at the facility was the development of "working tools," such as timber cruising methods and long-log scaling tables, for the anticipated new industry.[7]

The Forest Service's post-war vision for Southeast Alaska involved the construction of five or six major pulp mills, each with an allocation of timber thought sufficient for perpetual operation. In addition to a mill at or near Ketchikan, other possible locations were Sitka, Juneau, Wrangell, and the Petersburg-Thomas Bay area. Consideration was also given to establishing a second mill at Ketchikan.[8]

In 1945, the Forest Service presented some potential uses for Tongass timber ancillary to the expected manufacture of pulp. With the exception of western redcedar piling and poles, the policy of the Forest Service in Alaska was opposed to the export of round logs. Opportunities were seen in the establishment of a creosoting plant, the manufacture of veneer (to be shipped to finishing plants on Puget Sound), and in the manufacture of Alaska-cedar boat material. Shipments of Sitka spruce and western hemlock lumber direct from Alaska to the Orient were expected in the not-too-distant future.[9]

Except in the very short term, however, the supply situation for pulp and paper was not as dire as had been portrayed by the industry. Though there was a substantial backlog in consumer demand for paper, the cutting of pulpwood was expected to get back to normal about a year after the war ended. Moreover, imports, particularly from Canada, could provide an interim supply, and it was expected that once hostilities ceased, European pulp and paper producers would attempt to regain their markets in the United States. Sweden in particular, which had been neutral during the war, had built up an inventory of pulp and newsprint paper and was prepared to make substantial shipments to the United States. In November 1947, the Consumer Goods

Subcommittee of the President's Committee on Foreign Aid reported that "the world's pulp and paper supply-requirements picture is more out of balance than that of the United States." It predicted that by 1949 the United States "may be able to materially increase exports of paper and paper products." The lure of potentially expanding markets continued to encourage some government officials, such as Assistant Secretary of the Interior Warner Gardner, who reported to Congress that "it is a matter of utmost urgency and importance to start a pulp mill development in southeastern Alaska."[10]

Longer term, in anticipation of a post-war increase in demand, there were definite industrial plans for expansion of pulp and newsprint paper manufacturing capacity in the Pacific Northwest and the South. But it was Canada's industry that posed the greatest threat to the effort to establish a newsprint paper industry in Southeast Alaska. As had been the case for many years, the unparalleled Canadian capacity to produce newsprint paper was used largely to satisfy the American demand, and a robust postwar American demand made for good business in Canada. The Paper Trade Journal reported in 1946:

> Of all the peacetime Canadian manufacturing industries, pulp and paper stands first in employment; first in total wages paid; first in export values; first in the net value of production; and first in capital invested. In newsprint paper alone, Canada has a mill capacity of over 4¼ million tons a year, four times greater than that of any other country. Some 94 per cent of her newsprint output is exported. In the production of woodpulp, Canada is second only to the United States. In prewar years, Canada was responsible for two-thirds of the world's newsprint export trade. Canada supplied 3 of every 8 newspaper pages printed throughout the world. She now provides more than half the newsprint production of the world.[11]

Postwar Canadian pulp and paper mills maintained steady production levels near full capacity, though total output and efficiency were initially hindered by equipment that had not been upgraded since the Great Depression. By 1950, the United States was using six million tons of newsprint paper annually, 80 percent of which was supplied by Canada.[12]

Prince Rupert, British Columbia, is only 90 miles from Ketchikan. Plans for a pulp mill at Prince Rupert dated back to 1924, when the Prince Rupert

Sulphite Fiber Company purchased a sawmill and extensive timber rights, and prepared plans and ordered equipment for a 90-ton bleached sulfite plant. The establishment of a wood pulp industry at Prince Rupert was not realized, however, until after World War II, when the Columbia Cellulose Company (a subsidiary of the U.S.-owned Celanese Corporation of America) constructed a $27-million mill to produce not paper but pulp suitable for the manufacture of textiles and plastics.[13] The mill was dedicated on June 25, 1951, with a number of Alaskans present. According to the *New York Times*, the Alaskans admitted to "alarm and disappointment" that thus far no such plant had been erected in Southeast Alaska.

E. L. "Bob" Bartlett, Alaska's non-voting delegate to Congress, was among those at the dedication. He confessed to not knowing precisely why pulp mills had never been built in Alaska, but he was willing to speculate: "This might be one of those propositions where the water seems to be cold and everyone is afraid to plunge in. Some hearty boy will do it one of these days and will discover that the water is really fine. Then everyone will jump in after him."[14]

Some, ironically, saw possible environmental benefits for the lower 48 states in the development of a newsprint paper industry in Alaska. In the January 1947 issue of *Alaska Life*, Bob Callan wrote: "With many of the states in the U.S. facing the possible destruction of scenic and recreational areas to provide badly needed timber for paper manufacture, and with the present supply of paper at a low ebb, the opening of an Alaskan timber industry may mean the preservation of many of America's beauty spots, especially those in the Pacific Northwest." The National Park Service saw a potential benefit of a different sort, one that would benefit the tourism industry in Alaska. In a 1953 report on recreational opportunities in the territory, the agency suggested that a visit to a pulp mill "should prove to be an outstanding feature of a sight-seeing tour."[15]

Despite the support for establishing pulp mills in Southeast, and the dismay that it had not yet happened, there remained a major issue that needed to be addressed before Southeast Alaska's pulp-mill era could begin: Native land claims.

CHAPTER 16

Native Alaskans Claim Their Lands

THE FOREST SERVICE HAD BEEN PROMOTING SOUTHEAST ALASKA TO POTENTIAL PULP MANUFACTURERS AS A PACKAGE: CHEAP TIMBER UNDER GENEROUS TERMS, ADEQUATE WATERPOWER TO RUN MILLS, ECONOMICAL MARINE TRANSPORTATION, AND PRODUCTION THAT IN DOMESTIC MARKETS WAS SECURE FROM IMPORT OR EXPORT DUTIES AND EMBARGOES. WHAT WAS LEFT OUT OF THE EQUATION WAS ONE VERY IMPORTANT CONSIDERATION—SOMETHING THE AGENCY HAD BEEN AWARE OF FOR MANY YEARS:

> There are throughout the National Forest areas a number of old Indian claims and occupancies in all sorts of conditions from good to absolute nonuse and decay. Probably most of them are entirely abandoned. Ordinarily these claims give us no concern but if located on areas desired by industrial or public purposes, they are likely to spring to life. If valid it is essential that their interests be fully protected but if the claims have been abandoned for years they should not be allowed to interfere with legitimate use.[1]

Until Native claims on the Tongass National Forest were settled, the Forest Service did not possess clear title to the land and the timber that grew on it. Investors who were being courted to make expensive, long-term commitments to operate in the region would need to know the timber supply upon which they might base their business was secure.

Native land rights were not explicitly acknowledged in the treaty by which the United States purchased Alaska from Russia in 1867. They were, however, given limited recognition by the Alaska Organic Act (1884), which stated, "Indians or other persons in said district shall not be disturbed in the possession of any lands actually in their use or occupation or now claimed by them, but the terms under which such persons may acquire title to such lands is reserved for future legislation by Congress."[2]

In the years 1902 through 1925, when, as authorized by Congress, presidents Theodore Roosevelt and Calvin Coolidge made proclamations that

205

established and expanded the Tongass National Forest, Native claims in the region were ignored. The Tlingit and Haida people of Southeast Alaska were not compensated for their losses, despite ample precedent for doing so in the U.S. government's past dealings with other Native groups.[3]

The door for redress was opened with the Tlingit and Haida Jurisdictional Act of June 19, 1935. In this legislation Congress authorized the Tlingit and Haida Indians of Southeast Alaska to bring suit in the United States Court of Claims for "all claims, of whatever nature, legal or equitable...for lands of other tribal property or community property rights taken from them by the United States without compensation."[4]

The following year, an amendment to the Indian Reorganization Act of 1934 empowered the Secretary of the Interior, contingent upon a supportive vote by Indian or Eskimo residents, to designate certain specified lands within the territory of Alaska as Indian reservations. Among the lands available for such designation were those reserved by executive order and placed under the jurisdiction of the Department of the Interior, together with adjacent public lands, or any public lands that were actually occupied by Indians or Eskimos.[5]

The Secretary of the Interior at that time was Harold Ickes. A member of the progressive wing of the Republican Party, he had been brought into the cabinet of President Franklin Roosevelt for political considerations. "Honest Harold," despite shortcomings that characterized his personal life, had a sterling public record. Bold and outspoken, Ickes transformed a Department of the Interior tarnished by inefficiency and scandal into a reasonably efficient and nearly scandal-free agency.

Ickes had visited Southeast Alaska in 1938 and was sympathetic to the overall plight of the Native population. His desire to provide them with some redress dovetailed conveniently with an issue he was almost fanatical about: his belief that the Forest Service logically and rightfully belonged in his Department of the Interior, which he considered to be more conservation-oriented than the Department of Agriculture. Ickes hoped to rename his domain the "Department of Conservation." The establishment of Indian reservations in Southeast Alaska, as provided for in the Indian Reorganization Act, would afford a degree of economic justice for some of Southeast Alaska's Natives while, in effect, giving the Department of the Interior control over the use of national forest land that could be designated as Indian reservations.

To facilitate the establishment of Indian reservations, Ickes chose to use his authority to include in Alaska's commercial salmon fishing regulations a provision designed solely to encourage Native claims of aboriginal occupancy.

The U.S. Fish and Wildlife Service was the Interior agency responsible each year for writing the fishing regulations for Alaska, which were then subject to the approval of the Secretary of the Interior. In 1942, Ickes instructed Fish and Wildlife to include in the regulations the following provision:

> No [salmon] trap shall be established in any site in which any Alaskan native or natives has or have any rights of fishery, by virtue of any grant or by virtue of aboriginal occupancy, by any person other than such native or natives. Any native or natives claiming such rights may petition the Secretary of the Interior for a hearing with respect to the validity of such claim, and prior to any such determination such claimant and any interested parties desiring to appear in opposition to such claim shall have an opportunity to be heard.[6]

In 1944, pursuant to that provision, Native groups from the villages of Hydaburg, Kake, and Klawock petitioned the Secretary of the Interior for exclusive use of large tracts of land and water in Southeast Alaska, based on their aboriginal use and occupancy. The tracts encompassed much of the best timber in Southeast Alaska and totaled about 3.8 million acres, including Kuiu Island, the western half of Prince of Wales Island and most of the islands off its seaward coast, parts of Baranof and Chichagof islands, as well as a mainland sector.[7] To satisfy these claims, Ickes had full authority to create Indian reservations that would be administered by the Department of the Interior. Though the Tongass was a national forest and thus within the Department of Agriculture, federal lands there could only be designated as Indian reservations by the Secretary of the Interior, according to the amended Indian Reorganization Act of 1934.

The prospect of extensive Indian reservations administered by Harold Ickes was considered a major threat to the establishment of a wood pulp industry in Southeast Alaska. Ickes's reputation with the western lumber industry was soured in particular when the Roosevelt administration refused to open Olympic National Park to the logging of aero-grade Sitka spruce during World War II, and as *Pacific Pulp & Paper Industry* wrote, "The company proposing to build a pulp or pulp-paper mill would have to deal actually with Mr. Ickes personally, and to them that is an unpleasant prospect."[8]

The potential establishment of Indian reservations was also a serious threat to Southeast Alaska's salmon industry. In July 1944, Forest Service Chief

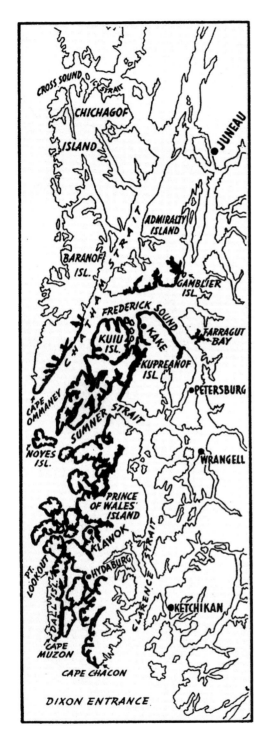

FIGURE 49. Coastal areas claimed by Kake, Klawock, and Hydaburg Natives as portrayed in *Pulp & Paper Industry*, August 1945.

Lyle Watts, his assistant, and Frank Heintzleman met to discuss the matter with H. L. Faulkner, a Juneau attorney representing the Alaska [salmon] Packers Association. In a subsequent letter to the Alaska Packers Association, Faulkner wrote,

> I found that they did not have a very clear understanding of the whole thing, and when I went over it with them, they were very much astonished and thought the whole scheme of the Interior Department was fantastic, but nevertheless very dangerous. However, when the matter comes to an issue these men are going to be prepared, and they are gathering all the data they can get, not only to be prepared against the creation of any Indian Reservations on the Forest Reserves, but also against the establishment of ancestral rights.

Faulkner added a postscript: "Please treat as confidential what I have said about the Forest Service officials, as I know they do not want to be quoted now."[9]

The Forest Service and the salmon industry were not alone in their opposition. Ernest Gruening, who had been appointed territorial governor by Franklin Roosevelt in 1939, did not get on well with Ickes and opposed the establishment of Indian reservations in Alaska, favoring instead the assimilation of Natives into Alaska's white-dominated society. Support for the reservations was also lacking within the Native community. The Alaska Native Brotherhood had opposed Indian reservations since its inception in 1912 and reaffirmed that position at its convention in 1947.[10]

To adjudicate the claims, Ickes appointed Judge Richard Hanna as examiner, and he detailed attorneys from the Department of the Interior's Office of Indian Affairs to represent Native claimants. Hanna, former chief justice of the New Mexico Supreme Court, was an expert on Indian law. In the fall of 1944, hearings were held at Hydaburg, Kake, and Klawock. In March 1945, Judge Hanna determined that aboriginal rights might exist on some uplands and recommended to Ickes that Congress be asked to authorize an extensive survey of the entire question.[11]

Ickes, however, considered the matter in ways different from Judge Hanna, and in January 1946 announced his controversial determination that the Indians at Hydaburg, Kake, and Klawock held aboriginal title to 273,000 acres of land inside the Tongass National Forest. Although the acreage did not include the best and most available timber in Southeast Alaska, *Pacific Pulp & Paper Industry* wrote, "No Forest Service official has made any comment, but there is no doubt that developments have made that department very unhappy, blocking its dreams of an Alaskan pulp industry."[12]

REMOVING THE OBSTACLES

Ickes resigned from his position as Secretary of the Interior in February 1946, shortly after Harry Truman became president and before any reservations were established or any land was transferred to the claimants. He was succeeded the following month by the decidedly more industry-friendly Julius Krug, who was eager for the issue to be settled in a manner favorable to the establishment of a wood pulp industry—something characterized by the *New York Times* in 1949 as being an "economically ripe" concept. Since 1946, the *Times* reported, "scads of people, including some big operators," had been looking into the Southeast Alaska pulptimber situation. One of those operators was the Timber Development Corporation of New York, which had an unconsummated contract to cut timber off lands in Southeast Alaska claimed by Natives.[13]

The last major obstacle to the establishment of the long-awaited wood pulp industry in Southeast Alaska was the cloud on the Forest Service's title to Tongass lands and timber. Until title was clarified, industrial interests would be unwilling to make investments.

Because of the purported acute shortage of paper pulp, time was of the essence. Secretary Krug was certain that "any legislation looking to a general settlement of the problem would be controversial and slow of enactment," and "It would take a considerable number of years to settle the matter by litigation." According to Krug, the development of a wood pulp industry in Southeast Alaska could not wait that long. As Senator Carl Hayden (D-Arizona), Chairman of the Joint Committee on Printing, told the *New York Times*, "We have to print the *Congressional Record*—and we have to have paper to do it."[14]

Krug suggested in a May 16, 1947, letter to House Speaker Joseph Martin (R-Massachusetts) that as an interim solution, the present Congress should pass legislation that would permit timber development to proceed. It would leave to a future Congress the "controversial question as to how the rights should be adjusted as between the Natives and the United States."[15] Krug and the pulp and paper industry knew that at least the outline of Native claims legislation favorable to the establishment of a wood pulp industry had been drafted even while Ickes was Secretary of the Interior, and they clearly expected it to become law, as was illustrated by a number of comments made in the early months of 1947.

In January of that year, a group of West Coast and New York financiers was described as "eager to submit a bid" on a Tongass pulptimber sale but was awaiting congressional action on pending legislation. Krug, in an interview with the *New York Times* two months later, predicted that "one or more of the [pulp] projects will go through before much time passes." He added, "Just one of the six mills proposed would support a greater population than the biggest town now in Alaska. One plant, with its employees and their families, would bring to Alaska 12,000 persons." The Association of Wood Pulp Consumers stated in April that Indian claims were "due to be settled soon."[16]

In his letter to Speaker Martin, Krug observed, "The question of native land titles in the Territory of Alaska has remained, in the large, unresolved throughout the history of that Territory," causing great uncertainty as to the amount and boundaries of the public domain in Alaska.[17] To press the point, the Alaska Territorial Legislature submitted to Congress the following memorial:

> Whereas every encouragement should be given to prospective settlers and investors to settle in and to develop the natural resources of Alaska; and
>
> Whereas every cloud on title to lands should be removed without delay; and
>
> Whereas the question of Indian or aboriginal title constitutes such a cloud on title to lands desired by homesteaders, trade and manufacturers, to pulp and paper investments; and
>
> Whereas economic and industrial progress and development of Alaska as a whole, and of the coastal areas in particular, are being retarded and hampered by the uncertainties of the Indian or aboriginal title controversy:
>
> Now, therefore, your memorialist respectfully urges the Congress of the United States to take prompt action to investigate and settle equitably to all parties in interest the question of Indian or aboriginal title to all lands in the Territory of Alaska.
>
> And your memorialists will ever pray.[18]

Some white Alaskans opposed sharing the territory's resource in any way. When the House Committee on Agriculture held hearings on a resolution authorizing the Secretary of Agriculture to sell Tongass timber, it received the following statement: "The Ketchikan Chamber of Commerce by unanimous approval is unalterably, repeat unalterably, opposed to giving Alaska Indian groups a 10-percent royalty on the natural resources in Alaska, including giving any such royalty on the timber resources of the Territory."[19]

Alaska's governor, Ernest Gruening, believed that the development of the pulp and paper industry would benefit Southeast Alaska's Natives more than continued pursuance of land claims. He conveyed to the committee the popular and official view of the expected effect of the expansion of the forest products industry on the Native population:

> I know of no one thing that will be more beneficial to the economy of the Indian population than the development of this pulp and paper industry…. It means a new day in the Indian economy. It means that instead of being obliged to subsist for 12 months on rather uncertain earnings of three or four months' fishing, they will have something that will keep them employed all year around, and I can think of nothing that will equal that in benefit.[20]

Secretary Krug's letter to Speaker Martin included draft legislation with a provision that all proceeds from the sale of Tongass National Forest timber and land would be held in a special account until the rights to the land and timber were determined. After little, mostly perfunctory deliberation and a slight modification (the elimination of commitment by Congress to enact legislation in the future to fully settle the matter), what became known at the Tongass Timber Act was passed by Congress in the last minutes of its session. It was signed into law by President Truman on August 8, 1947.[21]

The essence of the legislation reads,

> Sec. 2 (a) The Secretary of Agriculture, in contracts for the sale, or in the sale, of national forest timber under the provisions of the Act of June 4, 1897 (30 Stat. 11, 35), as amended, is authorized to include timber growing on any vacant, unappropriated, and unpatented lands within the exterior boundaries of the Tongass National Forest in Alaska, not withstanding any claim of possessory rights. All

such contracts and sales heretofore made are hereby validated.

(b) The Secretary of the Interior is authorized to appraise and sell such vacant, unappropriated, and unpatented lands, nothwithstanding any claim of possessory rights, within the exterior boundaries of the Tongass National Forest as, in the opinion of the Secretary of the Interior and the Secretary of Agriculture, are reasonably necessary in connection with or for the processing of timber from lands with such national forest, and upon such terms and conditions as they may impose.

(c) The purchaser shall have and exercise his rights under any patent issued or contract to sell or sale made under this section free and clear of all claims based upon possessory rights.

Sec. 3. (a) All receipts from the sale of timber or from the sale of lands under section 2 of this resolution shall be maintained in a special account in the Treasury until the rights to the land and timber are finally determined.

(b) Nothing in this resolution shall be construed as recognizing or denying the validity of any claims of possessory rights to lands or timber with the exterior boundaries of the Tongass National Forest.[22]

The Forest Service issued a statement on the legislation: "The Tongass Timber Act…neither denies or recognizes the validity of any [Native] claims of possessory rights. It was regarded as interim legislation designed to permit the immediate development of huge Alaska timber-using industries in which Alaskans, Indians as well as whites, stand to benefit greatly."[23]

Pulp & Paper reported, "The ink from President Truman's fountain pen was hardly dry on the bill authorizing the use of national forests in Alaska for private-owned pulp and paper industries, while relieving the operators of any threat of usurping Indian claims, before the U.S. Forest Service stepped out with its offer for sale of a reputedly 1,500,000,000 cu. ft. timber area."[24]

Harold Ickes, who was now retired from the government, was furious. According to Ickes, the Tongass Timber Act, referred to by some newspapers as the "Alaska Newsprint Bill," was, "one of the worst bills that has been passed in many a day," the result of a conspiracy between the Departments

FIGURE 50. In early 1947, it seemed to Alaskans that the pulp industry's time was close at hand. An article in this issue of *Alaska Life* optimistically touted the opportunities for the development of a newsprint paper industry in Southeast Alaska. (Alaska State Library, *Alaska Life*, Vol. 10, No. 1, January 1947)

of the Interior and Agriculture that was "intended to plunder the Indians ruthlessly for private profit." Ickes labeled Heintzleman as "probably the person chiefly responsible for this depredation of Indian property," though he heaped blame on many others as well.[25] (True to form, in the introduction to his

Autobiography of a Curmudgeon, Ickes wrote, "If in these pages, I have hurled an insult at anyone, be it known that such was my deliberate intent, and I may as well say flatly now that it will be useless and a waste of time to ask me to say I am sorry."[26])

Seeking compensation for the loss of hunting and fishing rights and the value of more than 20 million acres of land appropriated by the U.S. Government, some 7,000 Tlingit and Haida Indians sought redress in the U.S. Court of Claims. In 1959, the five-judge court unanimously agreed with their claim, and in 1969 the claimants were awarded $7.5 million from the Tongass National Forest Timber Fund.[27]

RESOURCES FOR SALE

Now that the Tongass Timber Act had become law, the Forest Service offered two 50-year timber sales with which it hoped to "facilitate the establishment of the new industry…on a commercially sound and permanently economical basis."

The first offering, in August 1947, was for 1.5 billion cubic feet of pulptimber located in the Thomas Bay area (near Petersburg), where a 1923 proposed Forest Service pulptimber sale had never been consummated. A minimum acceptable bid for pulptimber was set by the Forest Service at $0.85 per cubic foot. For sawtimber the prices were $3 per MBF for Sitka spruce (including high-grade logs of this species intended for pulp manufacture), $2 per MBF for western hemlock, and $1.50 per MBF for western redcedar and Alaska-cedar. The Forest Service called the Thomas Bay area offering "a major advance in the development of pioneer Alaska, with major implications for the economic security of the United States," and took the unusual step of calling for oral bids at an auction to be held at the Department of Agriculture offices in Washington, D.C. The successful bidder would be required to show proof of having $8 million for plant construction. It was anticipated that the successful bidder for the Thomas Bay timber would probably find it necessary to construct an entirely new town with housing for 1,000 persons.[28]

A like amount of timber was offered in the vicinity of Ketchikan on September 15, 1947. This offering, a variation of earlier offerings, consisted of timber mostly on Prince of Wales and Revillagigedo islands.

In the late summer of 1947, three members of the U.S. Senate Newsprint Subcommittee traveled to Southeast Alaska to examine firsthand the situation there. The group was led by Senator Homer Capehart (R-Indiana), chairman

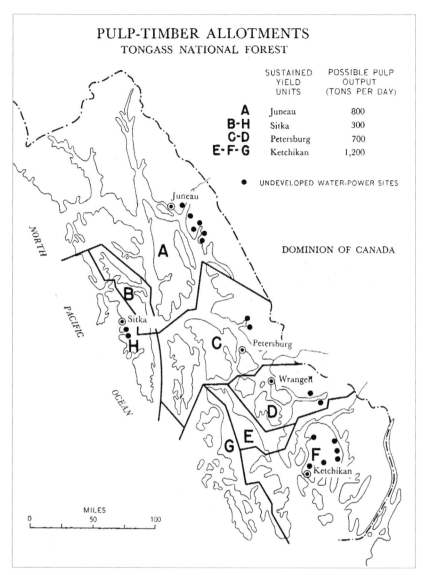

FIGURE 51. By 1947, the Forest Service had configured Southeast Alaska into eight sustained-yield pulp timber units. (USDA, *Trees, Yearbook of Agriculture, 1949*)

of the subcommittee, and included senators Harry Cain (R-Washington) and Allen Ellender (D-Louisiana). Capehart was particularly interested in the plight of small newspaper publishers, who were at the time somewhat squeezed for a supply of newsprint paper, and he wanted to consider Southeast Alaska's potential. It was not healthy to depend on a foreign country for newsprint

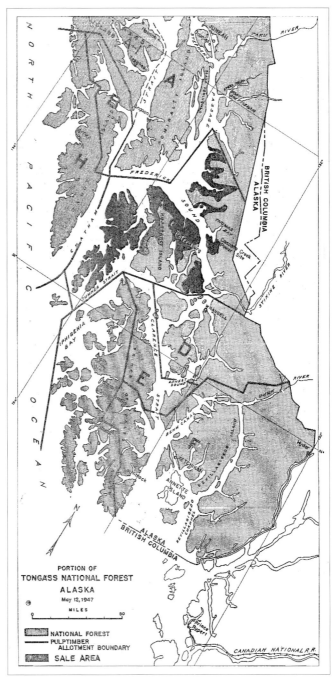

FIGURE 52. U.S. Forest Service–proposed Thomas Bay pulp timber offering, 1947. (*Alaska Life*, November 1947)

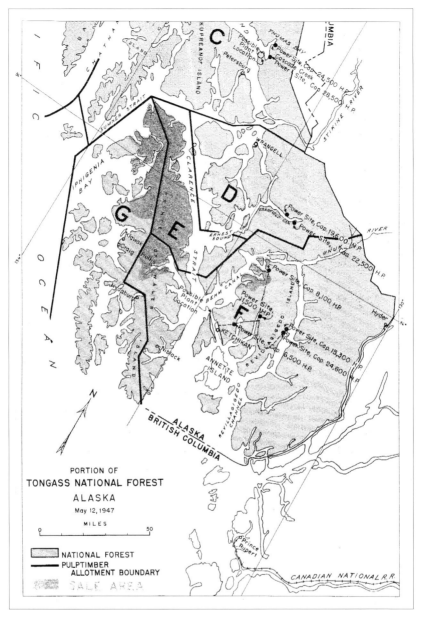

FIGURE 53. U.S. Forest Service–proposed Ketchikan area pulp timber offering, 1947. (Alaska State Library, Juneau Spruce Corporation Collection)

paper, he told his subcommittee, "when courageous publishers and editors are needed to combat communism."[29]

Capehart envisioned three ways in which the government could promote the development of a pulp industry in Alaska: "The development of power resources, the building of small cities which will be needed near the mills, and helping to solve Alaska's transportation problems." The editor of *Paper Trade Journal*, however, had earlier written an article titled "Alaska Dreamland," in which he said that "Congress, faced with the necessity of balancing a budget and of trimming taxes, will have to look with favor on other things than granting huge sums necessary to get paper production into full blast no matter how much it is needed."[30]

Another threat to the establishment of a large-scale wood pulp industry on the Tongass emerged that same year when Representative William Lemke (R-North Dakota) introduced legislation (H.R. 4059) that would have entitled war veterans to "homestead" claims in Alaska of up to 1,920 acres—three square miles. There were minimal requirements for "proving up" on such a homestead, which, if chosen from the better timberlands of the Tongass, could contain as much as 50 MMBF of timber.

The passage of this legislation would have been, in Chief Lyle Watts's words, "a deterrent to early development of a paper industry in the Territory." He added, "Transfer of these tremendously valuable public resources into a large number of small ownerships would lead to un-integrated piecemeal exploitation," with many of the tracts eventually coming into the hands of speculators; it would ultimately cause "the disintegration of the national forests of Alaska." Representative Lemke's legislation, derided in *American Forests* as a return to the "dark ages of American public land policy," passed the House by unanimous consent but was voted down in the Senate.[31]

Some suggested that American publishers pool their funds and launch an Alaska newsprint paper venture of their own, which is what George Cameron of the *San Francisco Chronicle* had attempted with the *Los Angeles Times* and *Portland Oregonian* in 1927. Cooperative newsprint paper ventures had met with success in the South, where the Southland Paper Mills, which began production at Lufkin, Texas, in 1940, counted 37 daily newspapers among its 144 original owners of common stock. Similarly, the Coosa River Newsprint Company, which began operations at Coosa Pines, Alabama, in 1950, was chartered with a provision that ownership of a share of common stock guaranteed annual delivery of 1,000 pounds of newsprint paper. A market for Coosa River Newsprint's entire production was backed by contracts with 115 newspapers.[32]

To facilitate the establishment of a wood pulp industry in Southeast Alaska, the Forest Service endeavored to establish and maintain operations on the Tongass in a position that was competitive with similar operations in the Puget Sound area. To accomplish this, the Forest Service would "do everything in our power and everything reasonable to fix prices, cutting requirements, and other features to enable the operation to proceed profitably." In 1947, the prevailing price for western hemlock sawlog stumpage on the Tongass was $1.50 per MBF, while western hemlock stumpage in western Washington was selling for around $2.50 to $3.50 per MBF, with some timber selling for as much as $9, now that war-related price caps for timber had been lifted. Additionally, the agency's administrative policies did not require that any specific kind or grade of product be manufactured.[33]

Frank Heintzleman estimated in 1947 that the establishment of a newsprint paper industry in Southeast Alaska would employ in the mills and forest combined a minimum of 6,000, and a maximum of 8,000 to 9,000 persons, about 80 percent of whom would not necessarily have to be skilled. The increased employment was expected to lead to a doubling of Southeast Alaska's population.[34]

Of opportunities for Southeast Alaska's Native people, the Forest Service wrote, "The native Indian population, too, can be expected to contribute greatly to the building of a new industry in Alaska. The Tlingit, Haida, and Tsimpsian Indians in this area are adaptable, intelligent people well able to take their place in modern industry. With basic plans for the industry's establishment put into effect, the Indians could be put to work immediately on such jobs as bunkering, constructing access roads, and building whole surrounding villages."[35]

But there were no bidders for the timber at either Thomas Bay or Ketchikan. Established paper manufactures in the East considered the Alaska projects, in the words of *Business Week*, "fantastically uneconomic."[36]

Disappointed but not completely deterred, the Forest Service in December 1947 extended the bid date on the Ketchikan timber, with an oral auction to be held in the chief's office in Washington, D.C., on April 14, 1948.[37] This timber seemed to have more potential than the Thomas Bay timber, and it was hoped that a buyer might be found among the West Coast producers who had surveyed the area. The general assumption was that because their operations were somewhat similar to what would be needed in Southeast Alaska, they would better understand the challenges.

But the Crown Zellerbach Company, the West Coast firm that had thoroughly assessed the practicality of establishing a newsprint paper plant in Southeast Alaska in the late 1920s and again in 1944, was pessimistic. "We found that it would be impossible to deliver the finished product [from Alaska] at a price competitive with that we could produce and deliver at our existing U.S. and Canadian mills," said James D. Zellerbach, president of the company.[38]

Though there were no buyers for the April 14 timber offering, there was at least one firm that did not share Zellerbach's low opinion of Southeast Alaska as a viable pulp producing region. Puget Sound Pulp & Timber Company, an established operator based in Bellingham, Washington, was among those who had been courted over the years by Frank Heintzleman. The company had researched the Tongass's potential and ultimately became convinced that it offered the opportunity for the profitable manufacture of pulp. The manifestation of that interest is the subject of Chapter 17.

CHAPTER 17

Ketchikan Pulp Takes the Plunge

PUGET SOUND PULP & TIMBER WAS ONE OF THE NATION'S LEADING PRODUCERS OF SULFITE PULP. FORMED BY THE MERGER OF A NUMBER OF EXISTING COMPANIES IN 1929, THE COMPANY DESCRIBED ITSELF AS "AN AGGRESSIVE ORGANIZATION IN SEARCH OF RAW MATERIALS AND [HAVING] A DESIRE FOR EXPANSION."[1]

In the summer of 1945, Puget Sound Pulp & Timber sent a logging engineer to Ketchikan to assess the opportunities on the Tongass National Forest. After studying various maps and reading reports on the Forest Service's Ketchikan pulptimber unit, the engineer related to Forest Service officials that his company was definitely interested in the prospect of manufacturing wood pulp in Southeast Alaska. That fall, Frank Heintzleman met in Seattle with Fred Stevenot, Puget Sound Pulp & Timber's president, for an exploratory talk and to discuss the terms of a timber contract to which the Forest Service might agree. The following summer, Puget Sound Pulp & Timber sent a three-man party consisting of its chief engineer, a forester, and a hydroelectric engineer to further investigate the situation.[2]

The men returned with a favorable report, and that fall a raft of logs from Prince of Wales Island was towed to the company pulp mill at Bellingham to assess their suitability. The logs were manufactured into pulp that was said to equal that manufactured from timber in Washington. Stevenot, meanwhile, had been looking into possible partnerships and sources of financing. His efforts bore little fruit in the short term, but in 1948 the American Viscose Corporation, of Philadelphia, sent a committee to investigate the potential of a pulp mill at Ketchikan. Founded in 1910, American Viscose was America's largest and oldest manufacturer of rayon, which is made largely from wood pulp. Shortly after the visit by American Viscose, Lawson Turcotte, Stevenot's successor at Puget Sound Pulp & Timber, negotiated a partnership agreement with the company.[3]

The Ketchikan Pulp & Paper Company, a joint venture of Puget Sound Pulp & Timber and American Viscose, was incorporated in the state of Washington in 1948. The purpose of the partnership was to bid on Tongass National Forest timber and to construct and operate a world-class pulp mill at Ketchikan. For

Puget Sound Pulp & Timber, an operation in Alaska offered the benefits of geographical expansion and diversity. For American Viscose, becoming a principal in the manufacture of wood pulp offered vertical integration as well as a guaranteed supply of the raw material critical to its operations.

In the Ketchikan Pulp & Paper Company the Forest Service had finally found a buyer for timber it had been unsuccessfully trying to sell, in its most recent configuration, for nearly a year. Preliminary award of the timber was made in Washington, D.C., on August 2, 1948. The *Wall Street Journal* heralded the event and its potential: "The opening of a little envelope in a government office here today will launch what may become a giant paper pulp industry for timber-rich Alaska." As soon as it received news via teletype that the contract had been signed, the *Ketchikan Alaska Chronicle* put out an extra edition announcing "Pulp Contract Let!" That day's regular edition of the paper headlined the contract's impact on local employment: "500-Ton Pulp Mill Assured to Employ About 1200 Men." And *Alaska Life*, a popular territorial periodical, enthusiastically proclaimed: "The Pulp Mill Boom Is On!"[4]

Negotiations over the final terms of the contract took almost three years. On July 27, 1951, Ketchikan Pulp & Paper was granted a certificate of necessity from the Defense Production Administration. The certificate—essential for the timely construction of the mill—qualified the company for priority in obtaining steel and other construction materials deemed critical during the Korean War.[5]

Timber Sale Contract No. A10FS-1042 between the Forest Service and the Ketchikan Pulp & Paper Company was signed on August 20, 1951. The contract provided for the cutting of 1.5 billion cubic feet (8.25 BBF), "more or less," of timber located on approximately 786,000 acres of Prince of Wales and Revillagigedo Islands, and full-scale operations until June 30, 2004 (50 years after the company's pulp mill was expected to begin operating). This was the largest sale of timber ever made by the Forest Service, and Ketchikan Pulp & Paper considered itself to be "assured of a perpetual supply of pulp timber." The 8.25 BBF of timber under contract provided for an average annual harvest of about 165 MMBF which, it should be noted, was expected to be supplemented with purchases through the normal bidding process of timber from outside the contract area. For its part, the company was obligated to construct a pulp manufacturing plant with a designed capacity of not less than 300 tons per day, with provision made for future expansion.[A] Soon after the

---

[A] Wood pulp is measured in air-dried short tons (ADST). One short tone equals 2,000 pounds.

contract was signed, the Forest Service gave Ketchikan Pulp & Paper permission to export up to 2 MMBF of Sitka spruce and western hemlock logs to Bellingham for experimental processing to determine the proper processes to be employed in the Ketchikan mill.[6]

The timber sale agreement characterized Ketchikan Pulp & Paper's venture as a "pioneering undertaking" that would entail "unusual risks due to many unknown conditions that may be encountered at the isolated site and during operations." But contemporary accounts lauded the step with great enthusiasm. In his 1948 annual report, Forest Service Chief Lyle Watts praised the inception of the new industry: "Providing a stable major industry with year-round operation and employment, the Ketchikan development marks the first step in opening up the territory's huge pulp-timber resources, the largest untapped resources of the kind on the continent. It is expected to play an important role in expanding the economy of Alaska on a sound and secure basis." Several years later, as the mill was about to come on-line, the *Ketchikan Alaska Chronicle* lauded the Forest Service's long effort to establish a wood pulp industry in Southeast Alaska: "The U.S. Forest Service in Alaska today said it recognizes the year 1952 as probably the greatest year in its history. It culminated over 45 years of field work in obtaining timber data, contacting timber users and publicizing the vast timber and water power resources of the Tongass National Forest so necessary for a wood pulp industry." In a 1954 issue that featured the Ketchikan pulp mill, *Pulp & Paper* wrote that "American industry…made one of the biggest leaps it ever made…when this massive cellulose plant was hewn out of the great Tongass forest." The publication praised the mill's owners, but with a probing undertone: "It took courage and faith to build this mill—if that is not so, then why did the government offers of low priced wood, with no carrying charges, go begging over 30 years?"[7]

For at least a short time after the preliminary award of the timber was granted, the Forest Service entertained the idea of a second pulp mill at Ketchikan. But before making an additional pulptimber sale, the agency wanted to conduct more precise cruises of timber and "have the benefit of some cutting performance" by Ketchikan Pulp & Paper. No such sale was ever offered, in large part due to pressure from the Ketchikan Pulp & Paper. Not long after the long-term contract was signed, the company questioned whether the volume of timber in its long-term cutting area was adequate to satisfy the terms of its contract. A subsequent, thorough Forest Service survey of the area revealed a timber volume that was more than adequate to satisfy the terms of the contract.[8]

PULP FOR RAYON

Though the need for newsprint paper had long been the economic rationale for the establishment of a wood pulp industry in Southeast Alaska, the manufacture of paper did not survive consideration at Ketchikan Pulp & Paper. The firm, which quickly shortened its name to Ketchikan Pulp Company, chose instead to construct a modern pulp mill to produce a highly-purified dissolving pulp that is particularly suitable for use in the manufacture of rayon and cellophane. Dissolving pulp, sometimes referred to as "chemical cellulose," owes its name to the fact that it can be dissolved by chemical action and reformed into a host of shapes, including fibers (rayon) and sheets (cellophane). Franklin Roosevelt's Natural Resources Planning Board had written in 1942 that the use of Alaska's forests for the manufacture of dissolving pulp would make them "even more valuable and important" than if they were used to make newsprint paper.[9]

Neither truly natural nor synthetic, rayon is sometimes referred to as "regenerated cellulose," and was commonly marketed in its early years as "fiber silk." It is used in numerous industrial and consumer products, including tires, rugs, industrial belting, clothing, and disposable diapers. The annual consumption of wood pulp for the manufacture of rayon grew sharply after rayon was first produced commercially in 1911, from about 5,000 tons in 1920 to more than 400,000 tons in 1948. As of 1950, the production of rayon was, after paper, the largest consumer of wood pulp in the United States. Rayon production accounted for about two-thirds of the dissolving pulp available domestically.[10]

As part of its partnership agreement with Puget Sound Pulp & Timber, American Viscose contracted with Ketchikan Pulp Company (KPC) for 100,000 tons of pulp annually for 20 years, or nearly all of KPC's production capacity prior to the mill's expansion in 1958. American Viscose had rayon manufacturing plants in Pennsylvania, Virginia, and West Virginia, and after World War II, the company had purchased the Sylvania Industrial Corporation, a manufacturer of cellophane.[11]

The rayon industry was, to quote an article from a 1950 edition of *Pulp and Paper Magazine of Canada*, "the most important factor in determining the future of the dissolving pulp industry," and many expected continued growth in the manufacture of dissolving pulp to supply a growing market for rayon. Others were less confident, for the price of dissolving pulp, which had showed a steady increase since 1939, began to decline in 1949, and some experts predicted a worldwide surplus of the material by 1955. They observed that, though the production and the consumption of dissolving pulp were both increasing in 1950, production capacity was on track to increase more rapidly than consumption. This was because the rayon industry's two dominant firms—American Viscose and DuPont—were in the process of constructing pulp mills capable of supplying their own needs as well as an additional quantity to sell on the open market.[12]

Thus the establishment at Ketchikan of a world-class pulp operation in itself carried an inherent risk: the amount of additional pulp that the facility would place on the market was internationally significant and, particularly if the market faltered, would affect prices not only in Ketchikan but around the world. The situation was candidly acknowledged in an article by the *Ketchikan Alaska Chronicle* written just prior to the mill's startup:

> The truth is that the Ketchikan mill may force the shutdown or curtailment of one or two other mills in the states or Canada, because there is no longer a shortage of pulp. A situation like that can mean a reduction in the price of pulp and closer operating margins for most or all of the industry. In such a case, the cost of operating in Ketchikan would be weighed against the costs elsewhere.[13]

There was also the risk that popular petroleum-based fibers such as nylon (developed in 1937), acrylic (1950), and polyester (1953) might substantially erode the market for rayon. But together, Puget Sound Pulp & Timber and American Viscose had a wealth of knowledge and experience in logging, the conversion of the timber into pulp, and the subsequent conversion of pulp into rayon and cellophane. Outside of its contract to purchase pulp, American Viscose was largely a silent partner. Logging operations and the operation of the pulp mill were managed by Puget Sound Pulp & Timber.[14]

Ground was broken for the $55 million pulp mill on June 1, 1951, at "Wacker City" (unincorporated), on Ward Cove, about eight miles north of Ketchikan. The 53-acre site was a part of a homestead that Eugene Wacker had proved up on in 1907.

Construction of the pulp mill was the territory's largest non-military construction project ever, with some 1,400 workers employed at its peak. Ketchikan's newspaper praised the effort and its result: "Mire and loose dirt, rock and slides could hamper, but not daunt the efforts of the many trained engineers and workmen that combined their many years of technical skill to produce the world's most modern pulp mill, a masterpiece of technology." A

FIGURE 54. Eugene Wacker. (Tongass Historical Museum, City of Ketchikan)

U.S. Department of Labor report described Ketchikan's pulp mill as "the nearest thing to automatic operation ever devised by the industry."[15] The high degree of automation helped make the project feasible in Southeast Alaska's high labor-cost environment.

Pulp was manufactured at Ketchikan using a new magnesium-base acid sulfite process (commonly referred to as the MgO process) pioneered by Weyerhaeuser Timber Company at its plant in Longview, Washington. In this process, magnesium bisulfite cooking liquor was prepared by burning sulfur to form sulfur dioxide, which was then united with flaked magnesium oxide and water. There were three major advantages to the MgO process as compared to the traditional sulfite process:

- most of the sulfur and magnesium oxide could be reclaimed for reuse. (Sulfur was reclaimed from boiler flue gasses; magnesium oxide from fly ash);

- organic solids recovered from spent cooking liquor could help fuel the boilers that powered the plant; and
- much of the pollution traditionally associated with sulfite pulp mills was eliminated.[16]

Though the Forest Service had touted Southeast Alaska's waterpower potential for decades, KPC generated its electricity not from hydroelectric power but with two steam-driven turbines totaling 28,000 horsepower. Sixty percent of the mill's steam was generated in specially-designed recovery boilers fueled with a concentrate of spent cooking liquid known as "heavy red liquor." The remaining 40 percent was generated in two boilers fueled with a 50-50 mixture of "hog fuel" (mill waste that included bark, fines, and sawdust) and heavy fuel oil.[17]

To supply the massive amounts of processing water required, an 85-foot-high concrete dam with an elevation of 250 feet above sea level was constructed at Lake Connell. A wood stave pipe capable of delivering more than 50 million gallons of water daily wound more than three miles from the lake to the Ward Cove mill. In its early years, KPC's pulp mill used 18 billion gallons of water annually, an amount equal to Seattle, Washington's annual domestic consumption.[18]

Although the company's long term contract required that Ketchikan Pulp "shall make such showing as may be required by the Chief, Forest Service, in respect to adequate measure for control of disposal of plant effluents," little provision was made to minimize the pollutants that were discharged by the mill. Liquid effluents from the pulp manufacturing and bleaching process that contained toxic and oxygen-depleting substances were, after only rudimentary treatment, discharged into Ward Cove. Though this was customary at the time, concern was rising in the state of Washington over the annual dumping by sulfite pulp mills of five billion gallons of toxic waste processing liquor into the state's waters. Smokestack emissions at KPC were primarily sulfur dioxide. Despite the company's sulfur reclamation process, about 40 pounds was still lost into the atmosphere for each ton of pulp manufactured.[19]

Formal dedication ceremonies for the mill were held on July 14, 1954, two months after the plant began operating. Among those present was the former chief of the Forest Service, William Greeley, who nearly a half century earlier had led the agency's initial effort to establish pulp mills in Southeast Alaska. His successor, Richard McArdle, called the mill's opening "the realization of a hope which the U.S. Forest Service has cherished for almost 50 years." The mayor of Ketchikan, George H. Beck, described the mill in grand terms: "It is

FIGURE 53. Ketchikan Pulp Company mill at Ward Cove, July 1954.
(Paul T. Saari photo, courtesy Tongass Historical Museum, City of Ketchikan)

a permanent monument to the men of vision—engineers, chemists, planners, builders—and back of them the men of finance whose confidence in Alaska made the mill possible."[20] And Frank Heintzleman, now territorial governor, made a confident prediction:

> From this year onward, July 14th will annually be celebrated by Alaskans to commemorate one of the outstanding important events or turning points in the history of their Territory—the dedication of the fifty million dollar mill of the Ketchikan Pulp Company at Ward Cove in 1954.... We now see the establishment of a major enterprise which will provide year 'round employment, based upon a renewable natural resource offering a supply of raw material in perpetuity under proper scientific management.[21]

In 1952, even before ground had been broken for the mill, Heintzleman had gone to Washington, D.C., to receive a superior service honor award from Secretary of Agriculture Charles F. Brannan. The citation was for "clear-sighted, persistent, and successful efforts in encouraging the beneficial utilization of forest resources in Alaska."[22]

The new pulp mill's annual timber requirement was more than 150 MMBF. "The vastness of such a quantity of logs," read a company annual report, "may be demonstrated by the fact that sufficient linear feet are hauled from the water in a single month to make a floating bridge to Juneau, more than 200 miles to the north."[23]

The process of turning logs into wood pulp began when the logs were sorted by species and drawn into the mill, where a circular saw 108 inches in diameter cut them to a maximum length of 22 feet. Logs more than 30 inches in diameter were also ripped lengthwise by a 10-foot band saw. The logs were then hydraulically debarked and sent to a 1,500-horsepower chipper capable of producing 80,000 cubic feet of chips per hour. The machine could reduce a log 20 feet long and 30 inches in diameter into chips in approximately 12 seconds.[24]

The recipe for making pulp was based on a wood fiber input of 82 percent western hemlock and 18 percent Sitka spruce, which reflected the approximate ratio of western hemlock to Sitka spruce found in Southeast Alaska's forest. The actual pulping process took place in six "digesters" (basically large pressure cookers) that were 58 feet tall and 17 feet in diameter. Each digester was filled with 55,000 gallons of cooking liquor and 100 to 110 tons of wood chips. The digester was then sealed, and the mixture cooked for five to seven hours at a temperature of approximately 300 degrees Fahrenheit (145 degrees Celsius) under a pressure of 110 to 120 pounds per square inch. The cooked pulp mixture was then transferred to a tank where the spent cooking liquor was separated out, and the raw pulp was routed through the plant to be filtered, bleached, and formed into sheets. Once dried, the pulp was wound into rolls weighing as much as 20 tons. As needed, it was cut to customer specifications and baled for shipment. Individual bales could weigh as much as 700 pounds.[25]

The mill's initial capacity was 300 to 350 tons per day, which was increased to 525 tons per day in 1958, and 650 tons per day by 1974. On some days, production was on the order of 800 tons. In 1982, the Forest Service considered the Ketchikan pulp mill to have a capacity of 200,000 tons per year.[26]

The pulp was marketed under the trade name Tongacell. The mill began producing pulp in May 1954, and the first commercial shipment, made in June, consisted of 2,000 tons destined for Buenos Aires. Within the next year, in addition to domestic shipments, pulp was also exported to Mexico, Germany, Japan, and Great Britain. As of 1974, about 75 percent of the plant's production was marketed domestically. Pulp destined for U.S. East Coast markets was loaded on railroad cars that were in turn transported on 20-car barges to Prince Rupert, then conveyed east on the Canadian National Railways.[27]

After KPC's first full year of operation, Lawson Turcotte wrote in his company's annual report that "the difficulties have exceeded any expectation," though he did not describe them. Turcotte even so expressed pride in the fact that his company had "successfully operated a relatively new and untried process in the manufacture of wood pulp."[28]

FAVORABLE TERMS

Following the pattern established in other parts of the U.S., government officials did their best to make conditions attractive to private investors. The establishment of the KPC mill at Ward Cove was the product of government incentives as much as capitalist entrepreneurship.

In a presentation given at the first Alaskan Science Conference, in November 1950 in Washington D.C., Frank Heintzleman suggested four public incentives that should be offered to the pulp industry:

1. Government financial assistance to establish barge and ferry service along the Inside Passage;
2. Liberal government loans to those who would establish pulp mills;
3. Positive assurance that territorial taxes on industries such as pulp mills will remain low during the critical first ten years of operation;
4. Improvements in living conditions in Southeast Alaska's cities.[29]

In 1954, Heintzleman wrote that the Forest Service "must be credited with making it possible for a concern to enter attractive long-term contracts for the purchase of timber."[30] In "Great Forest Industry Future Envisaged in State of Alaska," an article in a 1960 edition of *Pulp & Paper*, Heintzleman enumerated the benefits KPC had received at taxpayer expense, much of it in the form of groundwork conducted by various public agencies during the previous four decades:

The Forest Service as manager of the timber of the Tongass National Forest of southeastern Alaska cruised and mapped hundreds of thousands of acres of timberland. The Geological Survey investigated the available water supply and hydro sites of the region. The Bureau of Reclamation drilled dam sites to determine foundation conditions. The Bureau of Land Management made land surveys of areas to be occupied. The Bureau of Roads built highways to plant sites. The U.S. Congress enacted legislation to compensate Indians... The territorial government (of pre-statehood days) offered tax incentives for a 10-year period to large new industrial enterprises coming into the territory.[31]

The territorial tax incentive referred to by Heintzleman was a "new industrial enterprise" incentive enacted by Alaska's territorial legislature in 1953. The purpose of the legislation was "to encourage the establishment of new industry and the construction of new buildings and structures in the Territory which bring new payrolls, new settlers and, consequently, new wealth to the Territory." Under it, KPC was excused from paying half of its property taxes during its first ten years of operation.[32] KPC enjoyed a federal tax benefit as well. Under the certificate of necessity that was granted to the company by the Defense Production Administration in 1951, KPC was permitted over the next five years to amortize 65 percent of its plant cost in lieu of federal taxes.[33]

An additional benefit was a provision in KPC's Forest Service timber contract that was designed to help the company maintain a competitive position with similar enterprises in the Puget Sound region. The contract provided that, for the first 20 years of operation, timber would be provided such that the average cost of pulp logs delivered to the Ketchikan mill was 50 to 75 percent of the cost of hemlock logs of similar quality purchased by mills in the Puget Sound region. The actual percentage was determined by the regional forester and, for example, was established at 60 percent for the operating period ending June 30, 1964. Arthur Greeley (son of the former chief of the Forest Service), who became Alaska's regional forester in 1953, later referred to this as the "Puget Sound price gimmick." The gimmick seems to have worked. In 1954, Arthur Brooks, KPC's timber operations manager, wrote that one thing attracted his firm to Southeast Alaska: "The larger amount of timber available at relatively low stumpage costs."[34]

The contract prices of pulptimber, effective for the first 10-year operating period, were $0.85 per one hundred cubic feet (CCF) for all species, except that Sitka spruce logs 24 inches or larger in diameter were scaled and paid for as sawlogs. For material to be manufactured for sale or sold in other forms than pulp, the prices were set at $3 per MBF for Sitka spruce, including higher-grade spruce logs intended for use in pulp manufacture; $1.50 per MBF for western redcedar and Alaska-cedar; and $2 per MBF for western hemlock and other species. In addition to payments for stumpage, a "stand improvement" fee of $0.20 per MBF or $0.15 per CCF was charged to pay the cost of work on land cut over by the Ketchikan Pulp Company. Pursuant to the contract, logs were considered merchantable if not less than 16 feet long and at least six inches in diameter inside the bark at the small end, with not more than 50 percent defect, and containing not fewer than 20 board feet.

The Ketchikan Pulp Company, which also purchased logs from independent contractors, began logging operations at its first company logging camp in July 1953 at Hollis, on Prince of Wales Island's Kasaan Bay, some 42 water miles from Ketchikan. At Hollis the company planned to spend some ten years logging an estimated 400 MMBF of timber, much of it along Maybeso Creek. Logs were yarded with a high-lead system to cold decks and transported by truck to saltwater. The camp had about 150 employees. The general practice in Southeast Alaska logging camps at that time was to work 48 hours per week.[35]

ENVIRONMENTAL CONCERNS GROW

As the Forest Service after World War II geared up its efforts to attract pulp mill interests to Southeast Alaska, concerns about the possible environmental effects of logging and pulp mill operations began to be voiced.

Concern over the potential negative effects of extensive and intensive logging on Southeast Alaska's salmon resource, the backbone of the region's economy for more than a half century, was somewhat muted, though the fishing industry's West Coast trade journal, *Pacific Fisherman*, expressed a degree of unease—and of trust—in 1947, when its editors wrote:

> Unless the logging and manufacture of the timber is carefully planned and administered it can impair and perhaps destroy the priceless salmon fisheries of the region.

This is something which must not be permitted to occur.

The fisheries of Southeast Alaska are particularly vulnerable to logging, inasmuch as they are supported by spawning in short, steep streams whose entire watershed might be denuded and destroyed by operations not timed and planned and pursued with a view to protection of the fisheries.

There is good proof that logging of timber on a scheduled, selective basis need not be destructive of aquatic life....

The timber of Southeast Alaska can be manufactured into pulp without destroying the fisheries.

It *must* be.[36] (emphasis in original)

At least one federal fisheries official was less certain. C. Howard Baltzo, of the Department of the Interior's Fish and Wildlife Service, had lived in Southeast Alaska for four years and represented his agency at a congressional hearing on Alaska fisheries issues in Seattle, Washington, in the fall of 1949. Baltzo recognized that pulp mills in Puget Sound and elsewhere had been "considered a very large factor in the depletion of the local fisheries," but hoped that the same problem would not materialize in Alaska. He seemed to take at face value statements by pulp industry officials that "something" (Baltzo's word) would be done to eliminate pollution from the expected pulp mills.

Baltzo considered deforestation a greater threat to salmon than pulp mill pollution, and noted that there had already been specific instances in which logging had caused serious damage to Alaska's salmon streams. It was not, for example, unusual for "Cat" (tractor) loggers to use salmon streams as roadways on which to tow logs.

Fishermen in Southeast Alaska, Baltzo said, were "not entirely pleased" with the prospect of a pulp industry in the region, and would be "highly critical" of its "encroaching severely on their source of income." The "pulp people," according to Baltzo, had "no moral right to go up there and ruin the salmon industry."[37]

Concern for the salmon fisheries may have been greatest within the region's Native community, in which many relied on the fisheries for subsistence and their livelihood. In 1949, *Pacific Fisherman* quoted a "bitter" Native leader as saying: "The Forest Service says they will protect the streams

FIGURE 56. Tractor with arch towing logs in streambed. (Rasmuson Library, University of Alaska, Fairbanks, Angela Burke Janzen Collection)

from damage by logging, but why don't they do it *now*. Why should we expect them to do it *then*, when they have become still more powerful. We protest in writing against destruction of the streams by logging. What happens? Nothing."[38] (emphasis in original)

The commercial fishermen were not alone in their sentiments. In 1956—two years after the KPC plant opened—two writers for *Field & Stream*, America's premier sportsman's magazine, leveled a broadside at Southeast Alaska's budding pulp industry in a feature article titled "Lost Paradise." Corey Ford and Frank Dufresne had fished in Southeast Alaska in the 1930s and returned in 1956 to keep a long-planned rendezvous. The two were clearly unhappy about the arrival of the pulp industry in the region and did not mince words in communicating their displeasure. En route to Petersburg, they flew over what they described as the "stinking yellow plume" of the Ketchikan Pulp plant. The facility itself was, in their words, "a foul blot on a beautiful land." Likewise, clearcuts were "ugly bald spots, like patches of follicular mange." To Ford and Dufresne, the pulp industry was a "menace," a "destroyer of essential timbered watersheds, polluter of clean waters so necessary to the life cycle of the salmon, enemy of all wildlife," and "ruthless despoiler of a nation's recreational heritage." The writers concluded their piece by entreating readers to pressure their congressmen to protect Admiralty Island—the focus of the writers' concerns—by designating it as a "national game-management area."[39]

Almost ten years later, one of a pair of *Field & Stream* articles about logging on the Tongass National Forest acknowledged that Ford and Dufresne's entreaty was a "voice in the wilderness" that went unheeded.[40] Their piece, however, foreshadowed the controversy over the effects of clearcut logging on salmon and other wildlife in southeast Alaska.

Ketchikan Pulp's long-term timber contract did stipulate that the company's "operations shall not be permitted to interfere with the passage of salmon to their spawning grounds or to injure the spawning grounds in any way." In 1949, the Alaska Forest Research Center began monitoring salmon in the Harris River and Maybeso Creek, on Prince of Wales Island, in order to study the effects of clearcutting and high-lead yarding on salmon migration and spawning. Over the course of the next fifteen or so years, 20 percent of the Harris River and 25 percent of Maybeso Creek watersheds were clearcut. In a report published in 1969, the Forest Service concluded that "clearcutting apparently did not adversely affect the salmon spawning habitat" of either stream. Another Alaska Forest Research Center study, begun in 1950, examined twelve Southeast Alaska salmon streams where logging had occurred. Fifteen years later, it concluded that "no measurable harm was noted where…logging extended to the stream banks."[41] Time and better science would show, for both studies, the inaccuracy of their conclusions.

KPC'S EARLY LOGGING SHOWS

About 1955, the Ketchikan Pulp Company began logging operations at Neets Bay, on Revillagigedo Island, about 30 miles by water from its pulp mill at Ward Cove. The Neets Bay operation was based around the company's A-frame slackline skidder, at that time possibly the world's biggest logging machine. The *Lumberman* considered this skidder "the most spectacular piece of logging machinery in Alaska." It was able to efficiently "hot log" timber directly into saltwater at the rate of about 1.5 MMBF per month.[42]

It was no accident that Ketchikan Pulp began its logging operations at nearby Hollis and Neets Bay. In 1938, the Alaska Resources Committee, a subcommittee of President Franklin Roosevelt's National Resources Committee, had issued a plan for the development of Alaska that included this statement: "Initial sales within an allotment will ordinarily include those

timber units most accessible to tidewater, the more inaccessible units being left for later exploitation." The recommendation of the committee, one of whose members was Frank Heintzleman, was in contrast to what forest engineer Don Meldrum, who worked for the Cameron and Zellerbach interests, had suggested less than ten years earlier: The "better plan is one which contemplates logging a certain proportion of the difficult ground along with the favorable chances, thereby maintaining an average logging cost from year to year instead of the proposed low cost at the beginning of the operations."[43]

The timber near Hollis had been cruised in the late 1920s for the Cameron-Zellerbach interests and was known to be both excellent and accessible: "Too much stress cannot be put on the quality of the spruce found [at Maybeso Creek] as the trees are large and smooth and will produce first

FIGURE 58. Ketchikan Pulp Company logging at Harris River, Prince of Wales Island. (USDA Forest Service, Alaska Regional Office, Juneau)

quality timber or pulp."[44] Sitka spruce and western hemlock combined, the trees cut at Hollis were said to average 800 to 900 board feet per tree, approximately equal to ¾ ton of pulp. Timber volume harvested in one 700-acre unit that was representative of the area averaged 37 MBF per acre. An early Ketchikan Pulp timber manager was quoted as regularly saying, "Take the best [timber] first, and then after that you'll always have the best that is left."[45] This was a variation on the logger's adage: "When you're in the best, you'll always be in the best." Though the timber at Neets Bay was not as valuable as that at Hollis, it was good pulptimber and very accessible. For tugboats towing rafts of logs, the trip from Neets Bay to Ward Cove was short and in relatively protected waters.

The area being logged along Maybeso Creek was designated by the Forest Service as the Maybeso Experimental Forest in 1956, the purpose of which was to study the effects of clearcut logging that was on a much grander scale than had ever been experienced in Southeast Alaska. The clearcuts were contiguous and extensive: nearly all commercial timber for 4½ miles along both sides of the creek and up the valley walls was removed. One cut—the

FIGURE 59. Maybeso Experimental Forest. (USDA Forest Service, Alaska Regional Office, Juneau)

"mile-square unit" (actually 700 acres)—was designed especially large to study the amount and distance of seedfall. The mile-square unit study determined that such a large area would reseed naturally, and was used by the Forest Service to justify watershed-scale clearcutting elsewhere on the Tongass. Alder trees that survived clearcutting operations were poisoned to reduce the chance that the species would regenerate and become established where Sitka spruce and western hemlock were desired. Despite the evidence that clearcut areas would reseed naturally, for a number of years areas of the Tongass that met certain specified conditions were reseeded with Sitka spruce or western hemlock.[10]

EMPLOYMENT

The timber sale agreement signed in 1951 had specified that as much as possible, "labor for the conduct of logging operations, mills, and manufacturing plants conducted by the purchaser, its affiliates, subsidiaries, or contractors...will be recruited from residents of Southeast Alaska." In November 1954, the *Ketchikan Alaska Chronicle* recounted the changes wrought in Southeast Alaska over the four years since 1950, many of them no doubt due to the mill: population, up 37 percent; income, up 66 percent; employment, up 41 percent; school enrollment, up 51 percent; births, up 20 percent; number of automobiles, up 86 percent; and retail sales, up 21 percent.[47]

By 1960, there were approximately 407 hourly employees and 100 salaried employees at the pulp mill. The company's logging operations employed about 200 more, in addition to independent contractors.[48] It is clear that many of the expected economic benefits from the pulp mill were realized in Ketchikan and nearby communities. However, one area of concern was that many of the workers at the mill and in the woods came from the lower 48 states.

Logging in Southeast Alaska was said to be "a new and unpleasant experience to the experienced loggers from the states who form the bulk of our employees," according to a company official. In 1961, *The Timberman* noted an additional concern: "Alaska's reputation among loggers, largely untrue, that it's a seasonal logging country, has attracted a large number of drifters who travel from operation to operation and back and forth between Alaska and the Continental U.S. This high workman-turnover also increases costs."[49]

The year that the Ketchikan mill began producing pulp was later seen as a watershed. "The date 1954 was a significant one in the history of Alaska's national forests," wrote the author of a 1960 report on Alaska's national forests. "It marked the turning point between the custodial and management eras."[50]

CHAPTER 18

Alaska Lumber & Pulp Builds Its Mill

AS SUPREME COMMANDER OF THE ALLIED POWERS, GENERAL DOUGLAS MACARTHUR ADMINISTERED POSTWAR JAPAN, WHERE ONE OF THE MANY PROBLEMS WAS A SHORTAGE OF LUMBER AND WOOD PULP TO REBUILD THE DEFEATED NATION'S INFRASTRUCTURE AND ECONOMY. AMONG MACARTHUR'S OCCUPATION STAFF IN JAPAN WERE SOME 27 TRAINED FORESTERS.[1] JAPAN'S OWN FORESTS HAD BEEN SEVERELY OVERCUT DURING THE WAR, A SITUATION EXACERBATED BY THE LOSS OF SAKHALIN ISLAND AND ITS SUBSTANTIAL TIMBER RESOURCES.

Some 600 miles long, Sakhalin Island (*Karafuto* to the Japanese) is proximate to both Japan and Russia. Historically, control of Sakhalin Island alternated between Japan and Russia. It was controlled by Russia from 1875 until the Russo-Japanese War (1904–05), when a defeated Russia ceded the southern half of Sakhalin (below 50° N) to Japan.

Forty years later, the boundaries changed again. At the Yalta Conference, in February 1945, Josef Stalin negotiated a secret agreement with Franklin Roosevelt and Winston Churchill, part of which was that at war's end the southern half of Sakhalin Island would be retaken by the U.S.S.R. The loss of the island possession deprived Japan of what had been a very important domestic timber resource, much of which had been converted into pulp and used in the manufacture of rayon, an industry in which Japan had been a major factor before World War II. Faced with a massive postwar rebuilding project and little domestic timber, Japan now needed a major import program, but the situation was complicated by a shortage of foreign exchange with which to purchase timber on world markets. With the help of the International Bank for Reconstruction and Development (World Bank), Japan looked for a supply of softwood timber in three locations: eastern Asia, Brazil, and North America. Alaska seemed to be the most feasible.[2]

Japan had a long history of importing timber and wood pulp from Canada and, before World War II, it had been British Columbia's largest importer of wood pulp. In 1937, the Japanese purchased cutting rights to approximately 1.5 BBF of timber on Vancouver Island and the Queen Charlotte Islands, and unprocessed logs from these and other operations were shipped to Japan until World War II intervened and the Canadian government imposed an embargo. By 1949, shipments of Canadian wood pulp to Japan had resumed.[3]

In 1954, Japan began importing Soviet logs and lumber, initially on a barter basis in exchange for repairs on Soviet ships. A significant percentage of the imports came from Sakhalin Island. Boris Ivanov, the Soviet trade ambassador to Japan, wrote in 1964 that "the timber trade between Japan and the Soviet Union is one of the most important factors in the relationship between the two countries," with log imports from the U.S.S.R. to Japan amounting to about 670 MMBF annually.[4]

To secure the materials for rebuilding the nation and its economy, the Japanese government created four "national projects." The first of these was the Alaska Pulp Company, Ltd. (a.k.a. Alaska Pulp Corporation, APC) in Southeast Alaska, which was critical to supplying the Japanese with lumber and high-quality wood pulp for the manufacture of rayon. Other national projects involved steel from Brazil, and oil from the Middle East and Sumatra.[5]

The production of rayon had been a significant element of the pre–World War II Japanese export economy. Japan became a rayon producer and exporter in 1930 and by 1937 had become the world's leading producer of rayon; the following year it accounted for almost 30 percent of worldwide rayon production. The war and the destruction of Japanese rayon manufacturing plants reduced production in 1947 to about six percent of prewar levels. Under the direction of the Japanese government, the industry was quickly rebuilt, and by the early 1950s Japan was again a major factor in the production and export of rayon.[6]

The original objective of those who organized the Alaska Pulp Company was to establish a sawmill operation, the output of which would be shipped to Japan. The history behind the formation of the company was recounted in a 1953 letter from Donald C. Warner, of the Department of State, to O. F. Benecke, of the Juneau Chamber of Commerce. Warner wrote that in 1951 the Japanese had approached the Forest Service and sought to purchase Alaska timber. The agency turned down the request because the Japanese had proposed to use Japanese labor and construct their own logging facilities: the agency "considered the proposals contrary to established United States policy which is aimed at using national timber resources to develop Alaskan industry and population, and not merely as a source of logs and pulpwood for export."

In February of the following year, the Japanese government petitioned the supreme commander of the Allied Powers—now Lt. Gen. Matthew Ridgeway, who had succeeded Douglas MacArthur—to acquire softwood timber from Alaska's national forests, again proposing to send Japanese workmen to cut the trees. Warner, in his 1953 account, noted that:

> Before an answer was given to the Japanese, there was careful consideration by the Departments of Defense, Interior, Agriculture, Labor, Commerce and State. The Japanese Government was advised that the exportation of logs from Alaska, even into the continental United States, is prohibited…and that Japanese industries should not expect to obtain any softwood logs from Alaskan National Forests. It was told also that possibilities for employment of Japanese nationals in an Alaskan timber project were virtually non-existent.

In October 1952, the Japanese government and Japanese pulp and timber industries sent a mission to Washington to determine the feasibility of establishing a sawmill in Alaska; they now sought to import sawn lumber rather than raw logs. The petitioners also wanted to consider establishing a pulp mill at some later date. In response, the U.S. government described the conditions for the sale of Tongass timber and the existing marketplace for lumber from Alaska's sawmills. Any group—foreign or domestic—that wanted forest products from Alaska, Warner related, would have to demonstrate

> full compliance with the regulations which require that primary processing of timber harvested in the Alaska National Forests take place in the Territory.
>
> It was also pointed out to the Mission that any enterprise created for manufacture of forest products in Alaska for shipment to the United States, to Japan, or to any other destination, would have to be an American corporation, and that this corporation would have to bid successfully for National forest timber, and conform to all the applicable laws, regulations, and contract terms in the same manner as all its possible competitors.[7]

The *Seattle Post-Intelligencer* reported, in a conspiratorial tone, "What the Pacific Coast pulp industry now fears, according to its spokesman, is that after Japanese timber reserves were handed over to the Russians with the transfer of Sakhalin and Kurile Islands in secret conferences at Yalta, commitment was made to the Japanese that their timber deficiency would be made up by permitting them to draw on the American reserve in the Tongass National Forest."[8] But Regional Forester Arthur Greeley denied any secret arrangement and offered reassurance in a letter to the *Seattle Times*:

> I do think it is important that some people outside the Territory know that the Forest Service has not made any deal, and we have made only this commitment—that if the Japanese interests form a corporation under American law, and otherwise meet the necessary qualifications to bid for national forest timber, and if they submit a definite proposal which can be analyzed, if we approve their proposal we will advertise timber upon which they will have the opportunity to bid competitively.[9]

Timber within the Tongass National Forest, Greeley said, is for sale "under our conditions and on our terms."[10]

TERMS OF SALE

In January 1953, a Japanese technical group visited Southeast Alaska to assess the feasibility and scout a possible location for a sawmill and a pulp mill. The group was led by Takuji Oshima (who would later become executive director of Alaska Pulp Company) and included a representative of the Japanese Ministry of Agriculture & Forestry, two lumbermen, two pulp experts, and a paper expert. Though no firm decision was made, the group favored Sitka as the best location to build both a sawmill and a pulp mill. Their intent was to site the mills close to one another and integrate their operations to foster efficiency and ensure that timber would be used to best advantage.[11]

The prospect of a second sawmill in Sitka caused no end of alarm for Thomas Morgan, president of Columbia Lumber Company of Alaska, which owned the largest existing sawmill at Sitka. Fearing a strong local competitor and seeing an opportunity at the same time, Morgan initially proposed selling his mill outright to the Japanese. When his offer was refused, he proposed turning the mill's production over to the Japanese. He feared that "any other competitive arrangement would lead to a tremendous amount of confusion, and ruinous competitive bidding on every timber sale that might come up."[12] As they had his first, the Japanese declined Morgan's second proposition.

Strong objections to the Japanese plans were also voiced by Representative A. Walter Norblad (R-Oregon), who was among those who suspected that a secret deal had been struck at Yalta. Norblad believed that the establishment of a Japanese firm in Alaska set a bad precedent. And perhaps acting in his state's economic interests, Norblad made it clear that "the great timber-producing state of Oregon" could supply all the lumber the Japanese needed. The U.S. Department of State, meanwhile, seemed to endorse the project, which, it wrote, "would materially assist the Japanese economy and at the same time be of real benefit to the economic development of a self-sustaining economy in the Territory of Alaska."[13]

In June 1953, the Council for Integrated Counter-Measures for Forest Resources, a quasi-governmental Japanese organization that had been established in 1951, hosted a meeting in Japan of businessmen and government officials to discuss the possibility of establishing a sawmill and a pulp mill in Alaska. The Council's chairman was Tadao Sasayama, who after World War II had worked with a group tasked by General MacArthur to split up the large family-dominated industrial and banking combines (*zaibatsu*) that were considered a constraint to the rebuilding of Japan. The result of the meeting was a prospectus inviting potential investors to participate in an enterprise called the Alaska Pulp Company, Ltd.

The company was incorporated in Japan on August 15, 1953. Ninety-seven percent of its stock was owned by 76 Japanese companies, each of which had a direct interest in an additional source of wood pulp. Among the biggest holders of stock were Mitsubishi Rayon Company, Kokoku Rayon & Pulp Company, and Teikoku Rayon Company. There were no important producers of rayon fabrics in Japan that were not shareholders in Alaska Pulp Company. Tadao Sasayama was made president.[14]

Alaska Pulp Company's wholly owned American subsidiary, the Alaska Lumber & Pulp Company, was incorporated at Juneau on December 10, 1953, with an initial capitalization of $1 million. All officers and directors, save R. E. Robertson, a Juneau attorney, were Japanese. Tadao Sasayama was president of the subsidiary as well.[15] The Japanese investors hoped they would be able to make up for their shortage of foreign exchange by borrowing $30 million from the U.S. Export-Import Bank, but they were unable to do so because the bank was chartered to assist U.S. companies in export markets, not foreign corporations.

George Sundborg, general manager of the Alaska Development Board, was excited by the new interest in the exploitation of Southeast Alaska's forest: "We've got industrial prospects all over the world," he was quoted as saying in the *Wall Street Journal*. "Everybody seems interested in our timber." But those same prospects raised concerns by labor and industry organizations in the United States, particularly in the Pacific Northwest. Their anxiety focused on the perceived possibility of unfair competition.[16] As well, there were concerns that a Japanese company dependent on consumption of public timber in Alaska would jeopardize the future timber needs of American citizens, as voiced by Representative Russell Mack (R-Washington):

> This proposal offers such a serious threat to the long range prosperity and employment in many of our forest products industries and also would hasten the depletion of our diminished American forest resources that in my opinion, neither the State Department nor the U.S. Export and Import Bank should give the Japanese any encouragement that they will be permitted to exploit our federal forests in Alaska.
>
> Our American cut of timber will have to be substantially increased to take care of our own population's growing requirements. It, therefore, would seem to me to be the utmost folly to allow Japanese corporations to cut government-owned timber in Alaska and, after a small amount of primary processing, in Alaska, to export it overseas.[17]

To blunt some of the criticism, the Japanese indicated their willingness to enter into a joint venture with an American company. They found no takers.

The Forest Service, of course, was enthusiastic over the prospect of a second pulp mill in Southeast Alaska, and with its cooperation the endeavors of the Japanese progressed. In May 1954, Alaska Lumber & Pulp formally requested the Forest Service to prepare and advertise a 50-year contract for approximately 5.25 BBF of timber on Baranof and Chichagof islands. The sale boundaries were determined over the following months by Forest Service personnel working with Japanese government foresters "on loan" to Alaska Lumber & Pulp.[18] To ensure that there was sufficient timber to meet the terms of the contract, the Forest Service reserved Kuiu Island as a contingency sale area.

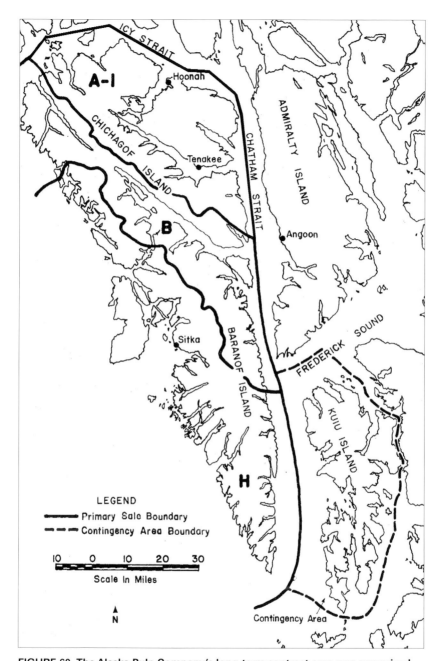

FIGURE 60. The Alaska Pulp Company's long-term contract area was comprised
of Allotments B and H, and, at the Forest Service's discretion, Allotment A-1.
(USFS-APC long-term contract)

The company initially planned to build a sawmill and soon thereafter begin construction of a pulp mill with a capacity of 100,000 tons per year. Until the mill became operational, wood waste from the sawmill would be compressed and shipped to pulp mills in Japan. At about the same time, the Japanese Kokuku Rayon & Pulp Company entertained the idea of combining wood waste from mills at Juneau, Sitka, and Whittier (in Prince William Sound) into monthly shipments to Japan for use in the production of pulp.[19]

On January 23, 1956, Alaska Lumber & Pulp Company, as the sole bidder, was granted the preliminary award of the timber it sought. In compliance with the preliminary award, plans were made for the construction of a pulp mill at Sitka with a capacity of 180,000 tons per year of high-alpha cellulose dissolving pulp to be used for the manufacture of rayon in Japan. The construction of a pulp mill at Sitka would be "an important step in the development of Alaska's resources," according to the company president, Tadao Sasayama. "It will also be of much importance for Japanese industry."[20]

On October 15, 1957, the parties entered into a long-term timber contract for an estimated 5.25 BBF of timber that would provide for the cutting of approximately 110 MMBF of timber annually until the year 2011, 50 years after the Alaska Lumber & Pulp Company's pulp mill was expected to start production. For its part, Alaska Lumber & Pulp Company was obligated to construct a pulp mill that would require 80 MMBF of logs annually. Of the timber under contract, an estimated 1.4 BBF was Sitka spruce, for which the initial stumpage rate was $2.90 per MBF; 3.6 BBF was western hemlock, for which the stumpage rate was $1.75 per MBF; and 210 MMBF was combined western redcedar and Alaska-cedar, for which the stumpage rate was $1.65 per MBF. As with the Ketchikan Pulp Company, Alaska Lumber & Pulp was expected to supplement its contracted timber with purchases through the normal bidding process of timber from outside the contract area. The utilization of logs as raw material for pulp mills was actually becoming an exception: of all the pulp mills constructed in the western U.S. in the late 1950s, the Sitka mill was the only one that did not rely exclusively on wood chips derived from lumber and veneer manufacturing residuals as raw material.[21] Chief administrator for the contract was Arthur Greeley, who had replaced Frank Heintzleman as Alaska regional forester upon Heintzleman's appointment as Alaska's territorial governor by President Eisenhower.

Both this contract and the original long-term contract with the Ketchikan Pulp Company permitted logs up to 40 feet in length to be scaled as one log and provided for a trimming allowance of eight inches. A major difference between the two contracts was the method used by the Forest Service to scale pulptimber. Under the contract with Ketchikan Pulp, pulptimber was measured in cubic feet on the basis of length and average middle diameter, inside the bark, taken to the nearest inch. The price was $0.85 per CCF. Under the Alaska Lumber & Pulp contract, pulptimber was scaled using the Scribner Decimal C log rule. Using this rule, a log is treated as a uniform cylinder, the diameter of which is average diameter, inside the bark, of the small end of the log, taken to the nearest inch. A table is then used to determine how many board feet the theoretical log contains. The price was $1.75 per MBF.

The cubic feet measurement used under the Ketchikan contract more accurately estimated the amount of pulpwood in a log than did the rule used in the Alaska Lumber & Pulp contract. Though this discrepancy could have been compensated for by adjusting prices, it was not, and the result of the different methods of measurement was that Alaska Lumber & Pulp paid considerably less for pulptimber than did Ketchikan Pulp. Under Alaska Lumber & Pulp's contract, a hypothetical evenly tapered, defect-free western hemlock log 38 feet long with a small-end diameter of 12 inches and a big-end diameter of 24 inches cost 33 cents (.19 MBF x $1.75/MBF). Under the Ketchikan Pulp contract, the same log was valued at 57 cents (.671 CCF x $0.85/CCF).[22]

That difference added up quickly with the millions of logs involved in each contract. The Ketchikan Pulp contract was eventually modified to specify that scaling be done the same as for Alaska Lumber & Pulp. The early reduction in the stumpage price for timber from the Tongass National Forest bespoke the Forest Service's desire to see both companies prosper, for the perceived success of the agency in fulfilling its agenda was inextricably tied to the success of the pulp industry that it had worked so long and hard to develop.

Though it may not have mattered much on the ground, another difference between the contracts involved the amount of control the Forest Service had over plans for logging operations. Under the Ketchikan contract, logging plans had to be approved by the regional forester, but the Forest Service had no formal participation in their formulation. The contract simply stated, "As far as may be deemed necessary for the protection of National Forest interests, the plan of logging operations on each of the logging units of this sale area or areas shall be approved by the Regional Forester." The Alaska Lumber & Pulp contract specifically required cooperation between the contract holder and the Forest Service in the formulation of logging plans: "The purchaser shall, before the start of logging operations within the sale area, and prior to the start of operations each year thereafter, join with the Forest Service in preparing a plan of operations which shall be followed except as modified by the purchaser and the Forest Service in writing."[23]

Shareholders of the Alaska Pulp Company contributed $8.5 million in capital stock to the venture and purchased $38.6 million in notes due January 1, 1977. Principal and interest on the bonds and senior notes were guaranteed by the Export-Import Bank of Japan, a government agency. In Japan, the line between where business ended and government began was somewhat blurred. Like the other national projects, the Alaska Pulp Company was a quasi-governmental enterprise. The fact that it possessed national project status was a clear signal to the Japanese banking industry that APC should receive special treatment in financial matters.[24]

Years later, the U.S. Court of Federal Claims described the company's privileged status: "The enterprise in Alaska was a symbol and an important component of the re-emergence of a strong Japanese economy. The company's ability to perform this timber contract was a matter of national pride and an essential component in restoring the economy of war-torn Japan."[25]

The goal of this pulp-making enterprise, along with the other national projects, was not to make money but to secure materials vital to the rebuilding of Japan. The Alaska Lumber & Pulp mill at Sitka would be a world-class producer of high-alpha wood pulp critical to the production of rayon. Despite the fundamental dangers inherent in financing a business in which profitability was a secondary goal, Japanese bankers were willing to assume the loans because their cooperation was enforced by the Japanese government through in an informal and extralegal carrot-and-stick manner. As noted in a court decision made many years later:

> [I]f companies go along with the government's wishes, they
> will be rewarded in the future, maybe through some license
> or permit or preferential treatment. Whereas if companies
> go against the government's wishes and "administrative
> guidance," they might at some point in the future face
> some retaliation by the government.[26]

On October 16, 1957, the day after the long-term timber contract with the Forest Service was signed, the New York City investment banking firm Dillon, Read & Company (which had also participated in the financing of the Ketchikan Pulp Company) negotiated the sale of $12 million in Alaska Lumber

& Pulp Company 6-percent first mortgage bonds to the Equitable Life Assurance Society of the United States (which had also purchased Ketchikan Pulp Company bonds), Prudential Insurance Company of America, and the General Electric Pension Trust. This was the first sale of securities of a privately owned Japanese corporation in the United States since well before World War II. In addition, some $7 million in loans were provided by eight U.S. firms who were major suppliers of equipment for the Sitka pulp mill. There was an unwritten understanding that the Export-Import Bank of Japan would, if necessary, stand behind loans made to AL&P by American concerns.[27]

PULP PRODUCTION AT SITKA

In the early fall of 1957, at Sawmill Cove on Silver Bay, where the Russian-American Company had once operated a crude sawmill, ground was broken for the construction of the pulp mill. The plant ultimately cost some $66 million, with much of the steel and cement coming from Japan on freighters that would return to Japan with a cargo of lumber from the company's subsidiary, the Wrangell Lumber Company mill.[28]

Utilizing the magnesium oxide process, the mill was similar in many respects to that of the Ketchikan Pulp Company. Logs were drawn into the mill and, if necessary, cut to length by a 108-inch-diameter cut-off saw. Logs over 32 inches in diameter were cut into cants by a 10-foot band saw.[29] The logs were then hydraulically debarked, chipped, and, in a highly automated process, manufactured into high-alpha cellulose pulp, which was marketed under the trade name Alapul. Production sold in Japan was transported aboard two specialized ships, the pulp carrier *Sitka Maru*, and the lumber and pulp carrier *Wrangell Maru*.

Electricity for the mill was produced on site by two 7,599-kilowatt steam turbines. As at the Ketchikan pulp mill, steam was provided by two special recovery boilers fueled with spent digester cooking liquor, and two others fueled with heavy fuel oil and hog fuel. Process water was piped from Blue Lake, a little more than a mile from the mill site. In 1961, the lake's level was raised nearly 150 feet by a dam constructed on Sawmill Creek by the City of Sitka to generate electricity and provide an additional water supply. The territory provided additional assistance: under the Alaska Industrial Incentive Act of 1957, Alaska Lumber & Pulp enjoyed a 10-year exemption from state

**FIGURE 61. Alaska Pulp Company
mill at Silver Bay.** (Southeast
Alaska Conservation Council)

and local taxes. The company, however,
did make payments to Sitka to cover
school costs.[30]

On November 18, 1959, the first wood pulp was produced at Sitka.
According to the local newspaper, it was pale brown "and resemble[d] rough
sponge, with a rather pungent wood odor." The following January, 2,260 tons
of pulp was shipped to Japan aboard the *Columbia Maru*, which was also
transporting a cargo of lumber from the company's sawmill at Wrangell. The
pulp was to be used for test purposes at 22 rayon mills.[31]

The Alaska Lumber & Pulp Company pulp mill was formally dedicated
on May 29, 1960, and initially employed about 440 workers, with an annual
payroll of $5.5 million. It was capable of producing 180,000 tons of pulp annually.
The Alaska Lumber & Pulp Company expected to meet the needs of its mills
by annually cutting about 105 MMBF of timber under its long-term timber
contract, and purchasing another 30 MMBF from independent suppliers. The
company employed about 75 men for its own logging operations, and about
150 worked for independent logging contractors.[32]

Pulptimber logging by Alaska Lumber & Pulp began at Katlian Bay on
Baranof Island, some 15 miles from its plant. There, the float-mounted

FIGURE 62. Alaska Pulp Company A-frame logging operation at Katlian Bay, Baranof Island. (*The Pulpzette*, Vol. 1, No. 10, June 1961)

Washington "Flyer" yarder, the largest logging machine ever built, began bringing logs to saltwater from more than a half mile away. Overall, it was expected that of the timber cut by Alaska Lumber & Pulp, about 25 percent would be logged directly into the water by A-frame equipment, 60 percent would be truck logged, and the remaining 15 percent would be handled by tractors (caterpillars) or rubber-tired machines.[33]

The *Marine Digest* praised Alaska Lumber & Pulp as a "monument to Japanese-American cooperation." According to Governor Heintzleman, all Alaskans saluted Alaska Lumber & Pulp, the second-largest industrial enterprise ever established in Alaska, "as an operation that will use yearly more than 125,000,000 ft. of timber now going to waste in our coastal forest."[34] Speaking for the Forest Service, Sitka District Ranger Ray Carr said that the company's successful entry into the production of pulp

> marks for us the completion of yet another major step forward in the orderly development of Alaska's national forest timber lands.... We can now start to concentrate on more intensive national forest management, which means that as timbered areas are opened up by logging, we will also manage them for the production of their recreation, wildlife, fisheries and watershed values as well as for timber production.[35]

CHAPTER 19

Interest in More Pulp Mills Wanes

AS THE PULP MILLS SWUNG INTO PRODUCTION, THE ECONOMY OF SOUTHEAST ALASKA, LONG IDENTIFIED PRIMARILY WITH COMMER-CIAL FISHING, QUICKLY BECAME DOMINATED BY FOREST PRODUCTS. DURING THE YEARS 1949–1953, JUST BEFORE THE KETCHIKAN PULP COMPANY'S MILL CAME ON LINE, FOREST PRODUCTS (F.O.B. MILL) ACCOUNTED FOR 9.3 PERCENT OF THE TOTAL AVERAGE VALUE OF SOUTHEAST ALASKA'S NATURAL RESOURCE PRODUCTS—A FIGURE THAT PALED BEFORE THE COMMERCIAL FISHERIES PRODUCTS' 86.7 PERCENT SHARE (WHOLESALE VALUE). THE SITUATION CHANGED DRAMATICALLY ONCE KETCHIKAN PULP BEGAN PRODUCTION: DURING THE YEARS 1954–1958, FOREST PRODUCTS' SHARE ROSE TO 41.9 PERCENT, WHILE THAT OF COMMERCIAL FISHERIES SLID TO 53.3 PERCENT. FOREST PRODUCTS' SHARE ECLIPSED THAT OF COMMERCIAL FISHERIES AFTER THE ALASKA PULP COMPANY'S PULP MILL BEGAN OPERATIONS IN LATE 1959. FOR THE YEARS 1959–1961, FOREST PRODUCTS SHARE WAS 53.6 PERCENT, WHILE THAT OF COMMERCIAL FISHERIES WAS 43.3 PERCENT. IN 1964, THE FOREST PRODUCTS INDUSTRY IN ALASKA EMPLOYED SOME 2,400 MEN AND WOMEN AND HAD AN ANNUAL PAYROLL OF $19.8 MILLION. (CONTRARY TO EARLIER EXPECTATIONS, VERY FEW NATIVES WERE EMPLOYED IN THE INDUSTRY.) THE ESTIMATED END VALUE OF FOREST PRODUCTS FROM SOUTHEAST ALASKA THAT YEAR WAS NEARLY $58 MILLION. BY 1975, SOUTHEAST ALASKA'S WOOD PRODUCT INDUSTRY HAD DEVELOPED TO BECOME, BEHIND OIL AND SEAFOOD, THE THIRD LARGEST INDUSTRY IN THE STATE.[1]

But the postwar Forest Service vision for Southeast Alaska went beyond the pulp mill operations in Ketchikan and Sitka; the agency targeted Juneau and the Wrangell-Petersburg area as well. Sustained-yield management, according to the Forest Service, would ensure the continuous availability of timber in "working circles" of dedicated timberlands proximate to each

FIGURE 63. Many loggers in Southeast Alaska lived in floating camps that could be moved to a new location once an area's timber was cut. (USDA Forest Service, Alaska Regional Office, Juneau)

location. The Ketchikan Pulp Company was already in operation when Regional Forester Arthur Greeley surmised in late 1954 that, "it is probable that logging experience now being gained in Southeast Alaska's pulp-timber stands, and that to come in years immediately ahead, will show that timber volume is economically available to support more than the four operations here mentioned."[2]

Indeed, other companies were seeking to enter the market. In 1953, the Georgia-Pacific Corporation announced its intention to construct near Juneau a facility to annually produce 96,000 tons of newsprint paper. In 1955, its subsidiary, the Georgia-Pacific Alaska Company, was given a preliminary award by the Forest Service of a 50-year contract for some 7.5 BBF of timber in northern Southeast Alaska. And on June 9, 1954, the Pacific Northern Timber Company was given a preliminary award of a 50-year contract for some three BBF of timber near Wrangell, contingent upon the construction and operation of a pulp mill, sawmill, and veneer mill.[3]

By early 1956, however, Arthur Greeley had moderated his expectations. The Forest Service, in addition to its firm obligations to the Ketchikan Pulp

Company, had contracted, at least conditionally, to provide timber to the pulp mills expected to be built at Sitka, Wrangell, and Juneau. It was time to garner solid information regarding the adequacy of the timber supply and the effects of the massive logging effort that would be required to supply the mills. In March of that year, Greeley, in addressing an Alaska Day luncheon in Portland, Oregon, analyzed the situation:

> Are there any more of these major pulp mill developments contemplated? Just now the answer is, "No, at least not yet." The Forest Service has gotten out ahead of its inventory information, and the technology of pulp production is changing so rapidly that adequate timber volume information is a routine essential. Further, the success of the operation now there, and those that will come, depends among other things on answers to some questions about the way our forests will behave under pulp-type cutting.

> So far, only theoretical answers are available for these questions, for there has been no comparable cutting heretofore. We think we know how to handle the timber stands, how to get reproduction, how to cope with problems of brush encroachment, how to handle logging so there will not be serious disruption to salmon streams, and also to especially scenic spots. But we need to check our assumptions through some experience on the more important of these questions....

> Until the figures are in and the analysis made, it will not be possible to determine if it is wise to encourage the substantial investment that would be needed for an additional pulp operation.

> Until more of the expected pulp mill capacity has been installed, I am not interested in seeing other types of forest industry move to southeast Alaska.[4]

By 1960, even Frank Heintzleman began to sound cautious. While he had written early in the year that "there should be no holding back in efforts to obtain, as quickly as economic conditions permit, the additional mills which the region can support," in late summer *The Timberman* quoted him as saying, "Alaskans are looking ahead toward more development in pulp, lumber, plywood and cedar, but pulp shouldn't be crowded too fast. Better information is needed on timber volume and growth. This takes time in this extensive forest area."[5]

Greeley's and Heintzleman's caution proved justified. Pacific Northern Timber Company built its sawmill in 1958 but decided not to proceed with the construction of its other two mills. Its timber contract was reduced to 693 MMBF over 15 years, and the entire operation was taken over by the Alaska Pulp Company in 1968. In 1958, citing a decrease in the demand for newsprint paper, Georgia-Pacific requested a five-year extension in which to comply with the provisions of the original award. The Forest Service allowed the company until July 1, 1961, to study the situation. In April 1961, *Pulp & Paper* quoted Georgia-Pacific regarding the possibility of establishing an operation in Southeast Alaska: "Market possibilities cast doubt on the economic feasibility of our making this commitment." In June 1961, the company made it official: It would abandon its Juneau paper mill project in favor of building a similar mill near Eureka, California, that would utilize lumber and plywood mill residuals as raw material. In 1963, however, Georgia-Pacific entered the pulp manufacturing business in Southeast Alaska when Puget Sound Pulp & Timber, with its half-ownership of the Ketchikan Pulp Company, was merged into Georgia-Pacific. Georgia-Pacific exited the pulp manufacturing business in Southeast Alaska in 1972, when, as part of a settlement with the Federal Trade Commission over allegations of monopoly, it agreed to divest itself of 20 percent of its assets. Those assets, which included KPC, became the Louisiana-Pacific Corporation.[6]

In 1965, the St. Regis Paper Company was given a preliminary award to the largest timber sale in the U.S. Forest Service's history. The sale was a modification of that rejected by Georgia-Pacific; it totaled 8.75 BBF in northern Southeast Alaska and near Yakutat. Two years later, however, the company opted not to establish operations in Southeast Alaska. The timber rejected by the St. Regis Paper Company was subsequently awarded on a preliminary basis to U.S. Plywood-Champion Papers in 1968. Like the earlier bidder, U.S. Plywood-Champion Papers opted not to establish operations in Southeast Alaska, and its contract with the Forest Service was cancelled by mutual consent in 1976.[7] This was the last serious consideration given by a large corporation to the establishment of an additional pulp mill operation in Southeast Alaska.

CHAPTER 20

The Heyday Passes

WHILE THE LONG-TERM CONTRACTS FOR THE KETCHIKAN AND SITKA MILLS WERE BEING CONSUMMATED IN THE 1950S, SUBTLE BUT IMPORTANT SOCIETAL CHANGES WERE DEVELOPING THAT WOULD AFFECT THE USE OF THE NATIONAL FORESTS NATIONWIDE AND EVENTU-ALLY THE LONG-TERM CONTRACTS SUPPORTING THE MILLS. AS WELL, MARKET FORCES, ECONOMIC CONDITIONS, AND INCREASING ENVIRONMENTAL REQUIRE-MENTS WOULD ADVERSELY AFFECT THE ABILITY OF THE PULP MILLS TO EARN A PROFIT. LEGISLATION—SOME OF WHICH CONTAINED TONGASS-SPECIFIC PROVI-SIONS— PLAYED NO SMALL PART IN ALTERING THE ECONOMIC ENVIRONMENT THAT THE PULP MILLS OPERATED IN.

MULTIPLE USES

In 1952, the Forest Service initiated a nationwide timber resource review. The extensive study, prepared with the assistance of other federal, state, and private organizations, was published in 1958 as *Timber Resources for America's Future*. The study found that the nation needed to grow more timber to meet expected demands. At the same time, due to increasing disposable income and leisure time, recreational demands on the national forests were increasing, more than tripling during the 1950s. Although the Forest Service's Organic Act cited only timber and water as valid uses of the forest, the agency recognized the realities of increased recreation and hunting, and the interest in wilderness, among other uses. These competing uses of the national forest had the potential to jeopardize the Forest Service's timber sale program. To establish a legal mandate for broader management of forest resources—and to guarantee that timber harvesting continued as a major use—the agency proposed that Congress enact legislation to recognize multiple uses of the national forests.[1]

In April 1960, *Pulp & Paper* quoted Forest Service Chief Richard McArdle as saying that if the policy of multiple use for public lands was not continued, "you [the pulp industry] are not going to get the timber you need—for you are a minority group." McArdle added, "There is an upsurge of demands for outdoor recreation lands. Earnest, dedicated people are in this battle."[2]

Two months after McArdle's statement, the Multiple-Use Sustained-Yield Act (MUSY) became law. The legislation, in the words of McArdle, "makes clear that no statutory priority is given to one resource over another." Though MUSY placed fish and wildlife, outdoor recreation, and other resource values on an equal statutory footing with timber, the legislation required only that the Forest Service give them equal consideration. There was no requirement that they be administered equally. On the Tongass National Forest, the agency expressed an "element of urgency" as it worked to develop an intensive multiple-use management program.[3] The eventual affect of MUSY was to put a general pressure on national forest lands available for timber production.

NATIVE CLAIMS

Of a more direct affect on the general Tongass National Forest timber supply was the Alaska Native Claims Settlement Act (ANCSA).[4] The 1971 legislation was spurred by the necessity of settling Native claims to land on the proposed route of the Trans-Alaska Pipeline. Engineered by Alaska Senator Ted Stevens, ANCSA addressed claims in all regions of Alaska. To settle the claims, for-profit Native corporations—both village and regional—were established and allowed to select some 44 million acres of government land and divide a cash payment of nearly a billion dollars. Southeast Alaska's 13 corporations have to date selected some 500,000 acres of mostly high-value timberlands from the Tongass National Forest—lands that represented perhaps 20 percent of the region's standing commercial timber volume. (Sealaska, Southeast Alaska's regional Native corporation, is entitled to select some 60,000–85,000 additional acres.) A small portion of the lands selected were within the long-term contract areas, but this did not materially affect the timber supply to the pulp mills because the Forest Service was able to provide make-up timber from contingency areas that had been previously set aside. Once in possession of their lands, the Native corporations quickly began implementing a financial strategy based on the liquidation of their timber resources to raise cash for investments and to pay dividends to shareholders. Unencumbered by primary processing restrictions, most of the timber was exported to Asia as round logs. After some three decades of logging, about 90 percent of the merchantable timber on Native corporation lands has been cut.

CLEARCUTTING

Public controversy over clearcut logging in the national forests grew with the increase in recreational activity. The legality of this practice was challenged, and in late summer of 1975 a federal appeals court upheld a lower court ruling that the 1897 Organic Administration Act of the Forest Service— the fundamental legal basis of the Forest Service's timber sale program— required live trees offered for sale to be fully mature and individually marked. What was referred to as the "Monongahela decision" effectively prohibited clearcut logging in the national forests and threatened to jeopardize the entire Forest Service timber sale program.[5] The court's reasoning was promptly adopted by the U.S. District Court in Alaska in *Zieske v. Butz* (1975), which involved a suit by the Alaska Conservation Society and residents of Point Baker, Alaska, over clearcut logging being planned on Prince of Wales Island under KPC's long-term contract.[6] Timber interests as well as the Forest Service wanted the law changed.

The National Forest Management Act (NFMA), which became law the following year, was an attempt by Congress to resolve the immediate problem created by the Monongahela decision and address the controversy over national forest timber policy in a substantive manner. It is today the primary statute governing the administration of the national forests. NFMA specifically permitted clearcutting in the national forests. It also directed the Secretary of Agriculture to develop management plans that "provide for multiple use and sustained yield of the products and services obtained [from the national forest system] in accordance with the Multiple-Use Sustained-Yield Act of 1960, and in particular, include coordination of outdoor recreation, range, timber, watershed, wildlife and fish, and wilderness."[7] Timber sale planning issues were addressed down to the forest level, interdisciplinary planning was mandated, and planning procedures and guidelines used by the Forest Service were converted into statutory and regulatory requirements. The agency was required to maintain updated plans, and to manage the national forests in accordance with them.

Specifically as it related to the Tongass National Forest, NFMA grand-fathered the long-term (50-year) timber contracts with Ketchikan Pulp and Alaska Pulp but limited future Forest Service timber contracts systemwide to 10 years.[8] NFMA also ordered the Forest Service to "revise the [50-year timber] contracts to make them consistent with the guidelines and standards" provided for in NFMA.[9] Under the general direction of the NMFA, the size of clearcuts on the Tongass National Forest was subsequently limited to 100 acres.

CLEAN AIR ACT AND CLEAN WATER ACT

Growing public awareness of and concern over air and water pollution resulted in the passage of the Clean Air Act in 1970 and the Clean Water Act in 1972.[10] Administered by the Environmental Protection Agency (EPA), the Clean Air Act and its subsequent amendments imposed increasingly strict limitations on air-borne pollutants emitted by Alaska's pulp mills, while the Clean Water Act and its revisions correspondingly established increasingly strict industrial wastewater standards applicable to the mills. Dioxin emissions, a result of using chlorine in the pulp bleaching process, were of particular concern. For Alaska's pulp mills, the costs of compliance were considerable, and only with great reluctance did they make any of the substantial expenditures necessary to reduce emissions.

Disputes between APC and federal pollution regulators began in the 1970s. The company's early strategy for dealing with the issue was to seek variances from the regulations. When the variances were denied, the company often made a calculated financial decision that it would be less expensive to pay fines than to make necessary upgrades.[11] This approach, needless to say, did not sit well with government regulators. As early as 1986, officials threatened to close the company's mill because of its failure to comply with federal clean water standards.[12] Alaska's Department of Environmental Conservation during the 1980s exerted additional pressure on APC (and KPC) for closer compliance with air emissions requirements but chose not to enforce its regulations aggressively against the two regionally important businesses, both of which claimed severe financial hardship. Despite the lack of aggressiveness, the Department of Environmental Conservation's efforts were incrementally successful, but more, a lot more, needed to be done to bring the mills up to acceptable standards.[13]

"NATIONAL INTEREST" LANDS

In 1980, the Alaska National Interest Lands Conservation Act (ANILCA) set aside more than 100 million acres of federal lands in Alaska for conservation purposes. In Southeast Alaska, the legislation reduced the region's timber base by prohibiting logging and certain other types of development on some 5.4 million acres of the Tongass National Forest, including most of Admiralty Island, which was designated a national monument.[14] To guarantee a continued adequate supply of economical timber to the pulp mills and sawmills of Southeast Alaska, a provision of the legislation inserted by Alaska Senator

Ted Stevens directed the Forest Service to supply Tongass timber at the rate of 4.5 billion board feet per decade and provided for an automatic $40 million annual appropriation for road construction and other timber-related operations.[15] The mandated annual timber supply was about equal to the average amount the industry had been using in the years leading up to ANILCA. Under the terms of the long-term contracts, the Forest Service was obligated to provide about 275 MMBF annually to the region's two pulp mills.

TONGASS TIMBER REFORM ACT

Despite the directive in the National Forest Management Act to modify the long-term timber sale contracts, only minor changes were actually made and many long-standing resource use issues were left unresolved. This was due in part to the fact that the vehicles used for the changes were the provisions in the original contracts that provided for bilateral modification of the contracts at five-year intervals. Under these provisions, all modifications had to be acceptable to both the pulp companies and the Forest Service.[16] It was a given that the pulp mills were not going to agree to anything that would undermine their position. Business as usual continued.

This set the stage for the Tongass Timber Reform Act (TTRA) of 1990, whereby Congress unilaterally modified the contracts to "enhance the balanced use of resources on the forest and promote fair competition within the Southeast Alaska timber industry."[17] Among TTRA's other provisions was the revocation of two of ANILCA's crucial Tongass-related provisions: the timber supply mandate and the automatic $40 million annual appropriation.[18] Additionally, the legislation prohibited logging on more than 1 million acres, mandated 100-foot buffer strips along fish-bearing streams, and constrained the Tongass National Forest timber program by forcing the agency to comprehensively consider non-timber forest resources when preparing timber sales. It also required the Forest Service to make stumpage rates paid by the long-term contract holders more comparable to independent sales.

The principal effects of the Tongass Timber Reform Act on the operations of the pulp mills were to reduce the region's timber base and increase the price of timber. The owners of the pulp mills viewed the legislation as an attempt by the conservation community to drive them out of business and, with the support of Alaska's congressional delegation, worked hard—but unsuccessfully—to overturn it.

THE MONOPOLISTS

Another problem for the APC and KPC was of their own making: beginning in the earliest years of their joint presence in Southeast Alaska, the two companies began colluding to eliminate competition and keep the price of timber down. In addition to timber cut from their contract areas, both KPC and APC purchased timber from independent loggers. As the only purchasers of pulptimber and the principal purchasers of sawtimber, the companies were able manipulate timber prices to their own advantage.

According to the *Use Book*, the pocket-size 1905 handbook that Gifford Pinchot prepared to explain the workings of the new national forests to both the public and his newly-hired cadre of forest rangers, monopoly over forest resources simply could not occur:

> The small man can buy a few thousand feet; the big man
> can buy many million feet, provided it is a good thing for all
> the people to let him purchase a large amount, but not
> otherwise. The local demand is always considered first.
> There is no chance for monopoly, because the Secretary of
> Agriculture must by law sell as much or as little as he
> thinks best, to whom and at whatever price he thinks will
> best serve the interests of all the people.[19]

Nevertheless, in 1947, an article in *International Woodworker*, the official organ of the International Woodworkers of America (IWA) labor organization, charged that "plans which will have a...monopoly effect with respect to Forest Service timber in Alaska are...afoot."[20] Nearly a quarter century later the Reid Brothers Logging Company, an independent logging firm based in Petersburg, and others filed antitrust claims in federal court against KPC and APC. The claims were investigated in 1970 by the Antitrust Division of the Department of Justice, but no action was taken. In March 1975, the Reid Brothers Logging Company again brought suit.[21] Among the Reid Brothers' allegations was that the pulp companies had conspired to monopolize the purchase of timber on Forest Service lands, the logging business, the sawmill and pulp mill business, and the selling of rough and dimensional lumber, pulp, and other wood products.[22] In June 1981, the federal district court ruled that the Ketchikan Pulp Company and Alaska Lumber & Pulp had during the years 1959 through 1975 violated the Sherman Antitrust Act by monopolizing the timber industry in Southeast Alaska.[23] The court found that the two companies had restricted and eliminated competition in all phases of the timber industry in Southeast Alaska; refrained concertedly from competing against others for timber or logs; kept would-be competitors out of Southeast Alaska by cutting off their timber supplies through preclusive bidding and other means; and attained and exercised monopoly power—that is, the power to set prices and exclude competition in the timber industry in Southeast Alaska.[24] Acting in collusion, KPC and APC had driven almost every independent logging company out of business and acquired or gained control of every substantial sawmill in the region.[A][25]

The defendants were ordered to pay some $10 million in damages to Reid Brothers Logging and three other companies. An appeal by Ketchikan Pulp and Alaska Lumber & Pulp was to no avail: the decision of the district court was affirmed by the Ninth Circuit Court of Appeals.[26]

Evidence obtained in the Reid Brothers litigation showed that, in addition to antitrust violations, the two companies had materially breached their long-term contracts by manipulating information that the Forest Service used to determine timber prices, and they had committed fraud against the U.S. government by making false statements of material fact, including false invoices.[27] The Forest Service estimated that total potential damages incurred by the government during the years 1959–1982 were between $76.5 million and $81.5 million.[28] In January 1983, however, the Department of Justice cited statute of limitations concerns in declining to prosecute KPC or APC for any civil or criminal offenses relating to fraud and breach of contract.[29]

PURCHASER ROAD CREDIT

The companies had actually achieved their dominant position partly with indirect help from the U.S. government. The National Forest Roads and Trails Act of 1964 authorized a program in which purchasers of national forest timber could convert a portion of the money spent on the construction of timber access roads into "purchaser road credit," which could then be used to pay the cost of stumpage. The amount of purchaser road credit that was available in a given sale was limited to the amount paid for timber that was over and above "base rates"—the lowest prices for which national forest

---

[A] Acquired or controlled sawmills were: Alaska Prince Timber Company (Metlakatla), Ketchikan Spruce Mills (Ketchikan), Schnabel Lumber Company (Haines), Alaska Forest Products (Haines), Alaska Timber Corporation (Klawock), Alaska Wood Products (Wrangell), Mitkof Lumber Company (Petersburg).

timber could be sold.[B][30] In 1975, purchaser road credit became "bankable": on a timber sale in which the amount of available purchaser road credit exceeded the cost of stumpage, unused credit could be transferred to other sales (including those in the future) within the same national forest.[31]

Also in 1975, provision was made during routine renegotiations of the long-term timber contracts between the Forest Service and Ketchikan Pulp and Alaska Pulp to allow the companies to convert into purchaser road credit all of their previous payments for stumpage purchased under the contracts, over and above base rates. In a sense, this was simply a change in accounting: prior to 1975 the Forest Service had simply deducted road building costs from the price of long-term contract stumpage. Yet because the companies had willingly paid more than base rates for much of this mostly long-since-gone timber, the refunds amounted to a windfall. The General Accounting Office reported in 1995 that the mills had used purchaser road credit to pay for about three-fourths of the timber cut under their long-term timber contracts.[32]

GLOBAL MARKETS

Like newsprint paper, wood pulp is a commodity, as is the rayon manufactured from it. Wood pulp is produced in various locations around the world and traded in international markets. It is subject to the same sort of market pressures and economic and political realities that precluded the production of newsprint paper in Southeast Alaska. As a structurally high-cost producer located far from markets, the wood pulp industry in Southeast Alaska was at a chronic disadvantage. The region's position worsened when competition from petroleum-based fibers such as nylon and polyester substantially eroded the demand for rayon and thus the demand for dissolving pulp. During the years 1977 through 1992, world rayon production declined at a rate of two percent per year, except for brief increases in 1988 and 1989. In Japan, the rate of decline was four percent per year. A notable decline in the domestic market for dissolving pulp occurred in 1989, when Avtex Fibers, the nation's largest rayon producer, shuttered its rayon manufacturing plant at Front Royal, Virginia, and declared bankruptcy.[33] The plant had been built by American Viscose, one of the original partners in the Ketchikan Pulp Company.

[B] Base rates are equal to the estimated forest regeneration costs (Knutson-Vandenberg fund) plus a mandatory $0.50/MBF deposit to the U.S. Treasury. Knutson-Vandenberg and the $0.50/MBF rule date from 1930.

CUTTING LABOR COSTS

Faced in the early 1980s with a difficult export market for their production because of a strong dollar, depressed pulp prices, and tough competition, the owners of Southeast Alaska's pulp mills successfully reduced their labor costs by ending union representation at their mills.

At Sitka, employees at the Alaska Pulp Company pulp mill were represented by the United Paperworkers International Union. In 1984, the company said it was losing money and negotiated concessions from the union, including an across-the-board pay cut of $2.40 per hour. In the summer of 1986, the company proposed that its workers accept an additional 25 percent pay cut. The union refused, and a bitter ten-month strike ensued. Replacement workers were hired to operate the mill, and in a subsequent election the workers, many of them newly hired, voted to decertify the union.[34]

Employees at the Ketchikan Pulp Company mill were represented by the Association of Western Pulp and Paper Workers. In 1983, the Louisiana-Pacific Corporation, parent company of the Ketchikan Pulp, offered to sell controlling interest of its Alaska assets to an employee group. Arrangements were made with Seattle First National Bank to finance the purchase by establishing a $45 million employee stock ownership trust. The financing was contingent upon a reduction of about $6.7 million in annual operating costs, largely through a rollback of employee wages and compensation to the 1981 level and "labor peace." Employees rejected the offer in June 1984, and shortly thereafter Louisiana-Pacific withdrew its sale offer and announced that the mill would be shut down for annual maintenance.[35] Unlike previous maintenance closures, however, managers instructed employees to take home all of their personal belongings, even their coffee cups, and warned that unless market conditions improved, the mill might never reopen.[36] The closure ultimately lasted six months and was viewed by workers as a lockout. Ketchikan Pulp reopened the mill after its contract with the union expired, and its members worked thereafter for reduced wages and benefits under an implemented offer by the company.

ANTIQUATED PULP MILLS

At the time they were built, the pulp mills constructed by KPC and APC were ultra-modern, state-of-the-art facilities. Forty years later they were antiquated, run-down, and hemorrhaging money. When building their plants, neither company likely anticipated the anti-pollution legislation that Congress would pass in the 1970s. Retrofitting the mills to comply with vastly higher pollution standards that had come into effect promised to be an extremely expensive proposition. In the early 1990s, APC officials estimated that an expenditure of $100 million was needed to bring operations into full compliance with pollution control laws.[37] In 1996, Mark Suwyn, CEO of Louisiana-Pacific, KPC's parent company, estimated the costs of necessary environmental and productivity upgrades to the Ketchikan mill would total $200 million. He pegged operating losses at that time to be $5 million to $10 million per month.[38]

SOUTHEAST ALASKA'S PULP MILL ERA COMES TO AN END

In the fall of 1993, Alaska Pulp Company shuttered its pulp mill "indefi-nitely." The stated reason for the closure, as reported by *Pulp & Paper*, was "because of the difficulty of obtaining economically priced quality timber."[39] Changes made by the Tongass Timber Reform Act were the focus of the company's complaint.

In April 1994, the Forest Service declared Alaska Pulp Company to be in material breach of its long-term timber contract, which it then terminated. In turn, the company sued the U.S. government for breach of contract and damages that ranged from several hundred million to nearly $9 billion. The U.S. Court of Federal Claims in January 2004 ruled that the Forest Service had breached its contract with APC when it implemented provisions of the Tongass Timber Reform Act (TTRA). It also concluded, however, that the TTRA was not a factor in the closure of the pulp mill, that the mill had "no reasonable prospect of profitability" and was entitled to no damages.[40]

Like APC, KPC blamed the Forest Service and the Tongass Timber Reform Act for most of its problems, and in 1994 and 1995 the company filed a series of suits against the U.S. government for damages, claiming modifications made to its long-term timber contract by the Tongass Timber Reform Act amounted to a breach of contract that was costing the company hundreds of millions of dollars.[41] The company also contended it needed a 15-year extension to its contract to obtain financing for the needed upgrades to its pulp mill. In September 1996, however, Ketchikan Pulp proposed a settlement that would put the entire issue to rest: in exchange for certain considerations, the company would drop its claims and close its pulp mill.[42] Negotiations involving Ketchikan Pulp, Louisiana-Pacific, the Forest Service, the Department of Justice, the White House, and Alaska's governor and congressional delegation commenced, and in March 1997 a settlement was reached: the long-term timber contract would be terminated, and the company would close its pulp mill and withdraw its claim for damages. Among the considerations Ketchikan Pulp would receive in exchange were $147 million and 300 MMBF of timber. The timber was considered sufficient to run the company's sawmills at Ward Cove and Annette Island for three years.[43] Marking the end of an era, the last wood pulp produced in Alaska was rolled out of the Ketchikan mill on March 25, 1997.

Southeast Alaska was not alone in experiencing the effects of the changed world supply of and demand for wood pulp. In 1997, Rayonier shuttered its dissolving pulp mill at Port Angeles, Washington, and Skeena Cellulose declared bankruptcy and shuttered its Columbia Cellulose pulp mill at Prince Rupert, British Columbia.[c] The Columbia Cellulose mill had been the envy of Alaskans when it began producing sulfite pulp in 1951.[44] Louisiana-Pacific exited the pulp business entirely after selling its Samoa, California, mill in 2001 and its Chetwynd, British Columbia, mill in 2002.

But the demise of the industry in Southeast Alaska seems ironic, given the confident predictions of then-Chief William Greeley in 1920:

> The paper mills that are established in Alaska forests will stay put, because permanency is the keynote of the National Forest policy, and of the technical standards and practice of the Forest Service. There will be no gutting of public forests in Alaska, no specter of timber depletion, no paper mills wondering where their raw material is to come from, no communities suddenly losing their principal industry.[45]

Calculated from numbers provided by the Forest Service, Southeast Alaska suffered the loss of some 2,200 direct jobs between the years 1990 and 2004 as a result of the dramatic decrease in the amount of Tongass National Forest-related logging and timber processing.[D] Adding losses in indirect and

---

[c] In 1976, to stem an earnings drain, the mill was converted to the production of kraft paper.
[D] The Forest Service estimates that there were a total of about 3,100 direct jobs lost during the period, but not all could be attributed to the Tongass situation. Timber harvest declined steeply on the Tongass, but there was also a contemporaneous steep decline in the harvest of Native corporation timber, which accounted for some 49 percent of the decline on the combined Tongass National Forest and Native lands in Southeast Alaska. Assuming logging is about equally labor intensive on Forest Service and Native land, the Tongass National Forest component of the 1,842 logging jobs lost is about 939. Since Native timber is exported as unprocessed logs, the entire loss of wood processing jobs (1,251) is attributable to the Tongass. The total direct job loss attributable to the reduced logging and processing Tongass National Forest timber, therefore, is approximately 2,190.

induced employment substantially increased the number of jobs lost. The effect of the job loss on the regional economy was mitigated, however, by the fact that some 35 percent of the workers in the industry were non-residents, many of whom simply left the state.[46]

Though numerous businesses have come and gone in Southeast Alaska, none have done so as spectacularly and with such long-term impacts as the pulp mills that the Forest Service had worked so long and hard to establish and maintain. Though not completed, the contracts were, in the words of the U.S. House Committee on Interior and Insular Affairs, "unprecedented in timber volume, duration…and benefits granted to corporations making use of public resources."[47]

Future Opportunities
in a Forest Transformed

HINDSIGHT AND FORESIGHT

IN CONTRIBUTIONS TO A 1999 COLLECTION OF ESSAYS ON ALASKA PUBLIC POLICY ISSUES, TWO ALASKA ECONOMISTS CONFIRMED THE END OF THE PULP INDUSTRY BOOM IN SOUTHEAST ALASKA. GEORGE ROGERS, A PRACTICING ECONOMIST IN ALASKA SINCE 1945, NOTED IN A HISTORICAL OVERVIEW THAT THE TIMBER INDUSTRY "HAD ITS BOOM AND TODAY APPEARS TO BE ON THE VERGE OF EXTINCTION." DAVID REAUME, AN ECONOMIST IN ALASKA SINCE 1977, WAS MORE SPECIFIC. IN A SUMMATION OF ALASKA'S PROSPECTS FOR FUTURE ECONOMIC DEVELOPMENT, REAUME WROTE: "WITH [KETCHIKAN PULP COMPANY'S] CLOSING, THE REMAINDER OF THE TIMBER PROCESSING INDUSTRY IN SOUTHEAST ALASKA STANDS A VERY GOOD CHANCE OF SIMPLY WITHERING AWAY…THERE IS NO REASONABLE EXPECTATION OF RENEWED FEDERAL SUBSIDIES…WITHOUT THE PULP MILLS, LARGE SCALE LOGGING AND SAWMILLING IN ALASKA IS NOT COST EFFECTIVE…THIS IS AN INDUSTRY THAT FOR ALL INTENTS AND PURPOSES HAS CEASED TO EXIST."[1]

Lawrence Rakestraw, author of *A History of the United States Forest Service in Alaska*, wrote in 1981 that "a major theme in the history of the Forest Service in Alaska is continuity of purpose. There has been more continuity than change—probably more than in any other Forest Service region."[2] Prior to the pulp mill era, "continuity of purpose" for the Forest Service in Southeast Alaska was its unrelenting effort to establish a world-scale wood pulp industry. As an institution, the agency was so heavily invested in this effort that once the pulp mills became established, "continuity of purpose" meant working to ensure their success, for the success or failure of the pulp mills became a direct measure of the Forest Service's success or failure to support economic development.

In 1975, David C. Smith, a University of Maine historian of the lumber and paper industries, had presciently summed up the Forest Service's saga in Southeast Alaska: "The story of these efforts is an excellent illustration of how promotional activities can carry people a long way down roads that have no turning."[3]

While the Forest Service had for many years been of one mind in its desire to establish and maintain large-scale timber operations in Southeast Alaska, since the closing of the pulp mills there has been serious debate within the agency over how and for what ends the forest should be managed. Some desire to see the reestablishment of pulp mill era-type timber operations, while others believe the Tongass should now be managed with broader goals in mind.

The logger's adage, "when you're in the best, you'll always be in the best," has application today. Though the best timber in Southeast Alaska is long since gone, what old-growth remains is, very simply and on a world scale, the best of what remains. There will doubtless continue to be pressures and opportunities to utilize this material, but the long-term prospects for a large-scale industry have dimmed. Although there is considerable local support for at least a limited amount of old-growth harvesting in Southeast Alaska, national support is lacking. Active opposition comes from three often overlapping fronts that present a formidable barrier to the continuation of extensive logging of old-growth timber on the Tongass National Forest:

- those opposed because of environmental concerns, especially the impacts of clearcut logging;
- those who consider their uses of the Tongass to be jeopardized by logging. This group includes tourism-based businesses, recreational users, hunters, both sport and commercial fishermen, and those engaged in subsistence activities; and
- those who desire an end to the taxpayer subsidies upon which the industry has long depended.

An additional impediment to the continued logging of old-growth timber on the Tongass is the increasing cost of access. There are today more than 5,000 miles of logging roads on the Tongass National Forest. These roads offer little benefit to potential large-scale loggers of old-growth timber because they were built not as a part of a program for long-term access to timber resources, but they were built directly to timber that was promptly harvested.[4] If there is to be large-scale old-growth logging in the future, new roads will be required. These roads will be expensive because the remaining timber is increasingly less accessible and because the Forest Service now requires that new roads be built to higher standards. The construction of new roads, however, has been strongly opposed by the environmental community and its allies, particularly in areas that were designated as "roadless" by the Clinton administration.[A]

FUTURE OPPORTUNITIES

Under what circumstances could an integrated forest products industry redevelop in Southeast Alaska? Two studies, one by the Juneau-based economic consulting firm McDowell Group (2004) and the other by the Forest Service (2006), addressed that question.[5] Both studies cited the fundamental requirement of a consistent, long-term supply of a relatively large amount of timber, but neither could say definitely how that supply could be guaranteed.[B]

Each study cited the need for a facility to utilize sawmill residuals and particularly the low-grade logs that generally comprise about 40 percent of a given old-growth timber sale. Perhaps best suited to this role was a medium-density fiberboard (MDF) plant.[C] Though there was uncertainty whether an MDF plant of any size could operate profitably in Southeast Alaska's high-cost environment, preliminary indications were that sustaining an industry based around a medium-sized MDF plant and ensuring that the region's existing sawmills operated near capacity would require the minimum annual harvest of approximately 200 MMBF of timber, while a more competitive, diversified, and efficient industry would require perhaps 350 MMBF. To reduce the cost of timber, a review of the efficacy of the Forest Service's standards and guidelines for logging and road-building operations was suggested. The premise is that it is possible that at least some of the standards and guidelines could be relaxed yet still provide adequate protection for fish and wildlife.[6]

[A] The January 2001 "Roadless Rule" prevented roadbuilding, except under special circumstances, in designated roadless areas of the national forest system. The Forest Service officially exempted the Tongass National Forest from this rule in December 2003, a decision that roused the ire of the environmental community and its allies. After the exemption, the Forest Service authorized several timber sales in roadless areas. All were challenged in court and none have been completed.

[B] The timber would necessarily come from the Tongass National Forest. Southeast Alaska's Native corporations, the other major holders of timberlands in Southeast Alaska, focus on the export of unprocessed logs.

[C] Medium-density fiberboard (MDF) is an engineered wood product made by combining wood fibers with wax and resin. The resulting material is then heated and pressed into panels. The principal use of MDF is as a substitute for solid wood and plywood in the manufacture of furniture. An advantage of an MDF plant is that—unlike a plant that manufactures oriented strand board (OSB) or laminated veneer lumber (LVL)—an MDF plant does not require roundwood as a raw material.

Less formally, others envision a far smaller harvest of mostly old growth to supply small mills that manufacture lumber and timbers for local consumption as well high value-added products. Currently, the Icy Straits Lumber & Milling Company at Hoonah (Chichagof Island) does just that. The mill, with a capacity of about 5 MMBF per year, cuts lumber and timbers as well a host of other "profiles" that include molding and trim material, siding, cardecking, and flooring. The company also manufactures timber-frame cabin and home packages. The largest sawmill operation in Southeast Alaska, Viking Lumber at Klawock (Prince of Wales Island), has a capacity of about 25 MMBF annually. Today it primarily cuts western hemlock stock that is shipped to a remanufacturing plant in Washington and Sitka spruce cants for the export market. Viking Lumber processes mill residuals and small logs into chips and then barges the chips to Puget Sound. The company also exports unprocessed logs. A number of small mills throughout the region cut rough dimension lumber and timbers for local consumption. A few also manufacture specialty products.

While the Forest Service and others ponder the scale and character of a future wood products industry in Southeast Alaska, there is one very simple sector of the region's timber industry that has been viable: the sector that exports unprocessed logs.[D] This has been the model for Native corporations for more than two decades, and has in recent years become a substantial component of the Forest Service's timber program on the Tongass. During the calendar years 1999 through 2005, an average of 18 MMBF of timber, representing 23 percent of the timber harvested on the Tongass, was annually exported as unprocessed logs. Most of the exported logs were Alaska-cedar and western redcedar, but considerable quantities of Sitka spruce and western hemlock were also exported.[7] Federal statute provides for the unrestricted export of Alaska-cedar logs from the Tongass National Forest. Western redcedar logs may be exported to the "lower 48" providing they are surplus to the needs of mills in Southeast Alaska.[8] The Forest Service has traditionally approved the export of Sitka spruce and western hemlock logs on a case-by-case basis, but, to foster timber utilization, the agency in early 2007 gave blanket approval—subject to grade and size restraints—for interstate shipment of more than fifty percent of the Sitka spruce and western hemlock sawlogs on any timber sale.[9]

[D] Exports include logs exported to other states as well as foreign nations.

The export sector of the timber industry has worked in large part because it does not require expensive long-term investments in facilities such as a modern sawmill or an MDF plant, and because it avoids the high operating costs associated with manufacturing wood products in Southeast Alaska. Lending institutions are willing to finance these lean and relatively short-term ventures, and the Forest Service has been fairly liberal in granting permits to export unprocessed logs. The downsides of exporting unprocessed logs are that it contributes relatively little to the regional economy, does nothing to build Southeast Alaska's pool of wood manufacturing skills, and further erodes the region's timber base. On the other hand, the inclusion of exportable material—particularly valuable Alaska-cedar—in a timber sale aimed at supplying a sawmill operation can make the sale more marketable.

SECOND-GROWTH TIMBER

The cutting of timber for commerce has occurred in Southeast Alaska for less than 200 years. During that time, the very best of the virgin timber, in terms of quality, value, and accessibility, has been largely cut.

Logging on the Tongass reached its apex during the 44 years (1954–1997) that comprised Southeast Alaska's pulp mill era. During this period, loggers felled 16 billion board feet of timber, mostly to supply the needs of the two pulp mills and their associated sawmills. The timber cut during the pulp mill era represented 86 percent of the total amount cut on the Tongass during the 101-year period 1909 through 2009.

A substantial portion of the area harvested in Southeast Alaska has been replaced with the "thrifty" forest (relatively decay-free and still growing) so desired by the Forest Service. More than 400,000 acres of the Tongass National Forest were clearcut during the pulp mill era (approximately, an additional 300,000 acres was harvested by Native corporations), and today at some locations the size and volume of timber in the second-growth (a.k.a. "young growth") forest is impressive. At other locations, however, regrowth is sparse or non-existent. Lacking at present is a comprehensive inventory. The Forest Service knows where the second-growth stands are, their acreage, and their age class, but lacks specifics such as data on tree size and volume per acre. This sort of information—and much more—will be necessary to plan any sort of broad utilization.

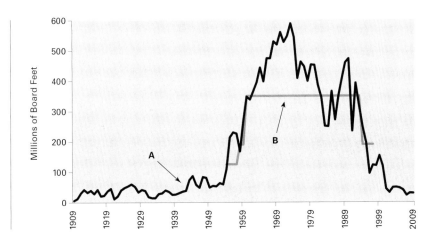

FIGURE 64. Recorded timber cut (A) and approximate pulp mill log capacity (B), Tongass National Forest, 1909–2009. (Compiled from Alaska Region Cut History spreadsheet at http://www.fs.fed.us/r10/ro/policy-reports/documents/Cut-History.xls. Accessed April 30, 2010)

YEAR	A	B	YEAR	A	B	YEAR	A	B
1909	6.2	—	1944	86.8	—	1979	453.2	350
1910	12.4	—	1945	58.3	—	1980	452.1	350
1911	30.2	—	1946	48.6	—	1981	385.7	350
1912	42.3	—	1947	83.4	—	1982	344.8	350
1913	32.9	—	1948	81.0	—	1983	251.2	350
1914	39.6	—	1949	49.2	—	1984	249.8	350
1915	28.7	—	1950	54.4	—	1985	365.3	350
1916	42.3	—	1951	52.9	—	1986	271.6	350
1917	41.0	—	1952	63.4	—	1987	351.5	350
1918	43.1	—	1953	59.2	—	1988	407.8	350
1919	37.4	—	1954	109.2	125	1989	460.5	350
1920	45.6	—	1955	213.8	125	1990	473.0	350
1921	11.7	—	1956	230.2	125	1991	234.8	350
1922	20.6	—	1957	226.4	125	1992	392.3	350
1923	40.5	—	1958	167.5	190	1993	317.5	350
1924	48.6	—	1959	266.6	190	1994	249.4	190
1925	53.7	—	1960	347.5	350	1995	197.5	190
1926	60.0	—	1961	338.2	350	1996	94.7	190
1927	52.0	—	1962	366.3	350	1997	122.1	190
1928	33.8	—	1963	395.1	350	1998	120.5	—
1929	42.0	—	1964	443.7	350	1999	153.2	—
1930	38.5	—	1965	397.6	350	2000	119.3	—
1931	18.2	—	1966	474.3	350	2001	44.1	—
1932	14.7	—	1967	474.3	350	2002	31.9	—
1933	14.7	—	1968	529.5	350	2003	48.1	—
1934	28.2	—	1969	519.3	350	2004	49.2	—
1935	30.5	—	1970	560.1	350	2005	46.6	—
1936	40.0	—	1971	527.7	350	2006	40.0	—
1937	35.3	—	1972	547.5	350	2007	22.5	—
1938	25.6	—	1973	588.5	350	2008	30.0	—
1939	26.5	—	1974	544.0	350	2009	28.3	—
1940	30.9	—	1975	408.4	350			
1941	35.8	—	1976	462.8	350	**Total**	**18,645.8**	
1942	38.5	—	1977	447.3	350			
1943	73.6	—	1978	398.7	350	**Pulp Era**	**16,118.5**	

FIGURE 65. Second-growth trees after clearcutting. This image shows a relatively common regeneration result following harvesting in Southeast Alaska, although not all sites are so well-stocked. (USDA Forest Service, Alaska Regional Office, Juneau)

Both the Forest Service and other interests in Southeast Alaska feel that this is very important information to secure. The Forest Service, for its part, is in the process of contracting with the Oregon-based Forest Biometrics Research Institute to construct a comprehensive model of the Tongass's second-growth timber stands to project timber availability in the future. The Southeast Conference is leading an effort to fund an on-the-ground cruise of selected second-growth timberlands in southern Southeast Alaska that, it says, will "effectively augment the USFS data and verify the present volume and stocking levels" and also determine the volume of merchantable second growth that would be expected to become available, by decade, in the future. The Forest Service also intends to initiate its own on-the-ground survey of second-growth timber resources.[10]

While the efforts toward a more comprehensive inventory proceed, the Washington office of the Forest Service has asked the Alaska Region to "explore ways to accelerate the transition of the timber management program on the

Tongass National Forest—and the timber industry in Southeast Alaska that is dependent on that program—away from its historical reliance on harvesting old growth forest stands, and towards a program and industry based on the harvest of young growth stands." The Forest Service in Alaska has acknowledged that the "harvest of old-growth trees has become increasingly controversial," and recognizes that the "interest in the management of young growth forest stands on the Tongass has increased dramatically in recent years."[11]

In response to the Washington office request, in May 2010, the agency formally chose to transition, in the words of Alaska Regional Forester Beth Pendleton, "quickly away from timber harvesting in roadless areas and old-growth forests." To ensure that a "bridge" exists for the remaining forest industry infrastructure to make the transition, the Forest Service will continue to offer a "limited number of old-growth sales in the near-term in roaded forest areas." The agency also intends to hire a "young growth coordinator" to help expedite projects in second-growth areas.[12] Overall, this decision to move rapidly from old-growth harvest to second-growth harvest probably represents the most significant change in direction ever for the U.S. Forest Service on the Tongass National Forest.

How quickly and successfully Southeast Alaska can transition to an industry based on second growth remains to be seen: A recent Forest Service study has determined that a 5- to 10-year transition period from old-growth to second-growth harvests would likely result in the complete collapse of the Tongass-dependent timber industry immediately after the transition occurred, in part because of the dearth of merchantable second-growth timber. The Forest Service estimated in 1997 that broad-scale logging of second-growth timber would not begin until around the year 2040.[13]

At the moment, however, the 2008 Tongass Land Management Plan (TLMP)—the document that establishes the framework for the management of the forest for the following 10 to 15 years—is still the management plan of record. The plan provided for an annual Allowable Sale Quantity (ceiling) of 267 MMBF.[14] In its planning process, the Forest Service had very optimistically anticipated a potential annual harvest level of 204 MMBF, about six times as much timber as has been harvested annually in recent years.[15] There is no second-growth component in the plan's timber harvest program.

The Forest Service's estimated forest rotation period on the Tongass has evolved with information obtained during more than a half century of clearcut harvesting operations. Since the late 1970s, the agency has had a program in which second-growth forests are pre-commercially thinned at 15 to 30 years of age. Such "properly tended" second-growth forests, according to the Forest Service, will "in just 90 to 120 years contain twice the volume with less defect and more uniform size" than the old-growth forests they replaced. At a cost of some $45 million, about 150,000 acres of the Tongass National Forest have been thinned by crews under contract with the Forest Service. The average cost was about $300 per acre. Currently, about 5,000 acres are thinned each year, and the Forest Service expects this schedule to continue.[16]

■ ■ ■

In an increasingly competitive world economy, lumber that might be manufactured from Southeast Alaska's second-growth timber will be expensive. The same limitations exist today for producing commodity wood products in Southeast Alaska as have hindered the development and operations of the wood products industry for more than a century: labor and operating costs are high, markets are distant, much of the terrain is severe, and the climate limits operations. Except on a local level, Southeast Alaska's second-growth timber will find it challenging to compete with the vast quantities of timber in the plantation forests of the Pacific Northwest's Douglas-fir region. Arguably the world's best timber-growing area, the Douglas-fir region is blessed with a mild climate, relatively gentle terrain, abundant labor, and a well-developed infrastructure. Modern mills in the region very efficiently churn out huge quantities of two-by-fours and two-by-sixes and other dimension lumber from relatively uniform second-growth logs.

For many of the same reasons that pertain to the manufacture of lumber, the prospects for the future manufacture of veneer or wood chips-based products in Southeast Alaska are equally grim. The situation for the manufacture of paper—the Forest Service's focus for decades—is similar, but the problems are exacerbated by two factors: the steep decline of newsprint paper consumption in the U.S. (a trend that is not expected to abate, in large part due to the rise in electronic communications), and the very substantial utilization of wood waste and recovered ("recycled") paper as raw material for the manufacture of paper.

■ ■ ■

In his 1960 book, *Alaska in Transition: the Southeast Region*, Alaska economist George Rogers documented Southeast Alaska's transition from a seasonal, fisheries-dominated economy to an economy dominated by a year-round wood products industry. Today, the pulp mills that anchored that industry for nearly a half century are gone, as are almost all of the region's sawmills. With the relatively sudden demise of all but small segments of its wood-processing industry, Southeast Alaska began another transition. The region was able to weather that transition without broad dislocations; it did so largely because the fishing industry held its own and employment in the tourism industry grew dramatically. A dozen years after the closure of Southeast Alaska's last pulp mill, the forest products industry is a relatively small player in the regional economy. It seems likely to remain so.

The Sawmills of Southeast Alaska

MANY OF SOUTHEAST ALASKA'S EARLY SAWMILLS WERE WATER-POWERED. SOME WERE LOCATED NEAR MAJOR STANDS OF MERCHANTABLE TIMBER. WATERPOWER IS SIMPLE, RELIABLE, AND INEXPENSIVE, BUT EVEN COMBINED WITH A NEARBY SUPPLY OF TIMBER, IT WAS FAR LESS IMPORTANT TO THE LONG-TERM SUCCESS OF A SAWMILL THAN WAS THE AVAILABILITY OF WORKERS. STEAM POWER ALLOWED SAWMILLS TO BE LOCATED IN THE REGION'S LARGER TOWNS, WHERE LABOR, SUCH AS IT WAS, WAS MOST AVAILABLE. ACCORDINGLY, THE LARGEST AND MOST SUCCESSFUL OF THE EARLY SAWMILLS IN SOUTHEAST ALASKA WERE LOCATED IN KETCHIKAN, JUNEAU, WRANGELL, SITKA, AND PETERSBURG.

The first regionally important sawmill in Southeast Alaska—based on the volume of material cut, the diversity of products produced, and the markets served—was the water-powered mill established at Shakan, on Kosciusko Island just west of Prince of Wales Island in 1879. Kosciusko Island was home to probably the finest Sitka spruce timber in Southeast Alaska. Lumber from this mill was used in the early construction of Wrangell, and was shipped as well to Sitka and Juneau, and to canneries in the region. This sawmill ceased operation prior to 1906.

The mantle of being the region's most important sawmill passed prior to 1900 to Willson & Sylvester's steam-powered sawmill at Wrangell, which was established in 1888. Though Harry Gartloy did an outstanding job of building and improving this business after the deaths of both Thomas Willson and Rufus Sylvester, he died in 1926, and after that the operation never regained the prominence it once enjoyed.

Sometime during World War I or shortly thereafter, Ketchikan Spruce Mills became the most important sawmill in the region. Continuity of ownership and a healthy dose of business acumen characterized this mill, which retained its position until the pulp mill era.

The following is a list (alphabetical, by location) and description of sawmills that operated in Southeast Alaska approximately between the years of 1867 and 1960.

Burnett Inlet (Etolin Island)

In 1930, the Forest Service's Ketchikan district ranger, J. M. Wyckoff, listed an unnamed sawmill at Burnett Inlet, where a cannery was also located.[1]

Coppermount (Prince of Wales Island)

The Alaska Copper Company operated a 5-MBF-per-day capacity water-powered sawmill to support its mining, milling, and smelting operations at Coppermount, on Copper Harbor (Hetta Inlet). The sawmill was in operation by 1902. Production in 1905 was about 600 MBF. The smelter with which the sawmill was associated was apparently last operated in early 1907.[2]

Craig (Prince of Wales Island)

Craig was founded in 1907 as a canning and cold storage site. The unincorporated town of Fish Egg was established in 1911 on nearby Fish Egg Island. Located at Fish Egg was a sawmill owned by an association of businessmen. In early 1912, the sawmill was sold to an individual from Seattle who expected to operate it that year.[3]

In 1917, the West Coast Lumber Company at Craig, with an annual cut of 500 MBF, was listed as one of the principal sawmills in southeast Alaska, even though it was still under construction. The mill was equipped with dry kilns and a box factory, and was said to be capable of cutting 50 MBF per day.[4]

The sawmill was purchased in late 1917 by F. J. Tromble of Hoquiam, Washington, who organized the Craig Lumber Company. Tromble's goal was to cut aero-grade Sitka spruce to be used in the World War I military aviation effort. He obtained financing from the fledgling Bank of Alaska (later the National Bank of Alaska) for ambitious and extensive improvements to the mill. In January 1918, Tromble secured a contract to supply the U.S. Army's Spruce Production Division with 300 MBF of aero-grade Sitka spruce per month. The lumber was to be shipped by water to Prince Rupert, and on to eastern manufacturing plants via the Grand Trunk Pacific Railway. To supply his sawmill, Tromble purchased 24 MMBF of Forest Service timber on 600 acres near Howkan, on Long Island.[5] The stand was almost pure Sitka spruce, and a Forest Service cruiser estimated 70 percent of it would grade out as clear. Tromble paid a dear price for the timber: $3.85 per MBF—about twice what Sitka spruce was usually selling for.

In the spring of 1918, Tromble leased the mill to Henry Shattuck, of Juneau. With about 40 employees, the steam-powered mill cut some 35 MBF of Sitka spruce lumber per day. The best material was set aside for the Army, and the rest used for the manufacture of salmon cases. Everything was said to be going along swimmingly, but this was hardly the situation. Two barge loads of what the mill considered aero-grade material was shipped to Washington in August. About half was rejected due to unsatisfactory quality. In all, only about 20 percent of the expensive timber at Howkan was in fact aero-grade. From the remaining 80 percent the Craig Lumber Company was manufacturing some very expensive salmon cases. By September, the company was unable to pay its bills. It ceased operations in November, the same month the armistice that ended World War I was signed. Aero-grade Sitka spruce suddenly became surplus, and government contracts for it were cancelled the following month.[6]

When it ceased operations the Craig Lumber Company had not paid for some 3 MMBF of timber it had contracted for with the McDonald Logging Company, most of which had been delivered to Craig. Bankruptcy ensued, and the Bank of Alaska faced possible collapse over the $100,000 it had loaned to the mill. In 1920, the bank purchased the Craig Lumber Company at auction for $16,000. It would spend the next eight years working to divest itself of its purchase.[7]

The Forest Service quickly reconfigured uncut timber at Howkan into a new 19 MMBF sale, which it sold in 1920 to the Craig sawmill's new operator, the O. P. Brown Box and Lumber Company. In the new sale the stumpage price was reduced to $2.60 per MBF.[8]

Still actually owned by the Bank of Alaska, the sawmill started cutting timber again in June 1920, and soon after began advertising spruce and hemlock lumber, salmon box shooks, and halibut boxes. The mill was equipped with a circular saw, employed about 20 men, and was capable of cutting about 25 MBF of lumber daily. The *Ketchikan Alaska Chronicle* reported that a shipment of clear Sitka spruce destined for England was made from Craig in 1922.[9] The sawmill at Craig was reported to be idle for the year 1924.

While the *Ketchikan Alaska Chronicle* opined at the time that "a competent mill at Craig could reap a good profit," the Bank of Alaska experienced only frustration with a series of sales and subsequent repossessions of the mill. In July 1924, the bank sold the mill to George L. Knapp, of Seattle. In February 1925, the *Petersburg Herald* reported the incorporation of the Craig Mill Company, which was planning to begin operations in early spring. Not long afterward, the *Ketchikan Alaska Chronicle* reported the mill to have been acquired by T. S. Steele, W. H. Rowe, and several others who planned to operate it as a cooperative. The men worked at improving the mill and cut some lumber for their own use, but never really went into operation. The Bank of Alaska finally divested itself of the sawmill in 1928, when it was sold to a local businessman for $1,500. It remained idle most of the time until 1945.[10]

In 1944, as the Alaska Spruce Log Program was winding down, the Walker Spruce Company, one of the nine gyppo logging operators contracted by the program, purchased the dilapidated mill, reportedly for about $800. With financing by Everett, Washington lumbermen, Walker Spruce began rehabilitating the mill, with the intention of cutting about 500 MBF of lumber per month. Lumber was urgently needed by the military for shipment offshore. At some point, the mill, which was equipped with a double circular saw headrig, was acquired by Frank V. Wagner, who, until he was declared insolvent in early 1947, had owned a sawmill at Wrangell. Wagner's fortune in Craig was similar that in Wrangell: In January 1949, his sawmill at Craig was offered at a marshal's sale. No bids were received. The mill may have been operated intermittently over the following few years, but gradually fell into disrepair and collapsed.[11]

Dolomi (Prince of Wales Island)

Ca. 1900, H. Z. Burkhart established a portable boiler sawmill at Dolomi, a now-abandoned mining community on Prince of Wales Island.[12] Production was largely mine timbers, and given as 287 MBF in 1901 and 338 MBF in 1902. Burkhart moved the mill's machinery to Ketchikan in 1903, where, with James J. Daly and C. M. Sommers, he established a company that later became the Ketchikan Spruce Mills.[13]

Excursion Inlet (mainland)

The Pacific American Fisheries Company constructed a cannery at Excursion Inlet in 1907, and the Astoria & Puget Sound Packing Company did the same the following year. Around 1910, Hoonah residents Oliver G. Hillman and Steve Kane constructed a small sawmill in a bay opposite the canneries. The bay later became known as Sawmill Bay. Powered by a Civil War-era Pelton wheel, the mill primarily cut lumber and timbers for the construction and rebuilding of fish traps. During the 1917–1919 cutting seasons, the mill was operated by Abraham Lincoln Parker, of nearby Strawberry Point (now Gustavus). Parker employed up to six men, usually Natives, in the operation, and figured his production was about 1,000 board feet per man per day. The mill last operated in 1928. Shortly thereafter it was partially dismantled, and the serviceable equipment was moved to Hoonah.[14]

■ ■ ■

The U.S. Army installed a sawmill at Excursion Inlet during World War II in conjunction with the construction of a large barge terminal. The sawmill was located at the head of the inlet, and was used to cut locally-logged timber into lumber for general construction. Much of the timber logged was along the Excursion River, in Glacier Bay National Monument.[15]

■ ■ ■

In 1953, Norman DesRosiers installed a gasoline-powered circular sawmill at Excursion Inlet and cut lumber part-time for local markets. The mill was capable of cutting 3 to 5 MBF per day, and was eventually equipped with a planer. The sawmill was converted to diesel power around 1971 and continued intermittently cutting lumber into the mid-1980s.[16]

Freshwater Bay (Chichagof Island)

In a 1917 interview with *The Timberman*, Tongass National Forest Supervisor W. G. Weigle noted that John Erickson had a sawmill at Freshwater Bay.[17]

Gustavus, a.k.a. Strawberry Point (mainland)

In about 1921, Abraham Lincoln Parker, likely with the help of three of his four sons, transported a steam-powered sawmill from Yankee Cove (north of Juneau) and installed it on his homestead along the Good River. A family enterprise, the Gustavus Lumber Mills was equipped with a planer and cut-off saw but operated only as needed.

The mill's production, as well as piling cut in logging operations, was mostly sold to the canneries at Excursion Inlet and Hoonah, with some additional lumber sold for local use at Gustavus and Hoonah. During the winter of 1933–1934, Gustavus Lumber Mills cut western hemlock timbers and planks for the construction of the long approach to the Gustavus dock.[18] The mill permanently ceased operations shortly thereafter.

Hadley (Prince of Wales Island)

Hadley was the location of three now-dormant copper mines. A steam-powered sawmill equipped with a band saw head rig, was constructed by the New York-Alaska Development Company about 1905 to furnish lumber with which to build a smelter. The sawmill may have been operated by the Alaska Smelting & Refining Company. The sawmill was shuttered once the smelter was completed, and remained idle for several years. In 1913, Foss Brothers and Company, headed by A.K. Foss, acquired the mill. Intent on making direct whole-cargo shipments of lumber to West Coast ports, the new owners quickly modernized and expanded the operation. The mill was operated as the Hadley Lumber Company and was said to have a capacity of 60 MBF per day. Advertised for sale in the mill's first year of operation were Sitka spruce, cedar and hemlock lumber, as well as halibut boxes, salmon boxes, and herring boxes. The company also sent 1.2 MMBF of Sitka spruce lumber to San Francisco aboard the steamer *Melville Dollar*. Though the weight of the lumber caused the Hadley Dock to collapse, its quality, according to the *(Ketchikan) Morning Mail*, so impressed buyers in San Francisco that all trade in Oregon and Washington spruce ceased, and the Foss Brothers soon had "sale for more spruce lumber than they can cut." In the spring of 1914, the firm shipped 180 MBF of Sitka spruce to San Francisco for use as finishing lumber in buildings being constructed for the Panama-Pacific Exposition.[19]

The operation was purchased in 1914 by the Alaska Lumber & Box Company. It continued to operate until May 29, 1915, when a spectacular fire destroyed the sawmill, planing mill, box factory, power plant, and 150 MBF of lumber. The mill was never rebuilt, and Hadley itself was deserted within several years.[20]

Haines (mainland)

There was reported to be a small "mom and pop" sawmill near Haines that about 1900 manufactured cases for nearby salmon canneries. This may have been the sawmill associated with the Alaska Fisheries Union cannery that operated at Chilkat Inlet from 1902 through 1905. In May 1909, the Combs Lumber Company advertised itself as "the largest saw and planing mill on Lynn Canal," and capable of cutting 25 MBF per day. A fire destroyed the mill in November 1912.[21]

■ ■ ■

In 1939, Frank Schnabel with members of his family purchased an old steam-powered circular sawmill for $500 and installed it at Jones Point, along the Chilkat River.[22] Their intention was to cut lumber for local use. The mill, known as the Haines Lumber Company, was capable of cutting about 10 MBF of lumber per day, but did very little during World War II. After the war, Schnabel increased the capacity of his sawmill with the installation of a double circular head saw, and converted the operation to diesel power. Haines Lumber also began trucking some lumber to Fairbanks, as well as to Whitehorse, Yukon Territories, Canada.

In the late 1940s, the mill was sold to Oliver and Lowell Colby, loggers from Oregon who operated the mill as the Colby Lumber Company. In 1956, the Colby mill employed seven men, was equipped with an edger and a planer, and had a daily production of 10 MBF to 16 MBF.[23]

Around 1957, the Colbys sold the mill to John Schnabel (son of Frank Schnabel), who operated it as the Schnabel Lumber Company. The uninsured sawmill burned to the ground in 1961. For the next several years Schnabel used a portable sawmill to cut mostly cants for export to Japan. In 1966, Schnabel began construction of a new 200-MBF-per-day mill at Lutak Inlet, on the waterfront east of Haines. The new Schnabel Lumber Company mill was used principally to cut cants, and reportedly was the first Alaska sawmill to export lumber to China, at least since the time of the Russian-American Company. Almost continually beset with financial difficulties, the mill was shuttered in 1977. It was re-opened in 1979, but again shuttered when Schnabel was forced into bankruptcy in 1983. The mill was taken over by Michael Chittick, who operated it as the Pacific Forest Products Company for about a year before going out of business. In 1986, the mill was purchased by Ed Lapeyri, owner of the Mitkof Lumber Company at Petersburg, in part to take advantage of deep-water port facilities at Lutak Inlet. Lapeyri shuttered his Petersburg mill and moved the operation to Haines. The Haines mill was renovated, including the installation of some equipment from Petersburg, and began cutting timber in the fall of 1987 as the Chilkoot Lumber Company. Market conditions forced a series of production curtailments in 1991, and the mill was permanently shuttered after a heavy snowfall collapsed a major part of the facility in 1993.[24]

■ ■ ■

In 1951, or within a few years thereafter, Frank Young, of the Chilkat Valley Development Company, established a small diesel-powered sawmill at Mile 28 of the Haines Highway. The mill employed six men and seasonally cut about 200 MBF of lumber per month for local markets. In 1956, the sawmill and logging operation was acquired by the Moose Valley Lumber Company (Ed Hosford, John Hoit, and Bill Chandler), which reportedly sold both rough and planed lumber on the local market. Additionally, the company trucked some lumber to Alaska's Interior. In about 1963, this sawmill was destroyed by fire. It was not rebuilt.[25]

Hidden Falls (Kasnyku Bay, Baranof Island)

The Hidden Falls Lumber Company was formed in 1927, and began cutting dock timbers, planking, and dimension lumber in 1929. The 25-MBF-per-day water-powered sawmill apparently ceased operating in the early 1930s, a victim of the Depression.[26] In 1940, the mill equipment was obtained by Sheldon Jackson School in Sitka, and used to replace equipment destroyed in a fire at the school's sawmill.

A water-powered sawmill was installed at Hidden Falls during World War II, and reportedly cut logs under a government contract. The mill was scheduled to receive logs from the Alaska Spruce Log Program. The mill itself was unique in that the planer was located a considerable distance from the head rig, and lumber was transported to it in a flume.[27]

Holbrook (Kosciusko Island)
Owned by the Alaska Coast Fish & Packing Company, the Holbrook Mill was built in 1902 and that year cut 72 MBF of lumber before permanently ceasing operations.[28]

Hoonah (Chichagof Island)
The small steam-powered "missionary sawmill" shipped to Presbyterian minister Sheldon Jackson at Sitka by his East Coast benefactors came through Hoonah in 1881, and was temporarily set up there to cut lumber for the construction of a school house and teacher's residence.[29]

The water-powered sawmill at Sawmill Bay in Excursion Inlet known as Kane's Sawmill was dismantled and brought to Hoonah in 1928. It was set up at a site just north of town (near the present-day ferry terminal) and was known as "Shotter's Sawmill," after its owner, Frank Shotter, the local postmaster. Shotter's sister, Louise, with her husband, Steve Kane, had owned the mill when it was at Excursion Inlet. The sawmill was powered by water routed in a flume from a small creek. Cutting was done only during daylight hours, as there was no electricity on the site. Shotter's sawmill intermittently cut rough lumber for the local market until the late 1930s.[30]

Howkan (village on Long Island, sawmill on Dall Island)
Alaska historian Robert De Armond wrote of a sawmill at Ham Cove, on Dall Island. The sawmill was apparently erected about 1882 by Presbyterian missionaries in conjunction with the mission station at Howkan, which was located about three miles away on Long Island. The 1889 Alaska governor's report lists a steam-powered sawmill operated by a Presbyterian mission "at or near Howkan." Two years later, however, the *Coast Pilot* reported a sawmill southeast of Ham Cove, but none at the cove itself.[31]

In 1902, George Emmons reported a small sawmill operating at Howkan. William Langille was more specific in 1906. He noted the water-powered Howkan Mill, at "Sinclaire's Cove," another name for Ham Cove. Langille listed the mill's capacity as 5 MBF per day. The mill was owned by Gould & Petty. It cut a variety of lumber, and was equipped with a machine for making barrel staves. The annual cut circa 1905 was 150 to 250 MBF.[32]

Hunter Bay (Prince of Wales Island)

Howard Kutchin, who examined Alaska fish processing facilities for the Treasury Department in 1900, reported that the Pacific Steam Whaling Company planned to construct a sawmill at Hunter Bay later that year. In 1906, William Langille reported the Northwest Fisheries Company as owning the 10-MBF-per-day steam-powered Hunter Bay mill, which cut lumber and box material. It had apparently closed, however, two years earlier.[33]

Hydaburg (Prince of Wales Island)

In 1917, a Forest Service official noted the existence of the Hydaburg Lumber Company, with an annual cut of 500 MBF. The *Wrangell Sentinel* reported the Hydaburg Co-operative Company had purchased a new sawmill in early 1919.[34] The author encountered no further information on either company.

Hyder (mainland)

Hyder, Alaska, was in established in large part because of the nearby Premier Mine, which was located in British Columbia. The town was served by the Hyder Lumber Company, which was probably established about 1910. The mill had a capacity of 15 MBF per day, and was capable of cutting logs six feet in diameter. The mill was particularly busy about 1920 in providing lumber for a 12-mile tramway constructed from saltwater to the Premier Mine. The mill's cut in 1923 was expected to be 3 MMBF.[35] The Premier Mine operated until 1948, but the sawmill probably closed much earlier.

Juneau (mainland) and Douglas (Douglas Island)

A water-powered sawmill was constructed at Sheep Creek by E. H. Boggs and D. H. Murphy in 1883, and began cutting lumber in July of that year. "Yank's Sawmill! Situated on Sheep Creek, four miles south of Juneau, is prepared to furnish the public with all kinds of rough lumber," read an advertisement in the April 12, 1888, *Juneau City Mining Record*.[36]

The Alaska M & M Company constructed a sawmill on the waterfront at Douglas in 1884. The mill was reported in 1894 to be owned by the Alaska Treadwell Gold Mining Company, and powered by an 80-horsepower boiler.[37]

The Alaska governor's report for 1889 listed two sawmills as operating near Juneau: a mill at Sheep Creek owned and/or operated by a Mr. Depue, and the steam-powered mill on Douglas Island. Two years later, Governor Lyman Knapp provided a partial listing of Juneau-Douglas sawmills that included the "Alaska M & M Co." (Douglas), the "Eastern M & M Co." (Douglas), and the Silverbow Basin Mining Company (Juneau).[38]

In 1899, W. T. Iliff, of Sheep Creek and Douglas Island, Alaska, advertised himself as a "manufacturer and dealer in lumber." In 1900, a Juneau sawmill operated by Charles Green produced 100 MBF. In 1901, Green entered into a partnership with Harry G. Slater in a mill at Douglas. They cut 180 MBF that year and 334 MBF in 1902. J. P. Jorgenson cut 800 MBF of lumber at Juneau in 1902. In 1911, the Alaska Supply Company, of Juneau, was reported as having an annual cut of 1 MMBF or less.[39]

George E. James & Company (a.k.a. the George E. James Mill Company and the George E. James Company) was the largest operator on Gastineau Channel, with a production in 1900 of 693 MBF at his Sheep Creek mill and another 115 MBF at a mill leased by the firm at Douglas. The Douglas mill was likely established in 1898. The next year, the Douglas mill alone was operated, with a cut of 2,016,000 board feet. In 1902, James entered into a partnership with B. M. Behrends, a prominent Juneau merchant and banker. Production that year was 1,868,000 board feet. In 1906, Assistant Forester F. E. Olmsted reported a Mr. James as leasing a mill at Douglas from the Alaska Treadwell Gold Mining Company. In 1911, the George E. James & Company of Douglas was reported to have an annual cut of 2 to 4 MMBF, and the Alaska Supply Company of Juneau to have an annual cut of 1 MMBF or less. Six years later, a Forest Service official noted that the George E. James Company had an annual cut of 6 MMBF and mentioned that it had just established a box plant at Sitka. The company's sawmill ceased operations in 1918.[40]

On the mainland side of Gastineau Channel, the "Juneau Sawmill" was built in 1905. Eight years later, Juneau businessman Henry Shattuck and Puget Sound lumberman H. S. Worthen purchased the mill and promised to construct a new sawmill that would be "the most modern and up-to-date in Alaska." The mill was incorporated as the Worthen Lumber Mills Company. The Forest Service expected the mill to cut about 9 MMBF in 1917. Worthen died in the spring of 1918 after being struck by a rolling log at his mill.[41]

Perhaps as a result, Fairbanks lumberman Roy Rutherford formed a corporation to purchase the Worthen sawmill in 1919. The name of the corporation was Juneau Lumber Mills, which in 1923 employed some 65 men and cut about 50 MBF per day. The total cut for that year was approximately 8 MMBF. In 1924, the operation was closed for improvements, including new dry kilns and a box plant. The main mill was rebuilt and completely electrified during the winter of 1926–27. The expanded mill, with a capacity of 100 MBF per day, was one of the finest and most completely up-to-date lumber plants in Alaska. Production for the first year of operation (1927) of the "new" plant was 14 MMBF of Sitka spruce lumber and more than 700,000 salmon cases. Some 500 MBF of clear Sitka spruce flitches were shipped to Great Britain. Also produced were about 40,000 western hemlock ties for the Alaska Railroad. During the mid-1930s, Juneau Lumber Mills employed about 90 men at the mill, with the normal operating year considered to be six months, six days per week, and eight hours per day. Juneau Lumber Mills did a considerable retail business, maintaining a lumber yards in Fairbanks and Anchorage as well as Juneau. Among the items stocked were shingles (from Wrangell) and western hemlock flooring.[42]

In the spring of 1947, a group of Oregon lumbermen purchased the company and changed its name to the Juneau Spruce Corporation. The men were largely affiliated with the Coos Bay Lumber Company, in Oregon, and director's meetings were sometimes held in the company's Oregon office. The following winter, they shut down the sawmill for repairs and improvements. When re-opened in April 1948, the Juneau Spruce Corporation sawmill was capable of cutting 100 MBF per eight-hour shift. In the spring of 1948, the Juneau Spruce Corporation shipped approximately 800 MBF of Sitka spruce lumber by barge to Prince Rupert, where it was loaded onto railroad cars for transport to various U.S. destinations. The company anticipated that such shipments could eventually reach 3 MMBF per month. But labor troubles ensued, forcing a work stoppage that caused the Juneau Spruce Corporation to shutter its mill less than a month after it had reopened. In the fall of 1948, the mill's owners decided to permanently cease mill operations. The mill burned to the ground on August 28, 1949, and several months later the company announced a decision to liquidate its Alaska interests.[43]

The Juneau Lumber Company, owned by B. C. "Curly" Canoles and Joe Murphy, was described as a marginal sawmill that operated for a short while on the site where the Juneau Spruce Corporation sawmill had burned. The Columbia Lumber Company purchased a controlling interest in the company in late 1950 or early 1951. In the spring of 1951, after the completion of an improvement program, the mill was expected to begin operations with a capacity of 60 MBF per 8-hour shift. In 1952, B. C. Canoles purchased the box making equipment from the Columbia Lumber Company and opened the West Juneau Box Factory. A year later, however, the factory was destroyed by fire.[44]

The Columbia Lumber Company, which continued to operate in Juneau after 1952, also had sawmills at Sitka and Whittier and a subsidiary plywood plant, Alaska Plywood Corporation, which was built at Juneau in 1952. In addition, it owned the Anchorage Lumber & Supply Company. The company's sawmill at Juneau was equipped with a circular saw, employed about 25 men, and had an annual capacity of 10 MMBF.[45] This sawmill was destroyed in an August 1959 fire that also destroyed the adjacent Alaska Plywood Corporation mill.

■ ■ ■

The federal government's Civilian Conservation Corps (CCC) constructed a sawmill along Montana Creek at Auke Bay in about 1937. The mill operated for about three years and cut lumber for government projects in the area. In 1953, Juneau resident Joe Smith constructed a diesel-powered sawmill on the same site and operated as the Auke Bay Lumber Company and the Smith Lumber Company. In 1954, the company opened a retail lumber yard at Juneau. As of 1956, the sawmill employed about nine men and was capable of cutting 20 MBF per day. One product manufactured was fish boxes in which frozen fish from the local cold storage were shipped. The operation was eventually transferred to the Mountain View Lumber Company. By 1960, the sawmill, now known as the Bay View Lumber Company had an annual capacity of 4 MMBF.[46]

In 1946 or 1947, Joe Smith—prior to his involvement with timber at Auke Bay—formed a company named Smith Enterprises and constructed a gasoline-powered sawmill on the south side of Lemon Creek. Smith had a contract to cut 50,000 ties for the Alaska Railroad, but after cutting about 5,000 realized his operation was too small to do the work. The contract was cancelled, after which Smith cut lumber for awhile for the Army. After about a year, this sawmill was dismantled, and a larger, diesel-powered sawmill built on the north side of the creek. The mill was originally equipped with a 60-inch circular saw head rig, but was later converted to a double circular saw. Probably in 1949, this sawmill was purchased by B. C. Canoles and Joe Murphy, who operated as the Duck Creek Lumber Company. In 1950, the men changed the company's name to Juneau Lumber Company and moved their sawmill to the site where the Juneau Spruce Corporation sawmill had burned. (See note above. Two years later, Canoles founded the West Juneau Box Factory). Canoles and Murphy had ambitious plans to expand operations, but the mill was shut down after a short while.[47]

■ ■ ■

The Alaska Yellow Cedar Company was incorporated in July 1949, and the following November it began operating a small electrically-powered sawmill in the carpenter shop of the defunct Gastineau Mining Company at Thane. The mill focused on cutting Alaska-cedar lumber for local boatbuilders as well as furniture-making concerns in Haines, Alaska, and Portland, Oregon. It was said to have operated successfully before being destroyed by fire in 1950.[48]

■ ■ ■

The manufacture of plywood in Alaska had been considered for some time. Thomas Morgan, the principal owner of the Columbia Lumber Company, became interested in expanding his firm's line of production to include Sitka spruce plywood shortly after he purchased the company. Plywood of Sitka spruce is about 25 percent lighter than that of Douglas-fir, and, as a specialty item, commanded a premium price in certain markets. The Forest Service supported the establishment of a plywood plant in Southeast Alaska and in 1951 granted Morgan permission to ship 50 MBF of logs to Puget Sound for testing in established plywood plants.[49]

Satisfied with the results of tests, Morgan formed the Alaska Plywood Corporation. The construction on Alaska's first and only plywood plant, built at a cost of $1.5 million, began in 1952. The lathe used to peel logs was secondhand, obtained from a Brazilian operation on the Amazon River.[50]

Morgan organized the Alaska Plywood Corporation as a producer's cooperative: priority in employment was given to those willing to invest in the company, and, as of May 1954, all of the company's 100 employees were shareholders. Peak employment at Alaska Plywood was about 115 men, who mostly came from outside Alaska.[51]

The plant required about 60 MBF of "peelers"—high-quality logs suitable for the manufacture of veneer—per day. The initial supply of timber for the plywood mill came from a 117 MMBF Forest Service sale on Kosciusko Island, some 220 water miles from Juneau. Plans for the construction of the plant had been contingent on getting this timber, for which the company was ultimately the sole bidder. Though timber was also sourced from Tuxekan Island and from Ketchikan Pulp Company as well as independent operators, little provision was made to secure a steady supply of suitable logs.[52]

On June 22, 1953, the first log ever to be peeled for plywood in Alaska went through a lathe at the Alaska Plywood Corporation. The plant, with an annual capacity of the equivalent of 30 million square feet of ⅜-inch exterior-grade Sitka spruce plywood, went into full operation the following month. The main thicknesses produced were ⅝-inch and ¾-inch, and the principal anticipated market was as concrete form material for military construction projects in Alaska.[53] Unlike most plywood manufacturers, Alaska Plywood used prime veneer for both the outside faces of the plywood as well as for the core layers.

A major production problem at the plant was that the lathe from Brazil was designed to produce 5-foot-by-10-foot veneer, not the 4-foot-by-8-foot U.S. industry standard. The company made little effort to locate a market for material of this size, and instead chose to have laborers cut the large sheets down to four feet by eight feet. The excess material was used to help fuel the operation's boiler.[54] The final product was so costly that even in Alaska it could not compete with plywood imported from the Pacific Northwest.

Alaska Plywood shut down for three months during the winter of 1953–54, complaining the percentage of peeler logs it had obtained from the Forest Service was disappointingly short of estimates. The actual reason for the plant's closure, however, was that employees were not receiving regular paychecks, and had refused to work. Short of cash and delinquent in repaying a government loan, Morgan resorted to "borrowing" federal retirement funds withheld from his employees to use as operating capital.[55]

Morgan bought timber at Yakutat, and attempted to have it towed in a raft across the Gulf of Alaska, apparently counting on the profits from this timber to repay the retirement funds he had illegally taken. The raft, however, broke up as it was being towed, and all the timber was lost. Soon thereafter agents of the U.S. Department of the Treasury padlocked the Alaska Plywood Corporation mill.[56] The mill was eventually reopened, but after sporadic production at less than planned output, the Alaska Plywood Corporation permanently ceased operations in 1955.

In June 1956, the company appealed to the U.S. District Court for protection under bankruptcy laws while it reorganized. The court appointed George Rogers, a Juneau economist, as trustee for the company. Rogers spent more than a year analyzing the situation, and in the course of his efforts found viable markets for the large sheets of Sitka spruce plywood.

The *Lumberman* wrote in 1960 that for Alaska Plywood, "the high cost of Alaskan labor more than overbalanced the comparatively low stumpage costs for logs," but Rogers, who became intimately familiar with the company's operation, wrote that "low log costs were sufficient to offset other high operating costs." In documents provided to the court, Rogers blamed undercapitalization and poor management for Alaska Plywood's problems.[57]

Though there was talk of reorganizing the Alaska Plywood Corporation into a true cooperative, foreclosure proceedings were initiated in January 1959. The following August, however, Thomas Morgan reached an agreement with the Department of the Treasury to regain control of the plant. At dawn several days later (August 14, 1959) the operation of Alaska Plywood's mill became a moot point when the plant burned to the ground. The cause of the fire was never determined, though Morgan blamed "some electrical defect."[58]

Kake (Kupreanof Island)

At Kake in 1920 was the Eureka Trading and Sawmill Company, a Native cooperative.[59]

Kasaan (Prince of Wales Island)

A sawmill was reported to have been built at New Kasaan in 1895. The Kasaan Bay Company owned a copper mine, a cannery, and, in the words of a government fisheries inspector in 1903, a "fine sawmill." The steam-powered sawmill was capable of cutting about 25 MBF per day, and cut, in the words of William Langille, "all kinds of lumber and box material." A shingle machine was attached to the mill. In 1901, the sawmill cut 1.3 MMBF of lumber. It was reported by George Emmons to have operated in 1902. The Kasaan Bay Company failed financially, and the cannery, sawmill, and 100 MBF of logs were purchased by a competing cannery interest at a receiver's sale in early 1904. The sawmill was idle in 1904 and no production was reported after that date. In 1930, the Forest Service's Ketchikan district ranger, J. M. Wyckoff reported this sawmill to have been abandoned for a long while.[60]

Ketchikan (Revillagigedo Island and Gravina Island)

Henry Imhoff was one of the earliest manufacturers of barrels in Southeast Alaska. He may have also been the last. Imhoff established a barrel manufacturing plant at Ward Cove in 1879 and operated it continuously until his death in 1928.[61]

Young Tsimshian men trained at the Presbyterian Mission School at Sitka established the Hamilton, Simpson & Company sawmill at Port Gravina (north of the present-day Ketchikan airport on Gravina Island), probably in 1892. The steam-powered mill had a capacity of 8 to 10 MBF per day. In 1896, company employed 25 workers in the sawmill and another 15 in logging operations. In 1898, the mill was manufacturing salmon cases for the canneries at Loring and Wrangell. Production in 1900 was 1,206,000 feet, and 1,050,000 feet the following year. Shortly afterward, the mill was reported to employ about 15 men, all Indian, and to have a capacity of 15 MBF per day. On July 5, 1904, while nearly all the residents of Port Gravina were in Ketchikan celebrating Independence Day, a fire destroyed the sawmill.[62] It was not rebuilt.

■ ■ ■

In 1900, William Harper began constructing a sawmill at Ward Cove. What was known as the Revilla mill was completed early the following year, and Harper began cutting lumber under contract for the construction of a new Alaska Packers Association cannery at Loring. By 1903, however, the so-called Revilla Lumber Company was idle and later that year the company went bankrupt.[63]

■ ■ ■

Edmond and Francis Verney built a steam-powered sawmill at North Saxman in 1900, and cut lumber the following year for the Fidalgo Island Company cannery. The mill's cut in 1900 was 400 MBF, and the following year was 700 MBF. A shingle mill was added in 1903. The Verney Brothers Lumber Company had a lumber yard at Ketchikan, and among the products advertised in 1904 were western redcedar shingles, kiln-dried finished lumber, beaded ceiling material, flooring, turned work, molding, and pickets. Financial difficulties led to the liquidation of the company's assets at a marshal's sale in 1907.[64]

■ ■ ■

The Alaska Packer's Association established a salmon hatchery on Revillagigedo Island about seven miles up the Naha River. The cost of trans-porting lumber to the site for its original construction was extraordinary: $65 per MBF. The hatchery was substantially expanded in 1903, and to save money the company erected its own sawmill nearby, which remained active only as long as the hatchery was being constructed.[65]

■ ■ ■

In 1904, "Ott" Inman, along with a man named Burt, constructed a shop along Ketchikan Creek to manufacture barrels. The business soon expanded to include the manufacture of shingles. In 1911, the company was known as the Ketchikan Shingle Mills and owned by Inman and A. J. Dunton. In 1913, the mill advertised shingles and "16 inch wood." It not long after became known as the Dunton & Inman Shingle Mill, and prior to 1917 was renamed the Red Cedar Shingle Company. The operation was moved to Charcoal Point (near the present-day Ketchikan ferry terminal) sometime before 1919, and its name changed to Ketchikan Lumber & Shingle Company. It was also referred to as "Fleming's Plant," after R. H. Fleming, who seems to have become the principal owner. In late 1923, Fleming sold his interest in the operation and left Ketchikan.[66]

■ ■ ■

The Ketchikan Power Company (motto: "We Cut Spruce") was incorporated in May 1903 under the leadership of Alaskan H. Z. Burkhart, who managed the company in its initial years.[67] Burkhart and his Alaska associates, C. M. Sommers and W. J. Hill, built and operated a lumber mill and a hydroelectric power plant. The power plant was sold off prior to November 1913.

The steam-powered sawmill began cutting lumber on March 17, 1904— Saint Patrick's Day. When questioned about the significance of the Saint Patrick's Day start up, Burkhart was said to have replied, "After all, we are cutting green lumber." Thereafter, it became the goal of the company to begin operations each year on Saint Patrick's Day. The principal products produced at the sawmill were salmon cases, rough and planed construction lumber, and "cedar furnishings" for local use.[68]

The mill was situated on 14 acres on Ketchikan's waterfront. It was capable of cutting 40 MBF per day of 10 hours and was equipped with a 3.5-MBF-per-day capacity dry kiln. In 1913, an 8-foot band saw head rig was installed and the mill's capacity was increased to 50 MBF per day. The mill was designed primarily to cut large Sitka spruce logs on a one-shift basis and was capable of cutting dimension stock up to 50 feet long. Salmon cases were also produced.[69]

James "J. J." Daly purchased Hill's interest in 1906 and assumed direction of the company shortly after the death of H. Z. Burkhart in 1909. James Daly died in 1919, and ownership soon passed to his sons Eldon and Milton. As of June 1, 1923, the name of the business was changed to Ketchikan Spruce Mills.[70]

Production in 1924 was 10.5 MMBF. The following winter, $150,000 was spent to modernize the mill and to expand its capacity to approximately 80 to 100 MBF per day, which made it the largest sawmill in the territory. Owing to its increased efficiency, the number of workers employed was not expected to increase greatly. One of the improvements was the installation of some 65 electric motors to power machinery previously powered by steam. Electricity was provided by a 1,000-horsepower steam turbo-generating unit fueled with sawdust and planing mill waste. Though the mill was not yet ready to resume regular operations, one log was run through its saws on the evening of March

17—St. Patrick's Day. Production for 1925 was approximately 14 MMBF, about a quarter of all the timber cut in the Territory that year. Three MMBF was exported to Europe.[71]

A fire in the fall of 1925 destroyed the planing mill and box factory at Ketchikan Spruce Mills, neither of which had been part of the previous year's renovations. The planing mill and box factory were promptly rebuilt to the standards of the renovated sawmill. Some of the timber cut by the mill was exceptionally large, and the company eventually installed a pond splitting machine for use on logs more than six feet in diameter.[72]

Lumber magnate J. P. Weyerhaeuser toured Ketchikan Spruce Mills in the summer of 1926 and assessed it as follows: "While I would not describe it as the finest mill of its kind that I have ever seen, it is nevertheless a splendid plant and a far better one than I had anticipated in Alaska. It is a plant that would be a credit to any community and its owners should be commended for the fine industry they have created and the thoroughly modern manner in which they operate."[73]

Production in 1927 was about 16 MMBF, including 600,000 salmon cases. That year, the company advertised "the largest variety of dry, square-edge finish, flooring, ceiling siding and mouldings in the Territory." Company employment at that time was about 100.[74]

Ketchikan Spruce Mills established a lumber yard at Fairbanks in 1938, at Anchorage in 1941, and at Palmer not long after World War II. Military construction in the region caused a particularly strong demand for lumber during the war. Dimension lumber was also marketed to canneries throughout Alaska, and cannery vessels journeying north before the salmon season often loaded lumber at Ketchikan.[75]

Eldon Daly died in 1944, which left his brother, Milton, as sole manager of the business. Milton's son, John, assisted him in running the operation. He was the third generation of the Daly family to be associated with Ketchikan Spruce Mills.

During the 1950s, Ketchikan Spruce Mills provided lumber for military projects in Korea. In its later years, the company switched to cutting cants for export to Japan. The first Japanese lumber ship was loaded at the plant in 1961.[76]

Ketchikan Spruce Mills was sold to the Georgia-Pacific Corporation (part-owner of the Ketchikan Pulp Company) in 1967, ending a 61-year span in which the business was largely owned and managed by the Daly family. The Louisiana-Pacific Corporation, which succeeded Georgia-Pacific as owner of Ketchikan Spruce Mills in 1972, shuttered the venerable facility in 1983.[77]

■ ■ ■

In 1917, the Beaver Falls Shingle Company was noted as being one of the two principal shingle mills in Southeast Alaska. Beaver Falls is located in George Inlet on Revillagigedo Island, about 10 miles northeast of Ketchikan. With both waterpower and a supply of western redcedar, Beaver Falls was an ideal location for a shingle mill. In 1924, the mill was reported to be operated or owned by a Mr. Silvas, and produced about 75,000 shingles.[78]

■ ■ ■

In 1919, the Ketchikan Lumber & Shingle Company advertised western redcedar, hemlock and spruce lumber, and western redcedar shingles. In 1921, the company advertised a specialty in western redcedar and Alaska-cedar, as well as "AAA-1 Red Cedar Shingles—Extra Heavy." The mill produced 2.9 million shingles in 1920, and in 1924 production at the plant was 3.7 MMBF of lumber. After the mid-1920s, the company was sold to J. E. Berg, and by 1932 was known as the Berg Lumber & Shingle Mill. By 1937, the company was acquired by John Bertoson and renamed the Alaska Lumber & Shingle Mill. The plant burned to the ground on April 17, 1951, and was not rebuilt.[79]

■ ■ ■

A small sawmill built by Fred Brown at Dall Bay during the 1940s was operated intermittently.[80]

■ ■ ■

The Ketchikan Shingle Company, owned by Olaf Johnson, operated during the 1940s. It was in operation by 1943 and remained at least until 1947. The company's products were sold both in Alaska and elsewhere.[81]

■ ■ ■

Lewis Johnson had a small sawmill on Pennock Island. It apparently operated for only a short time before it was destroyed by a fire on March 18, 1954.[82]

■ ■ ■

The Ketchikan Pulp Company completed construction of a sawmill at Ward Cove in 1973. The equipment was integrated into the woodrooms of the pulp mill, and was used to saw lumber from low-grade logs previously used for pulp. Its annual capacity was 50 MMBF.[83] By 1974, the company owned or leased four sawmills: Ketchikan Spruce Mills and Ketchikan Pulp Company Sawmill at Ketchikan; Annette Hemlock Mills at Metlakatla; and Alaska Timber Corporation at Klawock. Waste wood from each sawmill was chipped and sent to the pulp mill.

As the Ketchikan Pulp Company's business in Southeast Alaska grew and evolved, sawn products became increasingly important, with more than half of the company's production in 1975 being cants, flitches and lumber.[84]

"Killiswoo" (likely Killisnoo, near Angoon, on Admiralty Island)

A photograph in a 1921 article in an industry publication shows an apparent steam-powered sawmill, at least one outbuilding, and a large number of floating and beached logs. The illustration is titled "Sawmill and Log Pond, Killiswoo."[85] The sawmill may have operated in conjunction with a fish processing plant.

Klawock (Prince of Wales Island)

A sawmill was established at Klawock in 1878 in connection with Alaska's first cannery, the North Pacific Trading & Packing Company of San Francisco. The sawmill, which was steam-powered and had a capacity of 15 MBF per day, was used for making boxes and supplying local demands. The average yearly cut was only about 60 MBF. It was reported in 1897 that Native labor was exclusively employed for common work. Lumber from the Klawock sawmill was shipped to San Francisco, Honolulu, and Japan. In September 1899, the cannery and the sawmill burned. The cannery was quickly rebuilt, but the sawmill was never replaced.[86]

■ ■ ■

In addition to the above sawmill, the governor of Alaska's report for fiscal year 1889 lists a water-powered mill operated at Klawock by "Salmon Packing Company."[87] In 1902, a small sawmill was reported as operating at Tlevak, which was another name by which Klawock was known.[88]

■ ■ ■

In 1969, Edward Head, an experienced mill operator from California, incorporated what became the Alaska Timber Corporation and began setting up a sawmill at Klawock. Unable to secure bank financing for the completion of the mill, Head sold the partially-completed facility to the Ketchikan Pulp Company in an exchange that included a lease-back and purchase-back option. As well, he essentially entered into a 10-year partnership with Ketchikan Pulp in which he operated the sawmill but the pulp company supplied logs to the mill and marketed its production. The mill was completed in 1973 and began cutting mostly 3-sided cants for the export trade. When operating two shifts, the mill, sometimes cut more than a million board feet per week. About 1978, Head reacquired ownership of the sawmill. During the early 1980s, the Alaska Timber Products sawmill cut only about 5–6 MMBF per year, and it was shut down in September 1984 due to financial problems. The mill was re-opened the following year, but operated only very intermittently over the next several years. In 1987, Head entered into a partnership with the Weyerhaeuser Company, and the name of the business was changed to Klawock Timber Alaska, Inc. The mill began operating in 1988, but the partnership soured and the mill was shuttered in 1991. Before its closure, the mill had been cutting cants at a rate of about 48 MMBF per year. In 1994, the Klawock sawmill was purchased by the Dahlstrom family, which also operated a sawmill and a secondary wood processing plant at Hoquiam, Washington. The mill was quickly renovated and began operating as Viking Lumber. It has a capacity of about 95 MBF per 8-hour day and cuts both dimension lumber and cants. Viking Lumber has operated continuously since its inception, and is today the largest wood processing operation on the Tongass.[89]

Metlakatla (Annette Island)

Metlakatla was and remains a community of Tsimshian people. The present-day community began with a group that migrated from Canada under the leadership of Reverend William Duncan in 1887. Duncan worked fast, hard, and smartly to quickly develop a modern economy for his followers. Metlakatla's steam-powered sawmill, which went into operation in 1888, provided lumber for local construction. In addition, it was one of the first mills in Southeast Alaska to manufacture salmon cases. Some 16,000 salmon cases were manufactured in 1888.[90]

The sawmill burned to the ground in June 1889. Duncan promptly left for Portland, where with donations and credit purchased a new mill with about twice the capacity of the one that burned. In 1889, Metlakatla mill produced around 50,000 salmon cases. The sawmill (as well as the local cannery) was in 1900 owned by the community's principal business concern, the Metlakatla Industrial Company, Inc., which was in turn owned almost entirely by William Duncan. The mill, cut 72 MBF of lumber in 1900, 286 MBF in 1901 and 190 MBF in 1902.[91]

In December 1917, the Council of Annette Island Reserve sold to the Metlakatla Commercial Company "one old Pelton Wheel Sawmill Plant, situated on the northeast end of the town of Metlakatla, Alaska, together with its pieces of machinery, buildings, yards, wharves and grounds," for $800.[92]

During the Depression years, the sawmill at Metlakatla was operated for "home use." Preparations for World War II created a huge demand for lumber, and the small mill at Metlakatla worked throughout the winter of 1940–41, principally cutting material for an army landing field being constructed on Annette Island. Late in 1944, a new small sawmill known as the Metlakatla Town Sawmill was being installed at Metlakatla. It remained until 1947, if not later.[93]

The locally-owned Metlakatla Indian Community Enterprises began construction of a modern sawmill in about 1948. The mill was completed in 1953, and had a capacity of about 75 to 100 MBF per day.[94]

In January 1954, a fire completely destroyed the mill, but it was quickly rebuilt and leased to the Annette Island Spruce Company, an affiliate of the Walton Lumber Company of Everett, Washington. The mill produced spruce and hemlock cants and had an annual capacity of about 20 MMBF. The Annette Island sawmill was acquired by the Alaska Prince Timber Corporation, which sold it to the Louisiana-Pacific Corporation, owner of the Ketchikan Pulp Company in 1971. Under the ownership of Louisiana-Pacific, the company was known as Annette Hemlock Mills, and produced western hemlock lumber and cants for the domestic and export markets. Louisiana-Pacific modernized and increased the capacity of the mill to 60 MMBF annually.[95] Annette Hemlock Mills permanently ceased operations in 1999.

■ ■ ■

John Davis Sr. was a Tsimshian Indian who, in about 1900, established several Metlakatla enterprises, including a boat yard and marine ways. To provide lumber for his boat yard, Davis also operated a sawmill and had a small logging operation. Davis & Son became famous for their "Davis boats," the rowing/sailing skiffs that helped launch the commercial troll salmon fishery in Southeast Alaska. The boats ranged from 13 to 16 feet long, and were framed with oak and planked with western redcedar.

Moria Sound (Prince of Wales Island)

The Forest Service's Ketchikan district ranger, J. M. Wyckoff, reported a small shingle mill in part-time operation at Moria Sound in 1930.[96]

Pelican (Chichagof Island)

In 1938, the Pelican Cold Storage Company began to construct a cold storage facility and town at an uninhabited site at the confluence of Lisianski Inlet and Lisianski Strait. A year later, the company installed a small gasoline-powered sawmill that was later converted to diesel power. The mill cut rough lumber for the town's extensive boardwalks, as well as for the construction of a number of buildings and homes. A box factory, operated in conjunction with the sawmill, manufactured boxes used in shipping frozen salmon and halibut. The mill passed through a number of private ownerships, and in 1947 it was known as the Pelican City Sawmill. In 1955, the Pelican Lumber & Manufacturing Company was noted as a producer of lumber and fish boxes. The sawmill at Pelican ceased operating about 1957.[97]

Petersburg (Mitkof and Kupreanof Islands)

Petersburg was founded by Peter Buschmann, one of Southeast Alaska's pioneer cannerymen. Under his leadership, the Icy Strait Packing Company in 1899 purchased a 6-MBF-per-day steam-powered sawmill and installed it at a site in what is now downtown Petersburg. It was known locally as the "Charlie Norberg sawmill," after the mill's foreman and chief sawyer, who with an 8-man crew cut 500 MBF of lumber during the first year of operation. The lumber was used in the construction of two canneries, of a Petersburg wharf and various town buildings, and of salmon boxes and barrels. Due to a change in cannery ownerships, sawmill production dropped off after 1901 and almost certainly closed after 1906.[98]

In 1901, the "Buschmann-Thorpe interests" erected a steam-powered sawmill on a site at the mouth of Hammer Slough. The mill was not put into operation, and may have been the same sawmill offered for sale in the summer of 1903 by August Buschmann. Buschmann advertised land, a brand new sawmill with a capacity of 30 MBF per day, and 100 MBF of varied lumber. In 1906, the Buschmann-Thorpe sawmill was acquired by the Pacific Coast & Norway Packing Company, which that year moved its cannery to Petersburg from Tonka, on Wrangell Narrows several miles south of Petersburg. The sawmill was operated by O. P. Brown, employed 40 to 60 men, and had an annual cut of approximately 2 to 4 MMBF. In 1911, the company advertised itself as "Manufacturers of all kinds of Spruce and Cedar Lumber: Building and Mining Timber to order in large or small quantities. Boat Lumber and all kinds of Dressed Lumber and Boxes." The mill did not operated in 1913 and the following year on only an intermittent basis.[99]

In the fall of 1917, the sawmill was purchased by Olaf Arness, the owner of another Petersburg sawmill (see below), who operated his new acquisition as the Arness Lumber Company through 1919. The mill's capacity was 50 MBF per day. Arness advertised spruce and hemlock lumber, including kiln-dried finish material, and salmon boxes. Portland, Oregon, interests soon purchased the mill and began operating in 1920 as the Petersburg Lumber Company. Production for that year was 4 to 5 MMBF, with spruce and hemlock lumber and kiln-dried salmon boxes being advertised. This sawmill did not seem to have operated in 1922 or 1923, but in 1924 some 1.7 MMBF of lumber was reported to have been produced at the "Petersburg Mill Company.[100]

Despite the production, the mill experienced financial difficulties and was purchased at a marshal's sale early in 1924 by interests from Hoquiam, Washington, who operated it as the Shields-Sather Lumber & Box Company. After extensive renovations during the winter of 1924–1925, the mill began cutting lumber for local and export markets. Between 35 and 40 men were directly employed there. A fire in late May burned the underinsured mill to the ground.[101] It was not rebuilt.

■ ■ ■

Pete and Ole Knutsen began cutting lumber at their Knutsen Brothers Sawmill in about 1905. The small three-man, steam-powered operation was located on Kupreanof Island at the south side of the mouth of Petersburg Creek, and was equipped with a dry kiln. A barrel manufacturing plant was installed in 1918, but the demand for barrels slacked off not long after, and the plant was closed. In the fall of 1922, the mill was engaged in cutting street planking, and in 1925 it manufactured halibut boxes. In 1936, the mill cut 7 MBF per day; it specialized in Sitka spruce ladder stock, green clear Sitka spruce, and western redcedar and Alaska-cedar. The mill continued to operate, perhaps intermittently, until the late 1940s. Then, in the 1950s, it was resurrected by Vern Chaffee, who cut lumber for local markets until about 1959.[102]

■ ■ ■

Olaf Arness was born in Norway and settled in Petersburg in 1909. He began operating a sawmill near Three-mile Creek at Scow Bay (several miles south of Petersburg) in 1912. In 1915, the Olaf Arness Company manufactured rough and dressed lumber and barrels. By 1936, Arness's mill was known as the Lincoln Manufacturing Company, cutting construction lumber and planking for roads, and manufacturing fish boxes of various sorts and Sitka spruce oars. In 1941, he sold the mill to his son and daughter, Octor and Agnes. Octor eventually became sole owner of the mill, and entered into a temporary partnership with Lars Eide in about 1948. At some point, the mill became known as the Petersburg Lumber Company, which had a capacity of about 12 MBF per day and employed up to 20 men. In 1955, a fire destroyed the mill, which was not rebuilt.[103]

■ ■ ■

A mill known as the Hofsted [Hofstad] mill operated in or near Petersburg, but the mill—which apparently operated for only a short time—was reported in 1917 as being idle.[104]

■ ■ ■

"Old Man Ketchum's" water-powered sawmill was located on Kupreanof Island at Skoags Creek, some five miles south of Petersburg, and operated in the early 1900s. Ketchum specialized in the manufacture of boxes used to pack fresh halibut.[105]

■ ■ ■

In 1952, Lars Eide constructed the Mitkof Lumber Company sawmill at Scow Bay, about one-half mile south of the Petersburg Lumber Company's facility. The mill was equipped with a double circular saw head rig. Mitkof Lumber cut construction lumber, manufactured fish boxes, and eventually began cutting cants for the Japanese market. In the early 1970s, one of the mill's principal customers, Fred Tebb & Sons, of Tacoma, Washington, acquired an interest in the mill. In 1981, the Tebb interests with Ed Lapeyri became the mill's sole owners. The Tebb interests sold out to Lapeyri in 1987, and not long afterward Mitkof Lumber was shuttered, and some of its equipment was moved to Haines.[106]

■ ■ ■

The Northwest Mill Company sawmill was located near the Hume Cannery at Scow Bay and was in operation by November 1926. In 1936, the sawmill, managed by J. W. Weeks, had a capacity of 7 MBF per day. The mill, which was known locally as the Jim Weeks Sawmill, ceased operations about 1940.[107]

■ ■ ■

In 1923, Charles West established the West Lumber Company, with a small portable sawmill at the north end of Petersburg. West powered his mill with a Fordson tractor, and expected to produce air-dried western hemlock lumber. The mill was reported to have a capacity of 12 MBF per day.[108] It apparently operated only for a short time.

■ ■ ■

In 1905, William Langille visited a non-operational sawmill located on Wrangell Narrows on Kupreanof Island about four miles from Petersburg. He was unable to determine its ownership.[109]

Port Snettisham (mainland)

The Alaska Pulp & Paper Company had a small steam-powered mill installed at its pulp mill near the Speel River in the early 1920s.[110] The bulk of the lumber used in the construction of the facility, however, came from the Willson & Sylvester sawmill at Wrangell.

Rodman Bay (Baranof Island)

The Rodman Syndicate was incorporated in 1899 to develop gold claims at Rodman Bay, near the north end of Baranof Island. Construction began in the late summer of 1900, and continued, at least off and on, for the next several years. During construction, lumber was sawn on-site in a sawmill capable of cutting 15 MBF per day. Mining activity, and the accompanying sawmill operations, ceased in 1904, when the business failed.[111]

Shakan (Kosciusko Island)

Shakan was the location of one of Southeast Alaska's earliest sawmills. This water-powered mill was established in 1879 by Oliver Fontain and George Hamilton. Known as Hamilton's Mill, it operated at a site known as Oliver's Place. By 1887, the mill was being operated by a Mr. Sprague, who remained in business at least until 1889. In 1891, the Alaska Trading and Lumber Company operated the mill in Shakan. The Shakan Lumber & Trading Company, perhaps a new name for the 1891 firm, operated from 1893 until 1897.[112]

In 1899, C. W. Young and W. Finn purchased the sawmill and began upgrading it, including the installation of a planer and dry kiln. Production the following year was 1.1 MMBF. The mill may not have operated in 1901, but the following year was operated by the Alaska Fish & Lumber Company, with a production of 800 MBF. In addition to the sawmill, which was also equipped to manufacture shingles, the Alaska Fish & Lumber Company operated a saltery, general store and hotel. The company, at this time, was apparently far more interested in canning salmon than it was in cutting lumber. The company built a cannery, which began operating in 1902 but went out of business in late 1903. The cannery was taken over by the Shakan Salmon Company, which was comprised mostly of shareholders of the failed cannery.[113]

The sawmill's production in 1905 was small, being mostly box shooks manufactured from old lumber. In 1906, the company was reportedly bankrupt. It was acquired that year by Gorman & Company, which, under the Shakan Salmon Company name, advertised itself in 1909 as a canner of salmon, manufacturer of lumber, and purveyor of general merchandise.[114]

For the next several years, the sawmill operated intermittently to supply the needs of the cannery and to furnish lumber to Natives. In 1915, the operation was sold to Booth Fisheries, which operated as Northwestern Fisheries. The sawmill apparently closed for good soon afterwards.[115]

Sitka (Baranof Island)

In the fall of 1869, the local *Alaska Times* reported that "Mr. Stevens is rapidly progressing with his sawmill. The flume has been built from Indian River a distance of half a mile. He now has to place the water wheel close to the mill to be ready to turn out high grade yellow cedar lumber." Stevens was apparently planning to market his production in San Francisco, but found little demand. His sawmill was shuttered in 1870.[116]

■ ■ ■

The U.S. Army's Quartermaster's Department began erecting a steam-powered sawmill alongside Sitka's wharf in 1869. In April 1870, the *Alaska Times* reported the sawmill to be turning out "large quantities of excellent lumber." It is possible this was the former steam-powered floating sawmill that had been built by the Russian-American Company; a "Floating Steam Sawing Shop," considered to be public property, was reported to have been aground at Sitka in 1867.[117]

■ ■ ■

Little is known about "Mr. Rudolph's old sawmill," from which the equipment was being removed circa 1870. In the fall of 1870, a census was taken of Sitka. Of the 391 men, women, and children counted, the occupation of three was listed as "sawyer," and reference was made to the "Fuller Mill."[118]

■ ■ ■

In 1879, mining engineer George Pilz, with James Ring, leased and repaired the old Russian sawmill at Sitka's port. Pilz used the sawmill to cut lumber for his various operations, one of which was to provide lumber for the new gold mining camp on Gastineau Channel named after Joseph Juneau. He also, less successfully, tried shipping lumber to San Francisco. The sawmill was purchased by Theodore Haltern sometime in 1884. Haltern may have never actually operated it, but simply sold lumber left by Pilz and Ring. The mill's daily capacity was reported in 1889 to be 2.5 MBF on waterpower alone, and 5 MBF to 6 MBF when both water and steam power were employed.[119]

■ ■ ■

W. P. Mills purchased the old Russian sawmill in 1893, and put it to regular use in competition with John Brady, who also operated a sawmill in Sitka. Each mill annually produced about 200 MBF of lumber circa 1900.[120] As well as serving local needs, both mills shipped lumber to gold rush-era Skagway, and steamers going westward to the budding communities of Cordova, Valdez, and Seward often carried lumber as well.

In 1903, the W. P. Mills Company installed on the site a new steam-powered sawmill capable of cutting 10 MBF per day.[121] The company's annual cut was listed in 1911 as being 200 to 500 MBF or less. This sawmill was dismantled in 1914.[122]

W. P. Mills entered into a partnership in 1924, forming the Sitka Spruce Lumber & Box Company. The business operated intermittently over the course of three years. During its first year of operation, it reportedly cut 604 MBF of lumber. In May 1927, Mills repossessed the sawmill from his partners and continued to operate it for another decade. After Mills sold his sawmill (which was reported to have a capacity of 30 MBF per day) in 1937, it passed through a number of hands until the spring of 1938, when it was purchased by the Columbia Lumber Company of Alaska, which was headed by Thomas Morgan. Columbia Lumber also had a sawmill at Whittier, and later at Juneau, along with retail yards in Sitka, Juneau, Anchorage, and Fairbanks. Before long, improvements were made to the mill. More were planned, but the mill was totally destroyed by fire in April 1940. It was, however, rebuilt in an enlarged and improved form, and was back in production slightly more than a year later.[123]

The rebuilt mill, powered by steam and electricity, had a double circular head rig and had a capacity of 35 to 40 MBF per day, though with large logs the mill sometimes cut 50 MBF in a day. A box factory was operated in conjunction with the sawmill. The boxes manufactured were used by the local fish-processing plant to ship frozen fish. Beginning in early 1943 and continuing until nearly the end of the World War II, the Columbia Lumber Company sawmill ran two shifts per day. A second shift was again added in March 1947. Total employment at the mill was 70 to 80.[124]

Major improvements to the mill were completed by the summer of 1951, but financial problems forced its closure two years later. It did not reopen again until 1958, after which it operated only sporadically. In 1960, the sawmill had a listed annual capacity of 10 MMBF. The mill was purchased in 1964 by the City of Sitka and razed to make room for a boat harbor.[125]

■ ■ ■

When John Brady, a Presbyterian missionary who had worked in sawmills while attending college, arrived in Sitka in the late 1870s to minister with Sheldon Jackson, the water-powered sawmill built by the Russians was in a state of disrepair, and used only occasionally by particularly resourceful settlers.[126]

Brady got his start in Alaska lumbering in 1880, when a chief of the Tlingit people he was ministering told him that his people desired lumber to build houses. Before long, the "missionary sawmill" was on its way north. It was a small, steam-powered mill that came to Sitka by way of Hoonah, where it was temporarily set up to provide lumber for mission facilities.[127] It was with this mill that Brady began his transition from missionary to lumberman. Before long, the mission's lumber needs were being met by the small sawmill. Additionally, lumber from the mill was being marketed for local use.

In the spring of 1889, Brady erected a completely new steam-powered sawmill, which had a capacity of 6 MBF per day and was located in a cove north of Sitka. The business was incorporated in Portland, Oregon, as the Sitka Mill Company, and cut its first lumber in June.[128]

A furious gale leveled the mill in January 1891. It was quickly rebuilt, and later cut 100 MBF of rough and dressed lumber for the construction of two government buildings. Brady became the sawmill's sole owner that summer. During the Klondike Gold Rush, lumber cut at Brady's mill as well as the rival W. P. Mills operation was shipped to Skagway and then over White Pass and used to construct boats at Lake Bennett.[129]

The sawmill was completely overhauled in the fall of 1902, with its capacity increased to 10 MBF per day. The following spring Brady became one of the first loggers in Alaska to use a steam logging donkey to yard logs from the woods, which he did at DeGroff Bay on Krestoff Island.[130]

After his appointment in 1897 as governor of the District of Alaska, the mill was operated intermittently under a succession of individuals. By 1904, Brady's sawmill was virtually shut down, and in 1905 most of his sawmill machinery was shipped to the Valdez area.[131]

■ ■ ■

The Alaska governor's report for fiscal year 1889 listed a water-powered sawmill operated by the Little Mountain Mining Company at a site about five miles from Silver Bay, southeast of Sitka.[132]

■ ■ ■

Sheldon Jackson School constructed what was known as the "Beach-combed Sawmill" during the Depression years of 1934–1936. The mill's foundation was constructed of timbers handsawn from huge Douglas-fir timbers beachcombed on Kruzof Island. Machinery for the mill was largely salvaged from a sawmill abandoned near Sitka by a mining venture that had ceased operations in 1911. The mill was powered by a 25-horsepower diesel engine, obtained from an abandoned cannery at Hood Bay, on Admiralty Island. The mill was used to cut timber to meet the needs of Sheldon Jackson School as well as some custom cutting for local home and boat construction and repair.

A fire completely destroyed the Beachcombed Sawmill in January 1940, and the construction of a replacement mill was quickly begun. The new mill began cutting timber in January 1941, and operated for some 20 years before falling into disuse in the 1960s.[133]

Skagway and Dyea (mainland)

Gold was discovered in Canada's Klondike in 1896, and within the next decade perhaps 100,000 stampeders arrived in Skagway or Dyea, many of whom continued on toward the goldfields. In 1897, the towns were rapidly being transformed from hamlets to among the largest towns in Alaska. Wood was in such great demand that lumber purchased in Puget Sound for $6 per MBF was being sold for $35 per MBF at Skagway and Dyea. That demand also jump-started the establishment of sawmills in both towns, and it was said that boards coming from the towns' sawmills never touched the ground because they were loaded directly from the saws onto wagons for immediate delivery to customers.[134]

A number of small sawmills operated at Skagway and Dyea and vicinity during the gold rush. C. W. Young, of Juneau, operating as the C. W. Young Trading and Transportation Company, erected a store and sawmill at Sheep Camp along the Chilkoot Trail in 1897. In January 1898, a sawmill was constructed in less than a month's time at Dyea by the Chilkoot Railway and Transport Company, which was constructing a combination wagon road and tramway over Chilkoot Pass. The mill contracted for 600 MBF of logs, sawed timber for several months, and shut down permanently, a victim of a gold rush's ebb. Also, there are remains of a sawmill a short distance up the AB Mountain Trail. There may have been a sawmill operating around White Pass, as well as a portable sawmill that supplied the Brackett Wagon Road and the White Pass & Yukon Route Railroad. The Dyea Wood Company apparently operated about 1934.[135]

What was likely the first sawmill in Skagway, that of the Alaskan & North Western Territories Trading Company, began operating in the summer of 1897 to cut planking from local timber for a wharf the firm was constructing. The sawmill may have been known as the Virginia Lumber Company, and was leased by W. T. Iliff in 1899. In 1900, Iliff's sawmill had 15 employees and produced about 1 MMBF of lumber, including Sitka spruce molding. Some 250 MBF of spruce logs that were cut at the mill came from Rocky Pass, near Kake. Others were from near Angoon, on Admiralty Island, and from Icy Strait. Iliff died in 1901, and his bookkeeper, Stanley Bishoprick, took over operations of the mill. In 1902, the mill was capable of cutting about 25 MBF per day of ten hours, but production was only about 600 MBF. Bishoprick went into the business with the intention of marketing his production in Canada via the White Pass & Yukon Route Railroad, seemingly unaware that lumber from Southeast Alaska could not be legally exported. He described the situation in Skagway in early 1902 as "very flat," with his only market being the town and canneries in the immediate vicinity, and added that "we can cut enough in 90 days to more than supply the present demand." In November 1902, the Bishoprick Sawmill was destroyed by fire. It was not rebuilt.

■ ■ ■

In 1930, the Forest Service's Ketchikan district ranger, J. M. Wyckoff, listed an unnamed sawmill at Skagway.[136]

■ ■ ■

The Skagway Lumber Company was owned and operated by Edward Hosford. "Hosford's sawmill," as it was called, was located about three miles up the Taiya River valley from Dyea. The mill, which was equipped with a circular saw, was originally powered by a gasoline engine and later by a diesel engine. It operated from about 1949 until it was abandoned in 1956. While in operation it principally supplied the White Pass and Yukon Route Railroad with wharf piling, bridge timbers, and ties. Lumber, including shiplap, was also marketed in Skagway and Whitehorse.[137] Employing a tractor and arch, the company did its own logging, mostly for Sitka spruce in the lower Taiya River valley.[138]

Suloia Bay (Chichagof Island)

The George E. James Lumber & Box Company was a short-lived joint venture of Sitka lumberman W. P. Mills and George James, an individual who for many years had operated a sawmill at Douglas. In early 1918, Mills and James constructed a steam-powered sawmill at Suloia Bay, about 25 miles north of Sitka. The mill began sawing logs in early 1918, but ceased operations completely before the end of the year. Not long afterward, the operation was sold off piecemeal at a U.S. Marshal's sale.[139]

Warm Springs Bay, a.k.a. Baranof (Baranof Island)

A water-powered sawmill capable of cutting about 3 MBF per day was constructed in 1911. Lumber produced at the mill was used to construct the nearby village. In 1920, the Forest Service ranger at Petersburg mentioned a small sawmill at Baranof. The sawmill was operated in 1926, and a Forest Service photo dated September 1927 shows what appears to be an operating sawmill. The Warm Springs Bay sawmill was reported closed in 1930.[140]

Windham Bay (mainland)

Just after World War II, Stan Price, an established logger, set up a small sawmill at Windham Bay. Price produced rough-cut lumber, which was delivered to local canneries and salteries via the mailboat that served Windham Bay. The lumber was used to repair docks and boardwalks. Probably in the early 1950s, Stan Price's sawmill and equipment were seized by the Internal Revenue Service, ending his career as a lumberman.[141]

Wrangell (Wrangell Island and mainland)

In 1888, Thomas Willson and Rufus Sylvester, who had come to Alaska in 1880 and 1884, respectively, formed a partnership to build and operate the steam-powered Fort Wrangel Mills, which was to be financed by Sylvester and operated by Willson. That winter Willson went east to purchase machinery for the mill. The original intent of the enterprise, which eventually became the Willson & Sylvester Mill Company, Inc., was to manufacture salmon boxes. For a number of years the Willson & Sylvester sawmill, with a capacity of 35 MBF per day, was the largest sawmill in Alaska. In 1898, the mill advertised Alaska-cedar, western redcedar, and Sitka spruce lumber for sale, and added that material for flooring, ceiling, and "rustic" was available. That spring, the sawmill was employing almost 60 men and was barely able to keep up with orders. Lumber was fetching from $16 to $30 per MBF. About three years later, the mill employed some 30 to 50 men for eight or nine months of the year, with almost as many more employed cutting logs. Production in 1900 was 3.2 MMBF; in 1901, 4.2 MMBF; and in 1902, 3.9 MMBF.[142]

Thomas Willson died in 1900, and Rufus Sylvester three years later. Legal complications after their deaths caused the mill to be operated under the jurisdiction of the District Court of Alaska. In 1912, Harry Gartley organized a new company that retained the Willson & Sylvester name. Five years later, Gartley's sawmill was cutting about 6 MMBF of timber annually. The mainstay product was salmon cases, but the mill also cut considerable lumber for local construction as well as aero-grade Sitka spruce, with a shipment of clear boards 36 to 48 inches wide being exported in 1914.[143]

In March 1918, the Willson & Sylvester sawmill was destroyed by fire. Over the course of the following year, it was rebuilt to a greater capacity (40 to 50 MBF of lumber and 6,000 salmon cases per day) and higher efficiency. Some of the clear Sitka spruce lumber in stock in 1920 was reported to be 28 inches wide, 4 inches thick, 24 feet in length. In 1924, the mill cut a little over 8 MMBF of lumber, and the plant's capacity was increased to 65 MBF per day.[144]

In 1924, Harry Gartley obtained a 20-year franchise to supply city of Wrangell with electricity, and incorporated additional boilers and a 375-kilowatt generating plant with the sawmill. The following year, Gartley began selling electricity and changed the name of the business to the Wrangell Lumber & Power Company. His market for wood products included lumber for local construction, salmon cases, and clear lumber for shipment to Seattle and other points.[145]

During 1925, the mill cut more than 10 MMBF of lumber and manufactured nearly 700,000 salmon cases and 14,000 fresh fish boxes. Exclusive of its logging operations, the plant had approximately 100 employees. Except for key jobs, most of the mill hands were Natives.[146]

Harry Gartley died in 1926, the same year the *West Coast Lumberman* published an article about his sawmill titled "This Modern Alaska Mill Ranks With Best Coast Plants." Soon afterward, the business was reorganized as the Wrangell Lumber & Box Company, and in 1929 was leased by Nick Nussbaumer, who in addition to the company's normal business, had secured an order from the Alaska Railroad for 40,000 western hemlock ties.[147]

The replacement of wooden boxes by fiber cases in the canned salmon trade was severely felt at the Wrangell Lumber & Box Company, and in 1936 Nussbaumer changed the name of the business to Alaska Spruce & Cedar Products Company. Production that Depression year was only 3 MMBF.[148]

The sawmill was later operated as the Wrangell Lumber Mills by a Mr. Custard and a Mr. Meadows, who incorporated a shingle mill into the operation in 1941. In less than two years, however, the mill was shut down and in a state of disrepair. In 1943, H. M. Olsen, of Seattle, took over operation of the mill and made plans to quickly resume operations, but little, if any, milling occurred. In September 1945, Carl Edlund became the mill's principal owner, but his tenure, like Olsen's resulted in little, if any, lumber production.[149]

In the late spring of 1946, Frank V. Wagner and his son, Rodney, purchased Edlund's interest and announced that their sawmill would be known as the Alaska-Asiatic Lumber Mills. The Wagners then began an ambitious renovation and expansion project. The somewhat renovated mill resumed operations in 1946, but was closed down in April 1947 because of financial difficulties. In August 1947, Frank Wagner was declared insolvent, and the Alaska-Asiatic Lumber Mills was placed in the hands of a receiver.[150]

In February 1949, C. T. Takahashi, of Seattle, became the new owner of Alaska-Asiatic Lumber Mills. Takahashi changed the name of the company to Alaska Wrangell Mills. In 1952, Alaska Wrangell Mills made Wrangell's first full-cargo direct shipment of lumber to Japan. Despite this seemingly auspicious shipment, the mill operated little until 1954, when the Wrangell Lumber Company, a wholly-owned subsidiary of the Japanese-owned Alaska Lumber & Pulp Company, leased it for a period of five years. The sawmill, at this time, was considered to have a capacity of 35 MBF per day.[151]

Alaska Lumber & Pulp intended to use the Wrangell sawmill to cut lumber for export to Japan. The first shipload, which tallied 3.5 MMBF, was shipped in 1955. It was not well received. For the next three years, the mill mostly produced cants and flitches for resawing in U.S. millwork plants that produced musical instrument sounding boards, finish lumber for window and door stock, and molding. During this period, the mill was thoroughly modernized and its capacity increased to 130 MBF per day. The mill then switched to cutting cants for export to Japan. In 1958, a dispute between Alaska Lumber & Pulp and the Forest Service over the allowable maximum thickness of cants for the export trade resulted in the Japanese temporarily shutting down the mill. Considerable local pressure was brought on the Forest Service officials, and a compromise was reached that allowed the mill to resume production.[152]

The Wrangell Lumber Company produced some 50 MMBF of cants per year, all of which was shipped to Japan. Ever proud of the community's lumber industry, the *Wrangell Sentinel*'s masthead for a number of years proclaimed Wrangell to be "The Lumber Capital of Alaska."[153]

The historic but run-down Wrangell sawmill was shuttered in 1981, when production was switched to the nearby sawmill that had been purchased (as noted below) from the Pacific Northern Timber Company.

■ ■ ■

J. T. Milner, operating as the Stickeen Trading & Transportation Company, built the Stikeen Lumber Mills sawmill on the mainland northeast of Wrangell in 1898, along what was known as Water Falls Creek. The water-powered mill had a capacity of 20 MBF per day, and its owners planned to cut principally Sitka spruce, Alaska-cedar and western redcedar for the local and Stikine River trade. Some 20 men were employed between the mill and the logging camp. This sawmill seems to have been purchased by the Thlinket Packing Company, a Portland, Oregon, firm that constructed a salmon cannery on the site in 1899.[154]

■ ■ ■

During World War II, the Alaska Metals & Power Company was interested in mining molybdenum—for which there was a great demand at the time—in the Virginia Lake area near Wrangell. The company erected a small circular sawmill at Mill Creek (Eastern Passage) in 1944. Power for the mill was initially furnished by a steam donkey, but once enough lumber was cut, a 700-foot flume was constructed, and the mill was converted to hydropower.[155]

A substantial renovation of the mill's machinery was completed early in 1949, and in 1951 the company shipped 100 MBF of clear Sitka spruce to the Boeing aircraft plants in Seattle and Wichita and to Steinway & Sons, the famous New York City piano manufacturer.[156]

Sometime before or during 1953, the mill was leased to Roy Smith, who operated it under the name of Mill Creek Spruce Company. Smith cut a considerable amount of concrete form lumber in 1953 for the construction of the Ketchikan Pulp Company pulp mill. In January 1954, a fire destroyed the sawmill.[157] It was not rebuilt.

■ ■ ■

The Pacific Northern Timber Company (PNTC) was incorporated in 1953 under the leadership of Dean Johnson, of the C. D. Johnson Lumber Company of Toledo, Oregon. The company was later reorganized by C. Girard Davidson, a former Assistant Secretary of the Interior in the Truman administration, and Daniel L. Goldy, former Pacific Northwest regional administrator of the Bureau of Land Management.[158]

At the request of PNTC, the Forest Service in the spring of 1954 advertised a 50-year contract for three BBF of timber located on about 335,000 acres near Wrangell.[A] PNTC, which was the sole bidder, was awarded the contract that June. PNTC had grand plans, envisioning the establishment of a world-class integrated wood processing facility at Wrangell that included a $3.5 million sawmill. The terms of the original contract required the establishment of a sawmill by December 31, 1957, and the construction of a pulp mill of at least 80 tons daily capacity by December 31, 1962. The contract was later modified to include the construction of a veneer mill.[159] The company's plans, however, were not fully realized.

In the fall of 1956, citing financial difficulties and a lighter than anticipated demand for lumber, PNTC requested and received a 1-year extension of the period the Forest Service required for the construction of a sawmill. The company acquired its sawmill through George Reller, owner of the shuttered South Fork Lumber Company in Sweet Home, Oregon. Reller became a principal in PNTC, and in 1958 the equipment from his shuttered mill was shipped to Wrangell,

[A]The timber was located on Zarembo, Etolin, Wrangell, Woronkofski, and Mitkof Islands, and on the mainland in the vicinity of Bradfield Canal.

where a new sawmill was constructed on 40 acres of waterfront land at Shoemaker Cove (Zimovia Strait) that was purchased from the Forest Service, a transaction that required congressional approval. Plans for a pulp mill and a veneer mill were eventually dropped, and the company's Forest Service timber contract sale volume was reduced to 693 MMBF over 15 years.[160]

The PNTC sawmill, with an annual capacity of 40 MMBF (based on two shifts per day), was the first in Alaska specifically designed to cut western hemlock. The mill, operated by George Reller, went into operation in 1960, but quickly encountered financial problems. It was 1965 before any significant volume of lumber was produced. In 1968, PNTC and its Forest Service contract were taken over by the Alaska Wood Products Company (see above), an affiliate of the Alaska Lumber & Pulp Company, which was operating the old sawmill in downtown Wrangell. The new owner eventually cut western hemlock for a few years, then shut down the mill and rebuilt it to a greater capacity. In 1981, the Alaska Wood Products Company shuttered its downtown Wrangell sawmill to concentrate its sawmilling operations at the facility acquired from PNTC, which it operated until December 1994. The sawmill remained shuttered until April 1998, when it was reopened by Silver Bay Logging, an established timber operator in Southeast Alaska. The mill cut at the rate of about 40 MBF per day—about 10 MMBF per year—but was shut down in early 2003, after Silver Bay Logging filed for bankruptcy. The company successfully emerged from bankruptcy a year later and soon restarted its sawmill. The mill, however, was permanently shuttered in 2006.[161]

■ ■ ■

Early in 1949, the Wells Sawmill advertised building material and boat lumber, red and yellow cedar, spruce and hemlock. The mill also offered to cut logs to order. J. Hurn was given as a contact.[162]

■ ■ ■

The manufacture of western redcedar shingles was important to Wrangell, and the community earned and was proud of its reputation for high-quality shingles. Shingles were sold throughout Alaska as well as along the Stikine River.[163]

The steam-powered Wrangell Shingle Company mill was built on Wrangell's waterfront by J. W. Gano and several associates, and it turned out its first shingles on June 20, 1907. The mill had a capacity of 40,000 shingles per day. In 1911, after the death of Mr. Gano, the mill was operated by F. E. Smith. The shingle mill was put up for auction by the Gano Estate in September 1915, and was purchased by Marion McKinney, an employee. McKinney almost immediately made improvements to the mill. Production in 1920 was 2.2 million shingles. In 1921, the McKinney Shingle Company reportedly had a daily capacity of 25,000 shingles.[164] Fritz Angerman purchased the mill from McKinney in about 1931 and operated as the Wrangell Cedar Products Company. He sold the business in about 1939 to Charles Eslick, Neil French, and William Featherstone. In 1946, the owners new began construction on the same site of an all-electric shingle mill capable of producing 120 bundles of shingles per day. The production of shingles began in the fall of 1946. Wrangell Cedar Products marketed its shingles in Alaska. "Our prices are right. We sell locally every week day at our mill," read an advertisement in 1947. Not long after, however, the company ceased operations.[165]

Yakutat (mainland)

In 1903, the Stimson Lumber Company installed a sawmill at Yakutat that cut 20 to 35 MBF per day. The sawmill was part of a larger venture of the Yakutat & Southern Railroad Company (of which F. S. Stimson was president) that involved the construction of a salmon cannery as well as a railroad running east along the coast to stands of Sitka spruce timber and salmon-rich streams. Unique in the industry, salmon were to be transported by train to the cannery. Thirteen MMBF of material was cut in 1903—ties for the railroad and lumber for the construction of the salmon cannery, associated buildings, and a wharf. The sawmill was equipped with a 5 MBF-per-day planing mill and equipment to manufacture salmon cases, which were to be used by the owners as well as sold to other canneries. By 1904, the company's railroad reached nine miles to the Situk River, and the lumber cut at the sawmill was reported to be "a high quality of spruce." In 1908, Forest Supervisor William Langille reported that the sawmill did not operate that year or the previous year.[166]

■ ■ ■

In 1947, sources indicated the existence of the apparently short-lived Antone Novatney sawmill at Yakutat.[167]

Technology in Southeast Alaska's Forest

"THERE IS NO CHAPTER IN THE EPIC OF CIVILIZATION MORE ELOQUENT OF THE HARDIHOOD, COURAGE, AND INVENTIVENESS OF MAN IN SERVING HIS FELLOWS THAN THE HISTORY OF THE LUMBERMAN'S CONTEST WITH THE ELEMENTAL FORCES OF NATURE."

— E. T. Allen in *American Forestry*, 1916[1]

Essentially, there are two components to logging in Southeast Alaska: felling trees and getting logs into saltwater. Technological improvements made cutting faster and cheaper, but logging costs overall increased as timber handy to saltwater was exhausted.

CUTTING

At the time Russia owned Alaska, the technology for felling trees was the single-bitted iron axe. Such axes, "of good quality, well tempered, with strong butts and thick sides," were manufactured at Sitka.[2]

To deal with the large trees in the Pacific Northwest, a double-bitted axe with a long, finely-tapered head was introduced in the late 1870s. One bit was kept almost razor sharp for the task of cutting through the huge amounts of wood necessary to fell a large tree, while the other bit was filed more bluntly, and used for cruder work in order to preserve the falling bit.[3]

Loggers began to use one-man crosscut saws for felling timber in about 1880, with a longer, two-man crosscut saw—referred to in logger jargon as a "misery whip" or "Swedish fiddle"—coming into use a short while later. The saws were greatly improved about 1900 with the invention of raker teeth that cleared sawdust from the cut. "Come to me, go to you," was a chant sung by loggers while cutting through a log. In good conditions, two skilled men using a crosscut saw could fell a Sitka spruce tree five feet in diameter in about half an hour. Saws used for this type of timber were generally about 8 feet long.[4]

Crosscut saws were largely replaced by gasoline-powered chainsaws in the 1940s. Though early chainsaws were very heavy and cumbersome, it was said that a logger with one of the machines could "cut timber all day and dance all night."[5]

Timber fallers in Southeast Alaska usually used a chainsaw fitted with a 36-inch bar. When cutting particularly big timber, they used a larger saw fitted with a 72-inch bar. The combination could weigh more than 50 pounds, but with it a logger working efficiently under favorable conditions could fell a Sitka spruce tree six feet in diameter in a about a dozen minutes. Fallers generally owned two or three saws, which, depending on their size, cost $800–1,500 each. Properly maintaining the saws required a considerable amount of time. In the mid-1970s, a "bushler," a faller who was paid not by the hour but by how much timber he cut, was typically paid about $3.25 per MBF for timber felled, limbed, and "bucked" into logs, plus an additional $0.25 per MBF for the cost and maintenance of his saws. When in good timber, experienced fallers—the elite of a logging camp—earned about $300 per day.[6]

HANDLOGGING

"Nearly all of the timber, so far, which has been taken in southeastern Alaska has been put in the water by hand loggers, and has hardly ever been over a distance of 200 or 300 feet from salt water," wrote Alaska's Governor John Brady in 1899.[7] As well as a livelihood, handlogging was a lifestyle, and of those who logged Southeast Alaska's forests, the handloggers were the most romanticized.

Usually working in crews of two or three, these independent, resourceful, and hardworking men cut timber along Southeast Alaska's shores, usually under contract to a mill, cannery, or mining company. In the early years, they traveled to their logging sites by sloop; then, in the later years by "gas-boat."[8] Their tools were hand-powered and basic: axe, saw, steel-tipped springboards, wedges, sledge hammer, logging jacks, ropes, chains, bottle of kerosene (sometimes water) to clean pitch from and lubricate the saw, and the ubiquitous cans of Copenhagen chewing tobacco. Handloggers mostly cut timber located on slopes directly above the water, and because of the amount of physical effort involved, only cut the most valuable trees. Ideally, a felled tree would land in the water, where it could be limbed and boomed to await the arrival of a tugboat.

Many of the handloggers were Natives, who, according to a government official in 1903, "bring logs to the mills at such low rates that white men can hardly compete in this kind of work."[9] For some Natives, handlogging was

FIGURE 66. Annoying insects only made a hard job harder. (USDA Forest Service, Alaska Regional Office, Juneau)

an opportunity to earn some money during the fishing industry's off-season, which was somewhat curtailed with the advent of regulated cutting by the Forest Service during the years after 1910.

Regarding the basic economics of handlogging, the *Wrangell Sentinel* reported in 1911 that a handlogger could earn "from $3.00 per day up, many making $5.00 which is considered good pay." A Forest Service handlogging permit covered selective logging on five miles of shoreline. The permit could be transferred to a new location as often as was needed.[10]

Despite the increasing scarcity of workable timber and the increasing use of steam-powered logging equipment, as late as 1940 booms containing 100 MBF of timber cut by handloggers on the west coast of Prince of Wales Island were yielding as much as 65 percent clear lumber. The *Wrangell Sentinel* reported that for a few years just after World War II there were approximately a dozen handloggers operating in the Wrangell area.[11]

Logging is dangerous business wherever it takes place, but it was particularly so for handloggers working alone in remote locations in Southeast Alaska. Despite the danger, some handloggers preferred to work by themselves, finding experienced partners hard to come by and enjoying the satisfaction of working alone. For them, the search for a "stumper"—a tree that when felled would land in the water—was like a search for gold.[12]

The time it took to complete a raft of timber by handlogging was considerable, and the logger was intimately familiar with every stick in it. The depredations of teredos were a problem with Sitka spruce logs, while western hemlock logs presented a problem of their own: heavy to begin with, they occasionally became waterlogged and sank.[13]

Despite the respect accorded handloggers for their hard work and resourcefulness in a dangerous occupation, there was a dark side to hand-logging: the wastage of timber. Estimates were given that in Southeast Alaska handloggers as a class invariably wasted as much timber as they got into a raft. This involved timber that was cut but did not make it to the water, timber cut for skids or to clear the path, and blowdown or slides that resulted from handlogging operations.[14]

The government lamented the waste, but was initially sympathetic to the plight of the handloggers, with William Langille writing in 1905: "To prohibit this cutting, or to enforce payment for all the material cut for this purpose, is to prevent the hand logger from operating, and it would work a hardship and injustice on a class of independent laborers, who usually get the worst of it in any event."[15]

But by the late 1920s, as the Forest Service's vision of a massive pulp and paper industry in Southeast Alaska gained momentum, the agency became less sympathetic to the needs of handloggers. Chief Forester Greeley wrote that "it is not advisable to permit the 'front' of timber to be skimmed off by small logging jobs which lack the equipment necessary to operate the tract as a whole in an economical and workmanlike manner."[16]

OXEN AND HORSE LOGGING

Early loggers in Southeast Alaska used two forces to move logs into the water: gravity and manpower. In the "lower 48," both oxen and horses had been commonly used to skid logs, but they were used for only a relatively few years and in a few places in Southeast Alaska. Probably the biggest reason for not employing oxen and horses to skid timber in Southeast Alaska was simply that the region was not horse and cattle country, and it must have taken a considerable effort to keep the animals fed and sheltered and to transport them from place to place. By using oxen and horses, however, early

FIGURE 67. A single logger with a chainsaw can fell a spruce such as this in less than a quarter hour. (USDA Forest Service, Alaska Regional Office, Juneau)

loggers were able to log inland from the shore without incurring the major expense of a donkey engine.

The first apparent use of oxen to skid logs was on Douglas Island about 1895 by Captain Charles Sally. By about 1897, two logging companies, Bennett & Goodwin and the Barnes Brothers, started logging with teams of oxen near Wrangell. Two years later, oxen were also skidding logs cut on Kupreanof Island. About that time, however, Southeast Alaska's brief oxen-logging history seems to have come to an end.[17]

Horse logging, like logging with oxen, was used only in a few situations in Southeast Alaska, but continued on and off for a couple of decades. One of the first loggers to practice horse logging in the region may have been J. R. Reynolds, later part owner of Sawyer & Reynolds Logging Company. In 1910, Reynolds attempted to use an old line horse to skid timber, but the horse went wild after wolves chased it into the water several times. More successfully, William Paddock, of Tenakee, logged piling with a team of horses and a logging crew of nine men, probably in 1913 and probably on Admiralty and Chichagof islands. Paddock also logged piling about that time at Falls

FIGURE 68. Logging with oxen near Wrangell. (Alaska State Library, Early Prints of Alaska Photograph Collection)

Creek, near Gustavus. In the late 1920s, there were horse-logging operations on the Cleveland Peninsula, near Ketchikan, and at Salt Lake Bay, near Hoonah.[18] By 1930, however, horse logging seems to have ended in Southeast Alaska.

DEEP-WATER LOGGING ("BOAT LOGGING")

Where there was deep water close to shore, loggers sometimes used boats to yard logs into the water. From a heavy boat a cable was run ashore and fastened to a choked log. The boat would then make a short run to build momentum and take the slack out of the cable. If things worked right, once the cable came tight the log would be jerked from its resting place and come crashing out of the woods and into the water. Such operations were able to reach up to 400 feet on hillsides.[19]

Going into the winter of 1953–54, Southeast Alaska salmon fishermen were severely distressed after two consecutive disastrous fishing seasons.[A]

[A]President Eisenhower declared Southeast Alaska a disaster area in 1953 and 1954.

The Ketchikan Pulp Company was in the market for logs, and the Forest Service helped the fishermen by laying out timber sales particularly suited to boat logging. The Ketchikan Pulp Company agreed to locate a number of boat logging operations on its allotment, where it covered fishermen's stumpage costs. As of early November 1953, some 15 boats had obtained markets for logs and were either at work or getting their gear ready. Another 25 were out looking for suitable shoreline.[20]

STEAM DONKEY LOGGING

The steam logging "donkey" mechanized the chore of transporting logs out of the woods. Although the first steam logging donkey was patented in Michigan, the device was said to have been invented in California in 1881, where it was used to move massive redwood logs. By 1908, there were nearly 4,000 steam logging donkeys in use in Pacific coast forests.[21]

The donkey was a relatively large piece of machinery, consisting of a wood or oil-fueled boiler that powered a reciprocating steam engine, which in turn drove two drums (winches). In Southeast Alaska western hemlock was preferred for wood-fueled donkeys because it contained more BTUs and was less valuable than Sitka spruce. The main drum, for the actual pulling of logs, held heavy wire rope, usually 1 inch to $1\frac{1}{4}$ inches in diameter. The other drum held a lighter "haulback" cable, which was used to pull the main cable back to the timber. The donkey was usually mounted on a heavy log structure called a "sled," and could pull itself around by using its winch. With proper maintenance, steam donkeys were very dependable and long-lived. Because they were relatively unsophisticated machines, they lent themselves well to on-site repairs by loggers. By about 1930, however, gasoline and diesel engines had largely replaced the steam engines.

The term "skidding" refers to bringing logs in by dragging them over the ground, at times with one end suspended.[22] The early use of the donkey was for "ground yarding" or "ground skidding"—basically dragging logs over the ground to the machine. Hangups on stumps and other obstacles were an inevitable problem, which was sometimes rectified by the construction of a corduroy "skid road." Donkeys were often referred to as "yarders," and as "skidders" or "roaders."

Steam logging donkeys came into use in Southeast Alaska as good handlogging sites became scarce. John Brady, Alaska's fifth governor, was one of the first in Southeast Alaska to employ a donkey, which he used in

FIGURE 69. Large steam-logging donkey. (USDA Forest Service, Alaska Regional Office, Juneau)

1903 to log a stand of timber on Krestof Island, near Sitka. Another early steam donkey operation was circa 1900 at Farragut Bay (Frederick Sound). Sawyer & McKay, among the best of the region's pioneer loggers, began using a steam donkey at Ernest Sound in 1908. In the early 1900s, the size of a Southeast Alaska logging operation was commonly gauged by the number of donkeys employed. In 1921, there were, by one count, 20 steam logging donkeys employed in the region. The *Ketchikan Alaska Chronicle* in 1924 hailed the purchases by two local logging operators of the latest type of steam logging donkey as indicating "a spirit of progressiveness and of general confidence in the future of the industry."[23]

Ground yarding gave way "high-lead" logging, in which the donkey was used in conjunction with a spar tree. This system was first employed in Southeast Alaska in about 1916, when it was reportedly introduced by the McDonald Logging Company. There were about as many high-lead logging

FIGURE 70. "High climber" descending spar tree after limbing and topping.
(USDA Forest Service, Alaska Regional Office, Juneau)

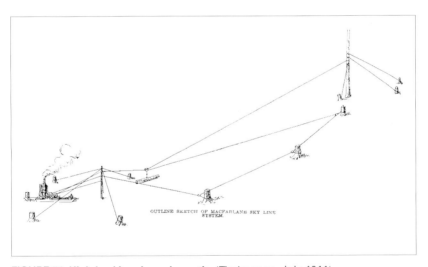

OUTLINE SKETCH OF MACFARLANE SKY LINE
SYSTEM.

FIGURE 71. High-lead logging schematic. (Timberman, July 1911)

configurations as there were operators and situations. By 1923, there had been one sky-line operation (logs suspended above the ground) in Southeast Alaska, but practically all steam donkey logging in the region remained ground-line work.[24]

FIGURE 72. A-frame logging, the equivalent of a floating spar tree. (USDA Forest Service, Alaska Regional Office, Juneau)

In the late 1950s, portable yarding towers —vehicle-mounted steel towers, some self-mobile and equipped with yarding machinery and able to be set up in a matter of hours— began to replace spar trees. The last use of a traditional spar tree in Southeast Alaska was about 1980.

A variation of high-lead logging that became widely used in Southeast Alaska involved a float-mounted A-frame/donkey engine combination. In this system, the A-frame replaced the spar tree, and logs were efficiently transported directly from the logging site into salt water. In the parlance of loggers, such a unit was a "slackline-skidder," and logging timber directly from the woods into saltwater was known as "hot logging."[25] Because they did not lend themselves well to working in rough water, the use of A-frame units was confined to relatively protected areas. Small A-frame units were sometimes powered by gasoline engines, but both gasoline and steam engines were eventually replaced by diesel engines.[26]

In 1938, with much of the high-quality sawtimber along Southeast Alaska's beaches suitable for A-frame operations having been logged off, truck logging was introduced to work previously inaccessible timber. Spar trees and sled-mounted skidders were used in conjunction with trucks. Logs were yarded to cold decks beneath spar trees, then transported by truck to saltwater.

With the establishment of the pulp mills in the 1950s, however, logging in Southeast Alaska focused for the first time on the massive cutting of pulp timber. Much of this comparatively low-grade material was located along the shorelines, and had been passed over by earlier operations. A-frame logging again came into practice, and some of the equipment used dwarfed that previously employed.

In 1954, the Washington Iron Works in Seattle custom built for the Ketchikan Pulp Company the largest skidder used to date in Alaska. The 550-horsepower diesel-powered machine weighed 132,000 pounds and was mounted on a log float that was 180 feet long and 130 feet wide. The construction of the float required some 750,000 board feet of logs, and the A-frame itself was 167 feet high. More than 33 tons of wire rope—7 miles worth—were used in its operation. The unit was capable of reaching 2,500 feet up the slopes, though the preferred operating distance was 1,400 to 1,800 feet. Production for this yarder was about 1.5 MMBF per month when yarding timber from the woods, more when yarding from cold decks. In an attempt to prevent teredos from ruining the float, arsenic-filled sacks were hung in the water at various locations beneath it. (Others operators tried to kill teredos by periodically detonating sticks of dynamite around their floats.)[27]

Eclipsing KPC's machine, in 1960 the Alaska Lumber & Pulp Company put what was reputed to be world's largest diesel-powered skidder, a Washington Iron Works "Flyer," in float-mounted operation at Katlian Bay, near Sitka. At 135,000 pounds, this skidder was slightly larger than the one used by the Ketchikan Pulp Company but had a significantly longer maximum reach—about 3,200 feet. The float on which the machine was mounted was innovative for the time. While the framework and deck were constructed traditionally of massive Sitka spruce logs, the floatation beneath was a new product—Styrofoam. The float's overall dimensions, including an area for "landing" logs, were 90 feet wide by 220 feet long. The A-frame sticks were 155 feet long.[28]

By 1970, however, after a long history in Southeast Alaska, A-frame operations accounted for only five percent of the timber yarded.[29]

RAILROAD LOGGING

In the Pacific Northwest, once the timber along the waterways was cut, railroads were often used to transport timber to mills. There were several proposals to build logging railroads in Southeast Alaska, though none came

FIGURE 73. Ketchikan Pulp Company's A-frame. (USDA Forest Service, Alaska Regional Office, Juneau)

to fruition. One proposal, circa 1928, considered the construction of a railroad to access and haul timber on Admiralty Island, and estimated the need for at least 200 miles of track to log just the timber on the west side of the island.[30]

The Yakutat & Southern Railroad, established in 1903, was the country's only railroad built specifically to haul fresh fish. Its first use, however, was to haul logs to the Stimson Lumber Company's sawmill at Yakutat. F. S. Stimson was an owner of the Yakutat & Southern Railroad, and lumber produced at his mill was used to construct the salmon cannery and other buildings owned by the railroad. The railroad ran east along the coast from Yakutat to the Situk River, a distance of nine miles.[31]

TRACTOR LOGGING

Tractor, or "caterpillar" or "Cat," logging was introduced in the U.S. after World War I, when modified surplus tanks were used to replace teams of horses or even railroads in transporting logs from the woods. Early tractors worked best on a 15 to 30 percent grade, which was considered too steep for railroads and too flat for horses, which usually pulled their load downhill.[32] The tractors were fitted with drum winches that could yard logs up to about 500 feet away.

The first tractor logging in Southeast Alaska was done in 1927 as an experiment. The experiment proved successful, and by 1930 there were six tractors operating in the region. The output per tractor was about 25 MBF for an 8-hour day.[33]

Developed in the 1930s and pulled by a tractor, the "tracked logging arch" was an un-powered tracked vehicle fitted with an arch-shaped frame and some stout pulleys. The equipment was used in both yarding and transporting logs.[34] When transporting logs, the front end of the logs were suspended just beneath the frame, which reduced the resistance of the load and the possibility of logs hanging up.

Tractor logging was employed in a few 1940s-era operations, specifically at Hoonah Sound, on Chichagof Island, and at Tuxekan, on Prince of Wales Island. When it began its logging operations in 1953, the Ketchikan Pulp Company "started out doing some tractor logging, but we dropped that in a hurry." Frank Heintzleman wrote in 1953, "The forest land [in the Tongass National Forest] is either too rough or too soft for tractor operation."[35]

TRUCK LOGGING

The introduction of truck logging was praised as a revolution in the western forests. Logging with trucks, which were initially used in combination with tractors, required a comparatively small initial investment, and allowed operators of limited means to enter the industry.

The first large-scale use of trucks for logging in the Pacific Northwest was by the U.S. Army's Spruce Production Division during World War I. The

FIGURE 74. Logging with
tractor and arch. (USDA
Forest Service, Alaska
Regional Office, Juneau)

success of truck logging afterward spread
throughout the Pacific Northwest, and by
1938 it was being seriously considered in
Southeast Alaska. The Forest Service's R. R.
Robinson wrote in 1938:

> For thirty years, the numerous logging companies that have
> worked in this section of Alaska have abided by the
> conventional forms of power logging which has entailed the
> use of steam donkeys on A-frame floats, and the various
> woods equipment needed in high-leading, ground-leading,
> slack-lining or combinations of these systems. However, the
> depletion of the desirable and easily accessible stands of
> timber (under present economic conditions) along the
> shores has caused the larger of the logging companies to
> give serious consideration to the advisability of introducing
> truck logging to this country.[36]

To that end, the McDonald Logging Company, an affiliate of the Ketchikan
Spruce Mills, probably in 1937 purchased some 19 MMBF of Forest Service
timber at Coning Inlet on Long Island with the express purpose of introducing
truck logging. The company also purchased timber on Prince of Wales Island,

and this timber, logged in 1938, was the first to be truck logged in Southeast Alaska. On the Coning Inlet sale, which was logged in 1941, McDonald Logging used two gasoline-powered trucks that operated over a plank road about 2½ miles in length, with grades up to 20 percent. The road was constructed of 6-inch thick planks laid on stringers, which required about 400 MBF of material per mile. The cost of constructing the road was about $10,000 per mile.[37]

During World War II, Edna Bay, on Kosciusko Island, became the main staging area for the Alaska Spruce Log Program (ASLP). A considerable amount of timber was logged by the ASLP on the island, and some was hauled by trucks over bulldozer-built roads. The ASLP was terminated in 1944, and virtually no truck logging was done again in Southeast Alaska until the Ketchikan Pulp Company began pulp timber logging operations at Hollis on Prince of Wales Island in 1953.[38] As it evolved in Southeast Alaska, truck logging was used in conjunction with high-lead logging. Logs were transported by the high-lead system to cold decks along the roads, and then loaded on trucks for transport to saltwater.

In 1958, eight miles of heavy-duty logging roads were constructed on the Tongass National Forest. The following year 14 miles were constructed, and that number increased to 33 in 1960. The total mileage of the system

FIGURE 76. Plank road,
logging truck, steam donkey.
(USDA Forest Service, Alaska
Regional Office, Juneau)

envisioned for Southeast Alaska was estimated in 1958 to be approximately 8,000.[39] To
encourage and facilitate logging, the Forest
Service, using its own funds, contracted for
the construction of roads into timbered areas.

A major boost to truck logging was the National Forest Roads and Trails
Act of 1964.[40] This legislation largely transferred the costs of building logging
roads to the government. By the year 2000, there were more than 5,000 miles
of forest development roads on the Tongass National Forest.

SURPLUS MILITARY EQUIPMENT

A large amount of surplus military equipment was sold by the government after World War II. Ever-resourceful and willing to improvise, Alaska
loggers were quick to put some of this equipment to use.

Logging donkeys had since the 1880s been slowly pulling themselves
around on wooden sleds. With the advent of truck logging, operators began
transporting their donkeys by truck, which were often used in tandem for the
heavy machines. It was a cumbersome and time-consuming process. Surplus
army tanks offered a solution. Fitted with a yarder, the resulting self-mobile
unit was attractive because it could negotiate fairly severe terrain yet travel

FIGURE 77. Loading logging truck with "hay rack." (USDA Forest Service, Alaska Regional Office, Juneau)

at 25 miles per hour on a level road. The disadvantages of using surplus tanks included a lack of replacement parts and "astronomical" fuel consumption.

In Southeast Alaska, the first mounting of a yarder on a military tracked vehicle was completed by Jim Campbell, a Ketchikan logging contractor who mounted a diesel-powered yarder on a surplus tracked gun carrier. His work preceded a larger project by the Ketchikan Pulp Company, which by 1961 owned three surplus tanks. Ketchikan Pulp's first mounting of a yarder on a tank chassis was deemed "an un-qualified success," but it seems that the overall effort with surplus tanks was something of an experiment that failed to endure.[41]

Logging, Lumbering, and Pulp-manufacturing Terminology

(as generally used, 1867–1960)

board foot. A unit of wood measuring 12" x 12" x 1".
- A thousand board feet is abbreviated *MBF*.
- A million board feet is abbreviated *MMBF*.
- A billion board feet is abbreviated *BBF*.[A]

board measure. Abbreviated *b.m.* or *B.M.*, simply denotes that a measurement is being given in board feet. Thus *10,000 feet b.m.*, for example, is equal to 10 MBF. Often this amount would simply be given as *10,000 feet*. Usually refers to lumber tally.

cant. A log that has been squared on two or three sides and is destined for further processing. The term is loosely used. *See* **flitch**.

CCF. 100 cubic feet. This unit of measurement is often used for pulp wood. One CCF of pulp wood is usually considered in Southeast Alaska to equal about 550 board feet. *See* **board foot**, **cord**.

cord. A unit of measurement often used for pulpwood. One cord represents 128 cubic feet (4'x4'x8') of stacked wood. One cord is today generally considered to equal 85 cubic feet of solid wood. Early Forest Service pulpwood sales in Southeast Alaska equated a cord to 100 cubic feet of solid wood unless or until the district forester and the purchaser of pulp timber, as a result of actual measurements, agreed upon a different ratio. The long-term timber contract between the U.S. Forest Service and the Ketchikan Pulp & Paper Company specified that one cord was to be considered equal to 90 cubic feet of solid wood. In his 1928 *Pulp-Timber Resources of Alaska*, Frank Heintzleman considered a cord to equal 600 board feet.[1] *See also* **CCF**.

cutoff saw. A large saw used to cut logs to a specific length before they enter a mill.[2]

decay. Disintegration of wood due to the action of fungi. *Incipient decay* is the early stage of decay, and is usually detected by a discoloration in wood that appears to be sound. *Advanced decay (rot)* is evidenced by a softening or breaking down of wood.[3]

dimension. Any lumber cut to size, as in 2 x 4s, etc.[4]

[A]A billion board feet of lumber is a very large amount. The average distance from the Earth to the Moon is about 239,000 miles, which is equal to 1.26 billion feet. Five billion board feet is about enough wood to build a one-inch-thick boardwalk four feet wide from the Earth to the Moon.

dry kiln. A chamber in which wood products are systematically dried through the controlled application of heat and withdrawal of moisture.

durability. Generally used to denote a wood's ability to resist decay.

economic timber. Defined by the U.S. Forest Service as: "A sale of timber wherein the average purchaser can meet all contractual obligations, harvest and transport the timber to the purchaser's site, and have a reasonable certainty of realizing a profit from the sale."[5]

edger. Sawmill machine that squares the edges of and rips cants into lumber.[6]

export. In the context used in this book, denotes the transport of logs, lumber, pulp and pulp products out of Alaska.

fell. In proper logger English, the act of cutting down a tree or trees. A *faller* is a workman who *fells* trees. Once a tree has been cut down, it is said to have been *felled*. In the woods, loggers often substitute *fall* for *fell*.

flitch. A log that has been squared on two or three sides and is destined for further processing. The term is loosely used. *See* **cant**.

forest types. *Old growth* is defined technically by the U.S. Forest Service timber managers as trees that have reached or passed the age of physiological maturity, assumed to be 150 years for coastal Alaska. In ecological terms, the agency generally defines old growth as an "uneven-aged or all-aged stand, usually exceeding 250 years of age."[7]

According to some forest scientists in the Pacific Northwest, a forest is not technically old growth unless it possesses four characteristics, three of which depend upon one's definition of "large." The characteristics are:

1. large, live trees;
2. large standing dead trees (snags) or stumps;
3. large logs lying on the forest floor, in various stages of decomposition;
4. a multi-species and multi-layered canopy.[8]

The terms *thrifty*, *mature*, and *overmature* are used by foresters in reference to whether the volume of merchantable timber in a stand of trees is growing, roughly static, or falling, while also denoting the amount of decay in the timber. Generally speaking:

- In a *thrifty* stand of timber merchantable trees are prevalent and still growing, and thus adding volume to the stand. There is very little decay present.
- In a *mature* stand of timber the trees have nearly quit growing, with growth roughly equaling decay, leaving the timber volume roughly static.
- In an *overmature* stand of trees decay exceeds growth and the timber volume is considered to be declining.

gyppo. A small, independent logging company.

head rig (head saw). The principal saw in a sawmill, which cuts logs into cants that are then sent on to other machinery for further processing.[9]

merchantable. In traditional use, as defined by the Society of American Foresters in 1918, timber that could be "manufactured and sold at not less than cost." The merchantability of timber varies with markets and the processing capabilities of mills.[10]

newsprint paper. Paper used in the printing of standard editions of newspapers. Newsprint paper is manufactured primarily of groundwood pulp to which some sulfite pulp has been added. The industry-standard price quote for newsprint paper is the contract price per ton delivered to New York City.[11]

pulp timber (pulpwood). Wood used to produce pulp. It is *usually* material that is too small, of substandard quality, or of a species not suitable for use in the manufacture of lumber or plywood.[12] *Contrast* **sawtimber**.

sawtimber (sawlogs). Regardless of their actual use, defined in *Terms of the Trade* as "logs of sufficient size and quality as to be suitable for conversion into lumber."[13] *Contrast* **pulp timber**.

scale. In the timber trade, the systematic estimation of the quantity of wood contained in a log. When a log is scaled as sawtimber, the goal is to obtain a reliable estimate of the amount of lumber, usually stated in board feet, the log can be expected to yield. Numerous different methods are employed in scaling sawtimber, with the table developed by John Scribner in 1846 and still bearing his name being the most common in America's west. It must be noted that the use of a given rule is due more to custom and agreement than to any consideration of its accuracy. The Forest Service in Southeast Alaska initially scaled pulp timber by the cord (1913), then by the board foot (circa 1920), then by the cubic foot (1921), and then back to board feet (1954).[14] *See also* **board foot**, **cord**.

stumpage. The value of standing timber or, more generally, the standing timber itself.[15]

teredo. A voracious marine wood borer of the family *Teredinidae*. Often thought to be a worm but actually a bivalve mollusk. May reach a length of 24 inches and have a diameter of ¾ inch. In Southeast Alaska, teredos have been known to completely ruin logs stored in salt water in four to five months.[16]

wood pulp. A spongy material made from wood that has been reduced to fibers through grinding or chipping and chemical treatment. Two types of wood pulp have been produced in Southeast Alaska:

- *groundwood pulp* is a type of *mechanical pulp* produced by forcing cleaned, debarked logs against a large, water-bathed revolving grindstone. This abrading process reduces the log to individual, relatively short wood fibers, many of which are broken. Lignin, the very complex natural glue that binds wood fibers together and is responsible for the strength of wood, is not removed in the manufacture of groundwood pulp, which causes the material to turn yellow with time. Groundwood pulp is used where strength and permanence are of minor importance. It is the primary component of newsprint paper and was manufactured at the Alaska Pulp & Paper Company mill at Speel River in the early 1920s.

- *sulfite (sulphite) pulp* is a type of *chemical pulp,* and is manufactured by cooking ("digesting") wood chips at high temperature in a large pressurized vessel containing an acidic liquor. The pulp mills at Ketchikan and Sitka were designed to process chips with a "fiber length" of ¾-inch.[B] The actual size of the chips were about ¾-inch x ¾-inch x ⅛-inch. In this process, the cellulose (wood fibers) is separated from the non-cellulose components of the wood, one of which is lignin. Refined pulps of different strengths and purity can be produced by adjusting the liquor and varying the cooking period. The yield of chemical pulp from a given amount of wood is less than that of mechanical pulp, but because wood fibers remain largely intact in this process chemical pulp is usually stronger than mechanical pulp. Sulfite pulp is a component of a range of paper products, from newsprint paper to high-grade paper. A highly-purified grade of sulfite pulp, known as *high-alpha* or *special-alpha* pulp, is used in the manufacture of fibers (rayon and acetate), films (cellophane) and plastics (cellulose acetate). The Ketchikan Pulp Company and the Alaska Lumber & Pulp Company employed the magnesium oxide (MgO) process to manufacture high-alpha pulp.[17]

[B] Also called *tracheids,* fibers are elongated wood cells, which in Sitka spruce and western hemlock are actually about ⅛-inch long.

Endnotes

CHAPTER 1

1. Arland S. Harris, et al., *Forest Ecosystem of Southeast Alaska: No. 1, The Setting*, Gen. Tech. Rept. PNW-12 (Portland, Oreg.: Pacific Northwest Forest and Range Experiment Station, 1974), 7; John N. Cobb, *Pacific Salmon Fisheries*, Bureau of Fisheries Doc. No. 1092, 4th ed. (Washington: GPO, 1930), 441; E. P. Stamm, A. J. F. Bromley, and P. S. Bonney, "Confidential Report to U.S. War Production Board on Plans and Feasibility of Logging and Transporting High Grade Sitka Spruce From Alaska to Puget Sound for Airplane Stock," February–March 1942, RG 95, NARA—Pacific Alaska Region, Anchorage.
2. H. E. Andersen, *Climate of Southeast Alaska in Relation to Tree Growth*, Station Paper No. 3 (Juneau: Alaska Forest Research Center, June 1955), 1.
3. *Ghost Trees: Measuring the Vanished Forests of Southeast Alaska* (Juneau: Southeast Alaska Conservation Council, 2000), 7; *Forests of Alaska* (Juneau: USFS, Alaska Region, July 1, 1944), 2–3; Leslie A. Viereck, and Elbert L. Little, *Alaska's Trees and Shrubs* (Washington: USFS, 1972), 60.
4. William A. Langille, "National Forest of Southeastern Alaska has 15,490,086 Acres," *Timberman* (September 1910): 27.
5. Greg Streveler (Southeast Alaska naturalist), personal communication with author, January 2004; A. E. Helmers, "Alaska Forestry—A Research Frontier," *Journal of Forestry* (June 1960): 469.
6. R. F. Taylor, "Site Prediction in Virgin Forests of Southeastern Alaska," *Journal of Forestry* (January 1933): 14; Henry S. Graves, "The Forests of Alaska," *American Forestry* (January 1916): 24; H. E. Andersen, "Clearcutting as a Silvicultural System in Converting Old Forests to New in Southeast Alaska," *Proceedings: Society of American Foresters Meeting, 1955* (Washington: Society of American Foresters, 1956), 61.
7. R. Kirk and J. Franklin, *The Olympic Rain Forest: An Ecological Web* (Seattle: University of Washington Press, 1992), 28–29.
8. C. Jeff Cederholm, et al., "Pacific Salmon Carcasses: Essential Contributions of Nutrients and Energy for Aquatic and Terrestrial Ecosystems," *Fisheries* (October 1999): 6–15.
9. Greg Streveler (Southeast Alaska naturalist), personal communication with author, October 2, 2004.
10. E. P. Stamm, A. J. F. Brandstrom, and P. S. Bonney, "Confidential Report to U.S. War Production Board on Plans and Feasibility of Logging and Transporting High Grade Sitka Spruce From Alaska to Puget Sound for Airplane Stock," February–March 1942, RG 95, NARA—Pacific Alaska Region, Anchorage.
11. J. M. Wyckoff, "History of Lumbering in Southeast Alaska," *Timberman* (June 1930): 40.
12. "Logging in Alaska," *Timberman* (April 1927): 176.
13. Royal S. Kellogg, *Forests of Alaska*, USFS Bull. No. 81 (Washington: GPO, 1910), 13; General Information Regarding the Territory of Alaska (Washington: U.S. Dept. of the Interior, 1917) 38; *Alaska, Outline of its History and a Summary of its Resources* (Juneau: Territory of Alaska, 1924), 39; B. Frank Heintzleman, *Pulp-Timber Resources of Southeastern Alaska*, USDA Misc. Pub. No. 41 (Washington: GPO, December 1928), 9; R. F. Taylor, *Yield of Second-Growth Western Hemlock-Sitka Spruce Stands in Southeastern Alaska*, USDA Tech. Bull. No. 412 (Washington: GPO, 1934), 2; John A. Sandor and John E. Weisgerber, *Timber Management Plan, Tongass National Forest, Alaska Region, 1958–1967*, 2 (Resource Summary revised May 22, 1962), O. Keith Hutchison and Vernon J. LaBau, *Forest Ecosystem of Southeast Alaska: No. 9, Timber Inventory, Harvesting, Marketing, and Trends*, Gen. Tech. Rept. PNW-34 (Portland, Oreg.: Pacific Northwest Forest and Range Experiment Station, 1975), abstract.
14. Heintzleman, *Pulp-Timber Resources of Southeastern Alaska*, 9; James W. Kimmey, *Cull Factors for Sitka Spruce, Western Hemlock and Western Redcedar in Southeast Alaska*, Station Paper No. 6 (Juneau: USFS, Alaska Forest Research Center, August 1956): 27.
15. *Western Hemlock*, USFS Silvical Leaflet 45, September 14, 1912: 3, 4; James W. Kimmey, *Cull Factors for Sitka Spruce, Western Hemlock and Western Redcedar in Southeast Alaska*, Station Paper No. 6 (Juneau: USFS, Alaska Forest Research Center, August 1956): 27.
16. Alexander L. Howard, *A Manual of the Timbers of the World: Their Characteristics and Uses*, 3rd ed. (London: Macmillan & Company, Ltd, 1948), 555; Frank H. Lamb, *Sagas of the Evergreens* (New York: W. W. Norton & Company, 1938), 249; "Hemlock Placed in Proper Light," *American Lumberman* (March 7, 1914): 34.
17. "Western Hemlock and Noble Fir," *Timberman* (September 1942): 22; "Answers to Correspondents," *Forester* (February 1898): 46; "Advisability of Changing the Name of Hemlock Discussed," *Timberman* (June 1921): 4; "The Future of Western Hemlock," *Timberman* (February 1930): 35; "The Hemlock Campaign," *Timberman* (April 1927): 35; "Western Hemlock Study," *Timberman* (August 1926): 196.
18. *Report of the Governor of the District of Alaska to the Secretary of the Interior, 1902* (Washington: GPO, 1902), 33.

19. L. J. Markwardt, *Distribution and Mechanical Properties of Alaska Woods*, USDA Tech. Bull. No. 226 (Washington: GPO, February 1931), 16; Edward T. Allen, *Western Hemlock*, USDA, Bureau of Forestry Bull. No. 33 (Washington: GPO, 1902), 23; W. Kendrick Hatt, *Second Progress Report on the Strength of Structural Timber*, USFS Circ. No. 115 (Washington: GPO, 1907), 18–19; O. P. M. Goss, *Mechanical Properties of Western Hemlock*, USFS Bull. No. 115 (Washington: GPO, 1913), 41–42.
20. Streveler, June 15, 2005 (related to Streveler by George Dalton Sr., of Hoonah, Alaska).
21. Advertisement, *Daily Alaska Empire* (Juneau), August 29, 1927.
22. C. R. Berry, "Western Hemlock," *Paper Trade Journal* (September 25, 1930): 40; R. P. A. Johnson, and W. H. Gibbons, *Properties of Western Hemlock and Their Relation to Uses of the Wood*, USDA Tech. Bull. No. 139 (Washington: GPO, December 1929), 56.
23. B. E. Hoffman, "Alaska Woods, Their Present and Prospective Uses," *Forestry Quarterly* (June 1913): 192.
24. "A Major Sitka Spruce Operation," *West Coast Lumberman* (June 1949): 66, 68; Donald J. Orth, *Dictionary of Alaska Place Names*, Geological Survey Professional Paper 567, (Washington: GPO, 1967), 880.
25. B. Frank Heintzleman, "Alaska's National Forests," *Timberman* (November 1923: 188.
26. J. E. Defebaugh, "Alaska and Its Timber," *American Lumberman* (September 5, 1903): 17.
27. Alaska Resources Committee, *Alaska—Its Resources and Development*, in: National Resources Committee, *Regional Planning*, Part VII (Washington: GPO, 1938), 96, 161; Clinton G. Smith, *Regional Development of Pulpwood Resources of the Tongass National Forest, Alaska*, USDA Bull. No. 950 (Washington: GPO, 1921), 8; J. P. Williams, "Timber Industry in Alaska," *Alaska Yearbook, 1926* (Seattle: *Alaska Weekly*, 1926), 15, 38; Heintzleman, "Alaska's National Forests," 188; J. M. Wyckoff, "Airplane Spruce Chances," November 5, 1940, RG 95, NARA—Pacific Alaska Region, Anchorage; Clinton G. Smith, *Regional Development of Pulpwood Resources of the Tongass National Forest, Alaska*, USDA Bull. No. 950 (Washington: GPO, 1921), 8.
28. "Tells Congress Alaska Has Only Scrubby Timber," *Wrangell Sentinel*, March 5, 1925. Note: reprinted from the *Alaska Pioneer*.
29. J. M. Wyckoff, "Lumber Industry on Tongass National Forest," *Pathfinder* (Valdez, Alaska) (October 1924): 4.
30. Heintzleman, *Pulp-Timber Resources of Southeastern Alaska*, 10; C. H. Forward, "Alaska Airplane Spruce," *West Coast Lumberman* (January 1940): 19; Ernest F. Burchard, *Marble Resources of Southeastern Alaska*, USGS Bull. No. 682, with a section on the Geography and the Geology by Theodore Chapin (Washington: GPO, 1920), 12–13; R. L. Furniss and Ivan H. Jones, "A Second Report Concerning the Bark Beetle Outbreak on Kosciusko Island," USFS (unpublished, 1946), 3.
31. Kellogg, *Forests of Alaska*, 15; Markwardt, *Distribution and Mechanical Properties of Alaska Woods*, 19; Greg Streveler (Southeast Alaska naturalist), personal communication with author, January 2004.
32. Heintzleman, *Pulp-Timber Resources of Southeastern Alaska*, 10; "An Alaska Big Stick," *Alaska Sentinel* (Wrangell), July 1, 1909.
33. H. S. Betts, *American Woods-Sitka Spruce* (USFS pamphlet series) (Washington: GPO, August 1944), 6; Markwardt, *Distribution and Mechanical Properties of Alaska Woods*, 19; Howard, *Manual of the Timbers of the World*, 557.
34. Arland S. Harris, "Sitka Spruce, Alaska's New State Tree," *American Forests* (August 1964): 34.
35. Brice P. Disque, *History of Spruce Production Division, United States Army and United States Spruce Production Corporation* (Portland, Oreg.: Kilham Stationary and Printing Company,1919), 3; *Distribution and Mechanical Properties of Alaska Woods*, 20.
36. H. S. Betts, *American Woods-Sitka Spruce* (USFS pamphlet series) (Washington: GPO, August 1944), 6; E. A. Sterling, "Flying on Wings of Spruce," *American Forestry* (March 1918): 133; "Spruce the Dominant Airplane Material," *West Coast Lumberman* (May 1928): 80.
37. Brice P. Disque, *History of Spruce Production Division, United States Army and United States Spruce Production Corporation* (Portland, Oreg.: Kilham Stationary and Printing Company, 1919), 3.
38. Armand Farmanian, former Locheed Corp. engineer, personal communication with author, September 2009; Lieutenant Colonel William D. Siuru Jr. "SLBM-The Navy's Contribution to Triad," http://www.tonyrogers.com/weapons/navy_slbm_triad.htm (accessed September 15, 2009).
39. *Sitka Spruce*, USFS Silvical Leaflet 6, October 7, 1907, 1; "An International Development Program For North Pacific Region," *Pacific Pulp & Paper Industry* (May 1943): 24; Nelson Courtlandt Brown, *Forest Products: Their Manufacture and Use* (New York: John Wiley & Sons, 1919), 24.
40. B. Frank Heintzleman, "Red and Alaska Cedar in Southeastern Alaska," *Timberman* (July 1923): 33; Edward S. Kotok, *Western Redcedar…an American Wood*, FS-261 (Washington: GPO, October 1973), 5; Arland S. Harris and Wilbur A. Farr, *Forest Ecosystem of Southeast Alaska: No. 7, Forest Ecology and Timber Management*, Gen. Tech. Rept. PNW-25 (Portland, Oreg.: Pacific Northwest Forest and Range Experiment Station, 1974), 18.
41. Kotok, *Western Redcedar*, 7; R. M. Ingram, "Western Red Cedar," *Timberman* (April 1926): 194; Markwardt, *Distribution and Mechanical Properties of Alaska Woods*, 11.
42. William L. Hall and Hu Maxwell, *Uses of Commercial Woods of the United States: I. Cedars, Cypresses, and Sequoias*, USFS Bull. 95 (Washington: GPO, 1911), 40; Wyckoff, "History of Lumbering in Southeast Alaska," 38; J. M. Walley, "Report on Market for Cedar Poles at Prince Rupert or Elsewhere and Alaskan Market for Shingles," USFS (unpublished), 1925, RG 95, NARA—Pacific Alaska Region, Anchorage.

43. R. A. Zeller, "Possibilities for Paper and Power Explained," *Ketchikan Alaska Chronicle*, July 4, 1920.

44. "Alaska Forest Product Manufacturers Face Numerous Problems," *Lumberman* (November 1960): 57; Arland S. Harris and Wilbur A. Farr, *Forest Ecosystem of Southeast Alaska: No. 7, Forest Ecology and Timber Management*, Gen. Tech. Rept. PNW-25 (Portland, Oreg.: Pacific Northwest Forest and Range Experiment Station, 1974), 18.

45. "Logging in Alaska is Tougher," *Lumberman* (November 1955): 135.

46. "Arland S. Harris, and Wilbur A. Farr, *Forest Ecosystem of Southeast Alaska: No. 7, Forest Ecology and Timber Management*, Gen. Tech. Rept. PNW-25 (Portland, Oreg.: Pacific Northwest Forest and Range Experiment Station, 1974), 18.

47. Streveler, January 2004; Heintzleman, "Red and Alaska Cedar in Southeastern Alaska," 33; Heintzleman, *Pulp Timber Resources of Southeastern Alaska*, 14; Langille, "National Forest of Southeastern Alaska," 27; William B. Ihlanfeldt (General Manager, Alaska Spruce Log Program) to Hugh P. Brady, October 26, 1943, RG 95, NARA—Pacific Alaska Region, Anchorage.

48. Warren D. Brush, "Alaska Yellow Cedar," *American Forests* (June 1950): 29; Arland S. Harris, *Alaska-Cedar…an American Wood*, FS-224 (Washington: GPO, Revised February 1984), 3; L. J. Markwardt, *Distribution and Mechanical Properties of Alaska Woods*, USDA Tech. Bull. No. 226 (Washington: GPO, February 1931), 9.

49. Paul E. Hennon and Charles G. Shaw III, "The Enigma of Yellow-Cedar Decline: What Is Killing These Long-Lived, Defensive Trees?" *Journal of Forestry* (December 1997): 5; Elizabeth Bluemink, "Warming Trends: Trees Fall to Climate Change," *Juneau Empire*, March 26, 2006.

50. "Properties and Uses of Alaska Cypress," *Timberman* (September 1929): 39–40.

51. Harris, *Alaska-Cedar*, 6; "In a Neglected Country," *New York Times*, October 5, 1884; William H. Dall, *Alaska and Its Resources* (Boston: Lee & Shepard, 1870), 453; John Muir, "Timber Resources of Alaska," *West Shore* (November 1879): 338.

52. William L. Hall and Hu Maxwell, *Uses of Commercial Woods of the United States: I. Cedars, Cypresses, and Sequoias*, USFS Bull. 95 (Washington: GPO, 1911), 35.

53. Charles Flory to Forester, April 1, 1922, RG 95, NARA, College Park, Md.

54. Harris and Farr, *Forest Ecosystem of Southeast Alaska*, 18.

CHAPTER 2

1. Ferdinand P. Wrangell, *Russian America: Statistical and Ethnographic Information*, trans. Mary Sadouski, ed. Richard A. Pierce (Kingston, Ont.: Limestone Press, 1980), 5.

2. Claus-M. Naske and Herman E. Slotnick, *Alaska, A History of the 49th State*, 2nd ed. (Norman: University of Oklahoma Press, 1987), 36.

3. P. A. Tikhmenev, *A History of the Russian-American Company*, trans. and ed. Richard A. Pierce and Alton Donnelly (Seattle: University of Washington Press, 1978), 421; Robert De Armond (Alaska historian), personal communication with author, April 10, 2005.

4. Katherine L. Arndt, *Supplemental Report: The Bishop's House as Documented by the Alaskan Russian Church Archives*, 2nd ed.,Prepared for Sitka National Historical Park, National Park Service, Sitka, Alaska, under Cooperative Agreement with the Department of Anthropology, University of Alaska, Fairbanks, 2004, 14; Svetlana Fedorova, *Russian Population in Alaska and California, Late 18th Century –1867*, trans. and edit. by Richard A. Pierce and Alton Donnelly (Kingston, Ont.: Limestone Press, 1973), 224; Katherine L. Arndt, Russell H Sackett, and James A Ketz, *Cultural Resource Overview of the Tongass National Forest, Alaska, Part 1: Overview* (Fairbanks: GDM, Inc., 1987), 188.

5. October 17, 1867, report of a correspondent before the *Alta California Newspaper* was forwarded to the Secretary of State, and reproduced in, 40th Cong., 2nd sess., Ho. Ex. Doc. 1339, Vol. 13, 1868, 69.

6. Katherine L. Arndt and Richard A. Pierce, *A Construction History of Sitka, Alaska, as Documented in the Records of the Russian-American Company*, 2nd ed. (Sitka, Alaska: National Park Service, 2003), 10, 19.

7. Kyrill T. Khlebnikov, *Colonial Russian America: Kyrill T. Khlebnikov's Reports* (Portland, Oreg.: Oregon Historical Society, 1976), 77.

8. Arndt and Pierce, *Construction History of Sitka*, 42, 172, 186; Tikhmenev, *History of the Russian-American Company*, 421; Harris and Farr, *Forest Ecosystem of Southeast Alaska*, 3.

9. Khlebnikov, *Colonial Russian America*, 77.

10. Arndt and Pierce, *Construction History of Sitka*, 226.

11. Fedorova, *Russian Population in Alaska and California*, 193; Tikhmenev, *History of the Russian-American Company*, 371, 420; Pavel N. Golovin, *The End of Russian America: Captain P. N. Golovin's Last Report, 1862*, trans. Basil Dmytryshyn and E. A. P. Crownhart-Vaughan (Portland, Oreg.: Oregon Historical Society, 1979), 90.

12. Golovin, *End of Russian America*, 89.

13. "Lumbering in Alaska 1811–1949," *Timberman* (October 1949): 112.

14. James R. Gibson, *Imperial Russia In Frontier America* (New York: Oxford University Press, 1976), 38.

15. Richard L. Williams, *The Loggers* (Alexandria, Va.: Time-Life Books, 1976), 37.

16. Arndt and Pierce, *Construction History of Sitka*, 62.

17. Ibid., 73, 77.

18. Ibid., 118.

19. Ibid., 113, 116

20. Ibid., 118.
21. Ibid., 98–99, 155.
22. C. L. Andrews, *Sitka, The Chief Factory of the Russian American Company*, 3rd. ed. (Caldwell, Idaho: Caxton Printers, Ltd., 1945), 85; Golovin, *End of Russian America*, 49; Robert De Armond, "Early Alaskan Sawmills," *Ketchikan Daily Alaska Fishing News*, February 5, 1946; Arndt and Pierce, *Construction History of Sitka*, 117, 156.
23. Robert De Armond, "Around & About Alaska," *Daily Sitka Sentinel*, July 27, 1994.
24. Arndt and Pierce, *Construction History of Sitka*, 123, 132, 160.
25. Ibid., 136, 150, 164, 170; Golovin, *End of Russian America*, 50.
26. Robert De Armond (Alaska historian), personal communication with author, April 10, 2005; Arndt and Pierce, *Construction History of Sitka*, 177, 224, 225, 228; Golovin, *End of Russian America*, 50.
27. Gibson, *Imperial Russia In Frontier America*, 38; Arndt and Pierce, *Construction History of Sitka*, 170.
28. Golovin, *End of Russian America*, 49; Arndt and Pierce, *Construction History of Sitka*, 236, 238, 249, 264, 265.
29. De Armond, "Around & About Alaska."
30. Robert De Armond, "Early Alaskan Sawmills," *Ketchikan Daily Alaska Fishing News*, February 5, 1946.
31. Arndt and Pierce, *Construction History of Sitka*, 200 228, 244; Arndt, Sackett, and Ketz, *Cultural Resource Overview*, 198.
32. Golovin, *End of Russian America*, 49, "An inventory of the public property in the City of New Archangel [Sitka] delivered to the United States of America, General Lovell H. Rousseau, United States Commissioner; by His Imperial Majesty the Emperor of Russia, Captain Alexis Pestchouroff, Russian Commissioner, on the 18th day of October, 1867, at New Archangel [Sitka]" in Archie W. Shiels, "The Purchase of Alaska," (unpublished, 1965), Archie Shiels Collection, Center for Pacific Northwest Studies, Bellingham, Wash.; Robert De Armond, "From Sitka's Past—No. 32," *Daily Sitka Sentinel*, May 6, 1986.
33. Khlebnikov, *Colonial Russian America*, 77.
34. Ibid.; C. L. Andrews, "Alaska Under the Russians—Industry, Trade and Social Life," *Washington Historical Quarterly* (October 1916): 285.
35. Tikhmenev, *History of the Russian-American Company*, 209, 360, 472; C. L. Andrews, "Alaska Under the Russians—Industry, Trade and Social Life," *Washington Historical Quarterly* (October 1916): 287.
36. Wrangell, *Russian America*, 8; Arndt and Pierce, *Construction History of Sitka*, 224.
37. Wrangell, *Russian America*, 9–10.
38. C. L. Andrews, "Russian Shipbuilding in the American Colonies," *Washington Historical Quarterly* 25 (1934), 9; Arndt and Pierce, *Construction History of Sitka*, 18.
39. Hubert Howe Bancroft, *History of Alaska, 1730–1885*, Vol. XXXIII (San Francisco: A. L. Bancroft & Company, 1886), 692.
40. Gibson, *Imperial Russia In Frontier America*, 38.
41. Kiril T. Khlebnikov, *Notes on Russian America, Part I: Novo-Archangel'sk*, comp. Svetlana G. Gedorova, trans. Serbe LeComte and Richard A. Pierce, ed. Richard A. Pierce (Kingston, Ont. and Fairbanks: Limestone Press, 1994), 100.
42. June Allen, *Spirit!* (Ketchikan: Lind Printing for Historic Ketchikan, Inc., 1992), 77; Harris, *Alaska-Cedar*, 6.
43. C. L. Andrews, "Alaska Under the Russians—Industry, Trade and Social Life," *Washington Historical Quarterly* (October 1916): 287; De Armond, "Early Alaskan Sawmills"; Arndt and Pierce, *Construction History of Sitka*, 240–241.
44. Golovin, *End of Russian America*, 10, 89–90.
45. Tikhmenev, *History of the Russian-American Company*, 335–336.
46. "Transfer of Alaska to the United States—Instructions from William H. Seward and report of General Rousseau," *Washington Historical Quarterly* (October 1908): 84.
47. De Armond, "Early Alaskan Sawmills."
48. *Report of Major John C. Tidball*, reprinted in *Letter from the Secretary of War, in relation to the Territory of Alaska*, 42nd Cong., 1st sess., Ho. Ex. Doc. No. 5, 1871, 4–5; William H. Dall, *Alaska and Its Resources* (Boston: Lee & Shepard, 1870), 439–41; De Armond, "Early Alaskan Sawmills."

CHAPTER 3

1. Nelson Courtlandt Brown, *The American Lumber Industry* (New York: John Wiley & Sons, 1923), 5; Nelson Courtlandt Brown, "America's Eternal Wealth…Growing Forests," *American Lumberman* (July 10, 1943): 35–36.
2. *Speech of Hon. Charles Sumner, of Massachusetts, on the cession of Russian America to the United States*, reprinted in *Russian America: Message from the President of the United States in answer to a resolution of the House of 19th of December last, transmitting correspondence in relation to Russian America*, 40th Cong., 2d. sess., Ho. Ex. Doc. No. 177, Pt. I, 1868, 188, 169.
3. Hector Chevigny, *Russian America* (New York: Viking Press, 1965), 255; Ernest Gruening, *The State of Alaska* (New York: Random House, 1954), contents.
4. Ellen Lloyd Trover, editor, *Chronology and Documentary Handbook of the State of Alaska* (Dobbs Ferry, N.Y.: Oceana Publications, Inc., 1972), 50.

5. George Davidson, *Coast Pilot of Alaska, from Southern Boundary to Cook's Inlet* (Washington: GPO, 1869), 30.
6. *Report of Major E. H. Ludington*, reprinted in *Letter from the Secretary of War, in relation to the Territory of Alaska*, 42d. Cong., 1st sess., Ho. Ex. Doc. No. 5, 1871, 7.
7. *Report of Major John C. Tidball*, 4–5.
8. *Letter from the Secretary of War, Transmitting Information in Relation to Alaska and Its Resources, the Alaska Commercial Company, the Conduct of Mr. Bryant of Saint Paul's and Saint George's Islands, and the Colonization of Icelanders* (year 1876 observations of Capt. J. W. White), 44th Cong., 1st sess., Sen. Ex. Doc. No. 48, 1876, 5.
9. Henry W. Elliot, *Report on the Conditions of Affairs in the Territory of Alaska* (Washington: GPO, 1875), 10, in *Seal Fisheries in Alaska*, 44th Cong., 1st sess., Ho. Ex. Doc. No. 83, 1876.
10. "North-West Possessions," *New York Times*, November 25, 1880.
11. Ivan Petroff, esq, special agent of the census, in *Letter from the Secretary of the Interior Transmitting a preliminary report upon the population, industry, and resources of Alaska* (January 15, 1881), 46th Cong., 3d. sess., Ho. Ex. Doc. 40, 1881, 73, 5; *Report of the Governor of Alaska for the Fiscal Year 1886* (Washington: GPO, 1886), 945.
12. Charles Hallock, *Our New Alaska; or, The Seward Purchase Vindicated* (New York: Forest and Stream Publishing, 1886), 54; "New Forest Chief on Saving Our Forests," *New York Times*, January 2, 1921; Royal S. Kellogg, *Timber Supply of the United States*, USFS Circular 97 (Washington: GPO, 1907).
13. Act of June 3, 1878 (20 Stat. 88).
14. General Mining Law of 1872, May 10, 1872, (17 Stat. 91); John Bufvers, *History of Mines and Prospects, Ketchikan District, Prior to 1952* (Juneau: State of Alaska, Dept. of Natural Resources, Div. of Mines and Minerals, 1967), 5.
15. William Gouverneur Morris, *Report Upon the Customs District, Public Service, and Resources of Alaska Territory*, 45th Cong., 3d. sess., Sen. Ex. Doc. 59, 1879, 107, 109.
16. Act Providing a Civil Government for Alaska, May 17, 1884 (23 Stat. 24).
17. Ted C. Hinkley, *Alaskan John G. Brady, Missionary, Businessman, Judge, and Governor, 1878–1918* (Columbus: Ohio State University Press, 1982), 144.
18. Lawrence Rakestraw, *A History of the U.S. Forest Service in Alaska* (Anchorage: Cooperative publication of the Alaska Historical Commission, Alaska Dept. of Education, and USFS, 1981), 7; "Cutting and Selling Timber in Alaska," *Alaskan* (Sitka), February 13, 1886.
19. Hallock, *Our New Alaska*, 56; *Alaskan* (Sitka), May 29, 1886; June 5, 1886; and March 30, 1889; Hinkley, *Alaskan John G. Brady*, 144; De Armond, "Around & About Alaska."
20. *Report of the Governor of Alaska for the Fiscal Year 1887* (Washington: GPO, 1887), 945.
21. *Report of the Secretary of the Interior, 1889*, 51st Cong., 1st sess., Ho. Ex. Doc.1, pt. 5, 1890, CVIII; *Report of the Governor of Alaska for the Fiscal Year 1890* (Washington: GPO, 1890), 29.
22. *Alaskan* (Sitka), August 31, 1889; *Alaskan* (Sitka), September 20, 1890; *Alaskan* (Sitka), December 20, 1890; *Report of the Governor of Alaska for the Fiscal Year 1891* (Washington: GPO, 1891), 32.
23. *Alaskan* (Sitka), September 12, 1891.
24. *Alaskan* (Sitka), April 4, 1891.
25. *Report of the Secretary of the Interior; Being Part of the Message and Documents Communicated to the Two Houses of Congress at the Beginning of the First Session of the Fifty-First Congress*, Vol. 1, 51st Cong. 1st sess., Ho. Ex. Doc. 1, Part 5, 1890, CX; *Report of the Governor of Alaska for the Fiscal Year 1889* (Washington: GPO, 1889), 9; *Report of the Governor of Alaska to the Secretary of the Interior, 1891* (Washington: GPO, 1891), 492.
26. Forest Reserve Act of March 3, 1891 (26 Stat. 1099), § 8.
27. Information published in government circular reproduced in *Decisions of the Department of the Interior and General Land Office in Cases Relating to Public Lands, From January 1, 1891, to June 30, 1891*, 7, Vol. XII (Washington: GPO, 1891), 456–459.
28. "Local News," *Alaskan* (Sitka), May 7, 1892.
29. "Big Federal Domain," *New York Times*, November 19, 1897.
30. "The Timber Law of Alaska," *Stickeen River Journal* (Wrangell), July 2, 1898.
31. Act of May 14, 1898 (30 Stat. 414), § 11.
32. Untitled, *Columbia River and Oregon Timberman* (May 1902): 25; Robert De Armond, "Abaft the Beam: Early Sawmills, III," *Ketchikan Daily Alaska Fishing News*, February 9, 1946.
33. F. E. Olmsted, "Inspection Report on the Alexander Archipelago Forest Reserve, 1906," USFS (unpublished), 33, Historical Collections, Alaska State Library, Juneau.
34. *Report of the Governor of the District of Alaska to the Secretary of the Interior, 1901* (Washington: GPO, 1901), 28; Robert De Armond, "Early Sawmills III."
35. "Free Use of Forest Reserve Timber," *Columbia River and Oregon Timberman* (March 1902): 33.
36. "Sale of Timber in Forest Reserves," (circular) General Land Office, January 22, 1902.
37. J. E. Defebaugh, "The Resources of Alaska: The Territory's Chief Executive on Its Natural Advantages and Needs," *American Lumberman* (September 5, 1903): 18–19.
38. William A. Langille, "Report on Alexander Archipelago Forest Reserve, 1906," USFS (unpublished), RG 95, NARA, College Park, Md.
39. William A. Langille, incomplete report, RG 95, NARA—Pacific Alaska Region, Anchorage.

40. Gifford Pinchot to W. A. Richards (General Land Office), December 18, 1903, RG 95, NARA—Pacific Alaska Region, Anchorage; Untitled, *Alaska Sentinel* (Wrangell), December 4, 1902; Olmsted, "Inspection Report on the Alexander Archipelago Forest Reserve, 1906," 7.
41. Langille, "Report on Alexander Archipelago Forest Reserve."
42. *Report to the Secretary of the Interior, 1901*, 28; "Stealing Logs," *Alaskan* (Sitka), April 23, 1904; Olmsted, "Inspection Report on the Alexander Archipelago Forest Reserve, 1906," 9.

CHAPTER 4

1. William B. Greeley, "Alaska—Uncle Sam's Coming Paper Factory," *World's Work*, November 1928, 85.
2. Act of March 3, 1891 (26 Stat. 1103), § 24.
3. Harold K. Steen, *The U.S. Forest Service: A History* (Seattle: University of Washington Press, 1976), 27–28; *Report of the Chief of the Forest Service, 1948* (Washington: GPO, 1948), 2; Rakestraw, *U.S. Forest Service in Alaska*, 10, 47.
4. Gifford Pinchot, *Breaking New Ground* (New York: Harcourt, Brace, and Company, 1947); (Washington: Island Press, 1998), 86. Citations are to the Island edition.
5. Sundry Civil Expenses Appropriations Act of June 4, 1897 (30 Stat. 35).
6. C. S. Sargent, "The Nation's Forests," *Garden and Forest* (January 30, 1889): 49.
7. "Cut Private Timber First," *Timberman* (September 1910): 19; "Pinchot's Timber Policy Wins," *Timberman* (March 1934): 7; "Strength of Alaskan Wood Tested," *American Lumberman* (September 4, 1915): 39.
8. George T. Emmons, "The Woodlands of Alaska," (unpublished, 1902).
9. Donald Craig Mitchell, *Sold American: The Story of Alaska Natives and Their Land, 1867–1959* (Hanover, N.H.: University Press of New England, 1997), 162.
10. Proclamation of August 20, 1902 (32 Stat. 2025).
11. *Report of the Governor of the District of Alaska to Secretary of the Interior* (Washington: GPO, 1902), 34–35.
12. *Report of the Governor of the District of Alaska to the Secretary of the Interior* (Washington: GPO, 1903), 19–20.
13. *Report of the Governor of the District of Alaska to the Secretary of the Interior* (Washington: GPO, 1904), 20–21. (reprint of Pinchot's August 16, 1904 letter to Governor Brady)
14. Overton Price to House Committee on Agriculture, January 8, 1903, quoted in: Pinchot, *Breaking New Ground*, 235.
15. "Transferred," *Forestry & Irrigation* (February 1905): 61; Steen, *U.S. Forest Service*, 74.
16. Merriam-Webster Online, http://www.m-w.com/cgi-bin/dictionary?book=Dictionary&va=forestry (accessed May 23, 2007); Theodore Roosevelt, "Forestry and Foresters," *Proceedings of the Society of American Foresters*, May 1905; Royal S. Kellogg, "R. S. Kellogg Speaks on Forestry," *Paper Trade Journal* (September 22, 1927): 46; Pinchot, *Breaking New Ground*, 1, 29.
17. Pinchot, *Breaking New Ground*, 260–261.
18. *Use of the National Forests* (Washington: GPO, 1905), 11–12.
19. "National Forest Reserve Rules," *Timberman* (December 1905): 32B.
20. John Muir had considerable firsthand knowledge of Southeast Alaska's forests. His article, "Timber Resources of Alaska," was published in *West Shore* in 1879. This frank and analytical piece was among the earliest on the subject, and contained none of the preservationist sentiment with which Muir is normally associated.
21. Roosevelt, "Forestry and Foresters."
22. U.S. Forest Service, *A National Plan for American Forestry*, 73rd Cong., 1st sess., Sen. Doc. 12, 1933, 1455.
23. *Alaska Sentinel* (Wrangell), June, 8, 1905.
24. "Alaskan Forest Reserve," *Forestry and Irrigation* (January 1904): 6.
25. *Field Programme for July, 1905* (Washington: USFS, 1905), 5; Olmsted, "Inspection Report on the Alexander Archipelago Forest Reserve, 1906," 18, *Field Program for April, 1907* (Washington: USFS, 1907), 10; "Alaska Forest Reserve," *Mining Journal* (Ketchikan), Illustrated Annual, January 1907.
26. *Field Program for November, 1906* (Washington: USFS, 1906), 16.
27. Rakestraw, *U.S. Forest Service in Alaska*, 30.
28. "Cut it Out," *Alaska Sentinel* (Wrangell), September 24, 1908; Olmsted, "Inspection Report on the Alexander Archipelago Forest Reserve, 1906," 1 in cover letter, 5 in text; "Alaska Reserve Created," *Daily Miner* (Ketchikan), September 14, 1907.
29. Olmsted, "Inspection Report on the Alexander Archipelago Forest Reserve, 1906," 3.
30. F. E. Olmsted to Forester, October 4, 1906, reprinted in *Juneau Empire*, August 18, 2002.
31. Olmsted, "Inspection Report on the Alexander Archipelago Forest Reserve, 1906," 3 in cover letter; 1, 30, 34, 36, 42 in text; 5, 20, 21 in "General Forest Conditions."
32. Excerpt from amendments passed by Congress, *Alaska Sentinel* (Wrangell), February 14, 1907.
33. "'National Forests' Hereafter," *Forestry and Irrigation* (April 1907): 168; B. Frank Heintzleman, "National Forest Administration in Alaska," *Pathfinder* (Valdez, Alaska) (October 1924): 1.
34. Proclamation of September 10, 1907 (35 Stat. 2152).

35. "Derivation of Forest Names," *Six Twenty-Six* (January 1921): 37; Orth, *Dictionary of Alaska Place Names*, 976; U.S. Coast and Geodetic Survey, *Pacific Coast Pilot: Alaska*, Part I, 3rd ed. (Washington: GPO, 1891), 79.
36. *Report of Governor of Alaska to the Secretary of the Interior, 1907* (Washington: GPO, 1907), 11.
37. Exec. Order No. 908 (July 2, 1908); Proclamation of February 16, 1909 (35 Stat. 2226); *Report of Governor of Alaska to the Secretary of the Interior, 1909* (Washington: GPO, 1909), 7.
38. Proclamation of June 10, 1925 (44 Stat. 2578).
39. *Annual Report of the Governor of Alaska to the Secretary of the Interior, 1930* (Washington: GPO, 1930): 68.
40. "National Forests of Alaska Now Are Separate District," *Ketchikan Alaska Chronicle*, July 8, 1921; "Forest District for Alaska is Created," *Wrangell Sentinel*, December 16, 1920; "Administration of Alaska Forests," *Science* (January 28, 1921): 86; *Report of the Forester*, 1930, 14; Rakestraw, *U.S. Forest Service in Alaska*, 91.

CHAPTER 5

1. Sawmill information in Chapter 5 is presented in greater detail, with citations, in Appendix A.
2. "Cong. Tawney and Party on Seattle," *Daily (Juneau) Alaska Dispatch*, August 14, 1903; J. E. Defebaugh, "Alaska and Its Timber," *American Lumberman* (September 1903): 18; Royal S. Kellogg, *Forests of Alaska*, USFS Bulletin No. 81 (Washington: GPO, 1910), 15–16; J. S. Baker, "Alaska's First Electrified Sawmill," *Timberman* (August 1925): 160.
3. B. E. Hoffman, "Alaska Woods, Their Present and Prospective Uses," *Forestry Quarterly* (June 1913): 187, 193.
4. Thomas R. Cox, *Mills And Markets* (Seattle: University of Washington Press, 1974), 234–236.
5. *1953 Annual Report, Governor of Alaska to the Secretary of the Interior* (Washington: GPO, 1953), 39.
6. "The Sawmill," *Alaska Sentinel* (Wrangell), February 4, 1909.
7. "Spruce Mills Started Again this Morning," *Ketchikan Alaska Chronicle*, March 23, 1925.
8. E. T. Allen, "Synopsis of Greeley Report," *Timberman* (February 1917): 30.
9. *Fisheries of Alaska in 1908*, Bureau of Fisheries Doc. No. 645 (Washington: GPO, 1909), 35.
10. "Territorial Law Regarding Saw Mill Wastes, Etc.," *Wrangell Sentinel*, April 29, 1915.
11. Barton Warren Evermann, *Alaska Fisheries and Fur Industries, 1913* (Washington: GPO, 1914), 39; E. Lester Jones, *Report of Alaska Investigations in 1914* (Washington: GPO, 1915), 104; "Territorial Law Regarding Saw Mill Wastes, Etc.," *Wrangell Sentinel*, April 29, 1915.
12. Refuse Act of 1899, August 3, 1899 (30 Stat. 1152), Ch. 425, § 13
13. *Report of the Governor of the District of Alaska to the Secretary of the Interior, 1902* (Washington: GPO, 1902), 33.
14. Olmsted, "Inspection Report on the Alexander Archipelago Forest Reserve, 1906," 5 in "General Forest Conditions."
15. "How Agriculture Department Is Aiding in Development of Huge Resources of Territory," *Ketchikan Alaska Chronicle*, June 27, 1923. (Statement by Henry Wallace prior to 1923 Alaska trip.)
16. *Report of the Governor of Alaska to the Secretary of the Interior* (Washington: GPO, 1913), 12.
17. W. G. Weigle, "National Forest Administration in Alaska Prior to 1920," *Yale Forest School News* (October 1953): 66; *Report of the Governor of Alaska to the Secretary of the Interior* (Washington: GPO, 1912), 16; Ray Taylor, "Early Forest Research in Alaska," unpublished report on file at USFS Forestry Sciences Laboratory, Juneau, 15.
18. "Seas Break Log Boom," *Wrangell Sentinel*, December 14, 1911; "Raft Breaks, Light Extinguished," *Wrangell Sentinel*, September 19, 1912; Taylor, "Early Forest Research in Alaska," unpublished.
19. John D. Guthrie, "The National Forest Resources of Alaska are for Use," *American Forestry* (January 1921): 12.
20. Heintzleman, "National Forest Administration in Alaska," 1.
21. William B. Greeley, "What is Wrong with Alaska," *American Forestry* (April 1921): 202.
22. R. A. Zeller, "Possibilities for Paper and Power Explained," *Pathfinder* (Valdez, Alaska) (August 1920): 30–31.
23. Henry S. Graves, "The Forests of Alaska," *American Forestry* (January 1916): 31–32.
24. "Stumpage Price Advances," *Petersburg Weekly Report*, July 15, 1916.
25. "National Forest Timber for Sale," *Petersburg Weekly Report*, August 19, 1916; "A Ray of Hope for Alaska Timber Industry," *Petersburg Weekly Report*, October 7, 1916.
26. "24,700,000 Feet National Forest Timber for Sale," (advertisement), *Ketchikan Alaska Chronicle*, October 25, 1923.
27. U.S. Forest Service, Eric Ness timber sale, March 26, 1914, RG 95, NARA—Pacific Alaska Region, Anchorage; J. L. Mackenchuil (USFS ranger at Petersburg) to Forest Supervisor, March 27, 1920, RG 95, NARA—Pacific Alaska Region, Anchorage.
28. *Wrangell Sentinel*, July 11, 1929.
29. B. Frank Heintzleman, "The Timber Industry of Alaska," *Cordova Daily Times All-Alaska Review for 1929* (Cordova, Alaska: *Cordova Daily Times*, 1929), 36–37.
30. "Is Rough Weather for Log Tow, Says Captain McDonald," *Ketchikan Alaska Chronicle*, October 5, 1923; "Log Boom Arrives," *Ketchikan Alaska Chronicle*, October 19, 1923.
31. Wyckoff, "History of Lumbering in Southeast Alaska," 40.

32. B. Frank Heintzleman, *Pulp Timber Resources of Southeastern Alaska* (Juneau: USFS, 1937), 25.
33. B. Frank Heintzleman, "Alaska National Forests," *Timberman* (November 1923): 188.
34. George H. Cecil, "Alaska Pulpwood Resources to be Developed," *Timberman* (June 1920): 37.
35. Heintzleman, "National Forest Administration in Alaska," 1.
36. B. Frank Heintzleman, "The Standing Timber Resources of Alaska," *West Coast Lumberman* (May 1, 1923): 102.
37. B. Frank Heintzleman, "Newsprint from the North," *Pacific Pulp & Paper Industry* (February 1927): 11.
38. B. Frank Heintzleman, "Large Paper Making Projects in View," *Cordova Daily Times All-Alaska Review for 1928* (Cordova, Alaska: *Cordova Daily Times*, 1928), 12–13; U.S. Forest Service, *A National Plan for American Forestry*, 1642.
39. C.R. S9592, August 21, 1937.
40. Alaska Resources Committee, *Alaska—Its Resources and Development*, in: National Resources Committee, *Regional Planning*, Part VII (Washington: GPO, 1938), 100.
41. B. Frank Heintzleman to files, May 18, 1946, RG 95, NARA—Pacific Alaska Region, Anchorage.
42. Milton J. Daly to B. Frank Heintzleman, June 12, 1947, Ketchikan Spruce Mills Collection, Tongass Historical Museum, Ketchikan.
43. Arthur W. Greeley to Milton Daly, March 10, 1954, Ketchikan Spruce Mills Collection, Tongass Historical Museum, Ketchikan.
44. Richard E. McArdle to Milton J. Daly, June 3, 1954, Ketchikan Spruce Mills Collection, Tongass Historical Museum, Ketchikan.
45. C. M. Archbold to Milton J. Daly, May 19, 1964, Ketchikan Spruce Mills Collection, Tongass Historical Museum, Ketchikan.

CHAPTER 6

1. B. E. Hoffman, "Alaska Woods, Their Present and Prospective Uses," *Forestry Quarterly* (June 1913): 187.
2. "Alaska Lumberman Looks For Good Season," *West Coast Lumberman* (March 1929): 20.
3. "Shipping Yellow Cedar," *Wrangell Sentinel*, March 24, 1910; "Yellow Cedar for Navy Yard," *Wrangell Sentinel*, May 2, 1910.
4. "Wrangell Is In The Very Front Rank," *Wrangell Sentinel*, March 23, 1911.
5. Graves, "The Forests of Alaska," 34; "Unlocking Alaska," *American Forestry* (July 1914): 485; B. E. Hoffman, "Alaska Woods, Their Present and Prospective Uses," *Forestry Quarterly* (June 1913): 187; "Alaska," *Timberman* (April 1914): 42–43; "Alaska Lumber Competing In State With Product of Washington for First Time," *Ketchikan Alaska Chronicle*, March 11, 1920; J. L. Mackenchuil (USFS ranger at Petersburg) to Forest Supervisor, March 27, 1920, RG 95, NARA—Pacific Alaska Region, Anchorage.
6. "Big Ship Arrives to Take on Lumber Cargo for New York Via Panama Canal," *Wrangell Sentinel*, September 9, 1926; "Alaska Spruce to Be Shipped East from Ketchikan," *Ketchikan Alaska Chronicle*, September 14, 1926.
7. "Alaska Spruce for the Manufacture of Airplanes," *Wrangell Sentinel*, June 21, 1917; "Alaska Lumber Shipped South Passed 2,000,000 Mark; Expect Increase," *Ketchikan Alaska Chronicle*, March 6, 1925.
8. Session Laws of Alaska, 1921, Chapter 31; Session Laws of Alaska, 1925, Chapter 70; Alaska Legislative Council, "Territorial Taxes in Alaska," Publication No. 22-4, June 1956, 5.
9. R. A. Zeller, "Some Points on Timber Removal on the Tongass National Forest, Alaska, and Their Bearing on the Character of the Future Stand," probably 1921 (unpublished), 19, RG 95, NARA—Pacific Alaska Region, Anchorage.
10. "Interdependence of Lumber Industry and the Fisheries," *Pacific Fisherman* (September 1925): 42.
11. *Fisheries of Alaska in 1907*, Bureau of Fisheries Doc. No. 632 (Washington: GPO, 1908), 42; "Pacific Salt Salmon Pack, 1921 Alaska," *Pacific Fisherman*, Yearbook (January 1922): 92.
12. Wyckoff, "History of Lumbering in Southeast Alaska," 38, 39.
13. "Russians Operated First Sawmill In Alaska With A Market In California," *Timberman* (June 1911): 32C; Jefferson F. Moser, *Salmon and Salmon Fisheries of Alaska* (Washington: GPO, 1899), 52, 110; "Pack of Canned Salmon in Alaska, by Districts, From Inception of the Industry," *Pacific Fisherman* (Yearbook, January, 1923): 56.
14. Jefferson F. Moser, *Salmon and Salmon Fisheries of Alaska* (Washington: GPO, 1899), 34; B. E. Hoffman, "Alaska Woods, Their Present and Prospective Uses," *Forestry Quarterly* (June 1913): 187; Kellogg, *Forests of Alaska*, 15; "Russians Operated First Sawmill in Alaska with a Market in California," *Timberman* (June 1911) 32C; "Alaska's Modern Sawmills," *Timberman*, 40.
15. "Specifications for Shipping Cases, *Pacific Fisherman* (Yearbook, January 1918): 89; Langille, "National Forest of Southeastern Alaska," *Timberman*, 27; "Interdependence of Lumber Industry and the Fisheries," *Pacific Fisherman* (September 1925): 42.
16. Hoffman, "Alaska Woods," *Forestry Quarterly*, 191; Kellogg, *Forests of Alaska*, 16; "Alaska Lumber Outlook," *Timberman* (January 1927): 131; "Modern Alaska Sawmill," *Timberman* (October 1927): 140; "Interdependence of Lumber Industry and the Fisheries," *Pacific Fisherman*, 43.
17. Jefferson F. Moser, *Salmon Investigations in 1900*, in *Bulletin of the United States Fish Commission, 1901*, Vol. XXI (Washington: GPO, 1902), 309.
18. Hoffman, "Alaska Woods," *Forestry Quarterly*, 191–192; "Fresh Fish at Seattle Last Month," *Wrangell Sentinel*, July 22, 1915; *Petersburg Weekly Report*, January 20, 1922, "Petersburg Sawmill May Be

Operated," *Petersburg Weekly Report*, February 9, 1923; *Fisheries of Alaska in 1907*, Bureau of Fisheries Doc. No. 632 (Washington: GPO, 1908), 49; *Fisheries of Alaska in 1910*, Bureau of Fisheries Doc. No. 746 (Washington: GPO, 1911), 43; "Review of Pacific Halibut Fishery in 1922," *Pacific Fisherman* (Yearbook, January 1923): 89.

19. Ward T. Bower and Harry Clifford Fassett, "The Fishery Industries of Alaska in 1913," *Pacific Fisherman* (Yearbook 1914): 61; *Fisheries of Alaska in 1907*, Bureau of Fisheries, 43.

20. Donald Nelson (Petersburg historian), personal communication with author, January 18, 2005.

21. "Alaska Timber Sales," *Timberman* (March 1926): 186; "Lumbering on Alaska Coast," *Timberman* (April 1917): 36; Canning Industry of Alaska, "The Cost of a Case of Salmon," (paid industry statement) *Ketchikan Alaska Chronicle*, January 10, 1925; "Protection for Alaska Made Boxes," *Timberman* (April 1931): 92; "Alaska Kills Proposed Fibre Container Tax," *Pacific Pulp & Paper Industry* (May 1931): 49; advertisement, Paraffine Companies, Inc., *Pacific Fisherman* (Yearbook, January 1919): 193; Canning Industry of Alaska, "A Problem Up to Alaska," (paid industry statement) *Ketchikan Alaska Chronicle*, February 6, 1925; "Lumber Mills are in Danger, Says Gardner," *(Juneau) Daily Alaska Empire*, October 2, 1930; "Alaska Cut 50,000,000 Feet," *Timberman* (November 1934): 80.

22. John A. Lee, "Why Wood Boxes are Preferred," *Timberman* (May 1927): 13; "Fibre Spells Doom of Wooden Box," *Pacific Pulp & Paper Industry* (February 1929): 36–37.

23. James A. Crutchfield and Giulio Pontecorvo, *The Pacific Salmon Fisheries: A Study of Irrational Conservation* (Baltimore: Johns Hopkins Press, 1969), 79; House Committee on Merchant Marine and Fisheries, *Fish Traps in Alaskan Waters* (hearings, January 15–16, 1936), 74th Cong., 2d sess., 1936, 211.

24. John N. Cobb, *Pacific Salmon Fisheries*, Bureau of Fisheries Doc. No. 1092, 4th ed. (Washington: GPO, 1930), 483; John N. Cobb, *Salmon Fisheries of the Pacific Coast*, Bureau of Fisheries Doc. No. 751 (Washington: GPO, 1911), 27.

25. U.S. Forest Service, "The Forests of Alaska," *Timberman* (July 1936): 17; Wyckoff, "Lumber Industry on Tongass National Forest," 4.

26. Adam Greenwald (former logger), personal communication with author, October 11, 2004 and July 19, 2005.

27. "National Forest Timber for Sale," *Ketchikan Alaska Chronicle*, January 28, 1920; U.S. Forest Service, Forest Description, S. J. Kane timber sale application, February 4, 1912, RG 95, NARA—Pacific Alaska Region, Anchorage.

28. Wyckoff, "History of Lumbering in Southeast Alaska," 40; Adam Greenwald (former logger), October 11, 2004.

29. Ernest F. Burchard, "*Marble Resources of Southeastern Alaska*, USGS Bull. No. 682, with a section on the Geography and the Geology by Theodore Chapin (Washington: GPO, 1920), 12, 13; Wyckoff, "Lumbering in Alaska," 8; Heintzleman, *Pulp-Timber Resources of Southeastern Alaska*, 31; "Tug Brings Big Tow of Piling Saturday," *Ketchikan Alaska Chronicle*, May 19, 1924; Ketchikan Trap Owners Association, *Fish Traps* (Ketchikan: Ketchikan Trap Owners Association, 1936): 15.

30. Cobb, *Pacific Salmon Fisheries*, 482; Steve Colt, *Salmon Fish Traps in Alaska* (January 11, 1999) www.iser.uaa.alaska.edu/publications/fishrep/fishtrap.pdf (accessed March 13, 2007); "Southeast Alaska Leads the World in Production of Canned Salmon," *Pacific Fisherman* (September 1923): 9; H. C. Scudder, *The Alaska Salmon Trap: Its Evolution, Conflicts, and Consequences*, Historical Monograph No. 1, 1970, 9, Historical Collections, Alaska State Library, Juneau.

31. Wyckoff, "Airplane Spruce Chances."

32. U.S. Forest Service, "Report of Timber Cut, Baldwin and Paddock, September 25, 1919," RG 95, NARA—Pacific Alaska Region, Anchorage; Wyckoff, "Lumber Industry on Tongass National Forest," 4; Wyckoff, "History of Lumbering in Southeast Alaska," 39, 40; Fish Traps: A Souvenir of the Ketchikan Industrial Fair, September 4–7, 1936 (Ketchikan: Ketchikan Trap Owners Association, 1936), 15.

33. "General Use of Fish Traps Barred in Alaska Salmon Fishery," DOI news release, March 9, 1959; Alaska Laws 1959, c. 17. As amended by id., c. 95.

34. "Tribute to Sitka Spruce," *Timberman* (August 1917): 2.

35. "Sitka Spruce for Airplanes," *American Lumberman* (October 8, 1927): 38; "Tromble Gets Big Spruce Contract," *Wrangell Sentinel*, January 17, 1918.

36. W. G. Weigle, "The Production of Airplane Spruce in Alaska," *Wrangell Sentinel*, September 5, 1918.

37. "Spruce Clears in Urgent Demand," *American Lumberman* (June 19, 1915): 31.

38. "Alaska Spruce for the Manufacture of Airplanes," *Wrangell Sentinel*, June 21, 1917; J. L. Mackenchuil (USFS ranger at Petersburg) to Forest Supervisor, March 27, 1920, RG 95, NARA—Pacific Alaska Region (Anchorage); "Sitka Spruce for Airplanes," *American Lumberman*, 38.

39. John Rustgard to Council of National Defense, Washington, D.C., March 2, 1918, RG 95, NARA—Pacific Alaska Region, Anchorage.

40. Executive Order No. 2854, May 2, 1918.

41. W. G. Weigle and Frank K. Andrews, unpublished report regarding the examination of Sitka spruce at Lituya Bay, June 22, 1918, RG 95, NARA—Pacific Alaska Region, Anchorage.

42. Weigle, "National Forest Administration in Alaska Prior to 1920," 68.

43. Executive Order No. 2938, August 16, 1918.

44. Proclamation No. 2330, April 18, 1939 (53 Stat. 2534); Public Land Order No. 551, 14 F.R. 695, February 4, 1949; B. Frank Heintzleman, approved notes, February 17, 1932, Glacier Bay submerged lands issue file, Glacier Bay National Park and Preserve, Gustavus, Alaska.

45. "Europe Lumber Trade Increasing," *Petersburg Weekly Report*, December 14, 1923; "Big Increase Alaska Spruce for Exporting," *Ketchikan Alaska Chronicle*, March 6, 1923; "Alaska's Modern Sawmills," *Timberman* (October 1930): 41; "Market Conditions—General Trade Outlook: Spruce," *Timberman* (April 1925): 42; "Spruce Mills 50 Years Young; Is Major Industry," *Ketchikan Alaska Chronicle*, May 21, 1953; USFS, "Alaska Lumber Shipments," November 30, 1923, RG 95, NARA (College Park, Md.); "Ketchikan Spruce Mills," *West Coast Lumberman* (November 1937): 57; "Spruce Plane is Proving Better Than Metal Ones," *Ketchikan Alaska Chronicle*, October 8, 1926.

46. *Report of the Governor of Alaska to the Secretary of the Interior* (Washington: GPO, 1920), 31.

47. "Petersburg Lumber Making Long Trip," *Petersburg Weekly Report*, August 5, 1921; "National Forester Tells of Alaskan Forest Possibilities," *Petersburg Weekly Report*, January 5, 1923.

48. "Ceiling Price Announced for Alaska Lumber, *Alaska Fishing News* (Ketchikan), February 10, 1942.

49. "Ships Alaskan Spruce to Australia," *American Lumberman* (July 7, 1923): 56.

50. "First Export Alaska Lumber Leaves Ketchikan," *Ketchikan Alaska Chronicle*, September 18, 1922; "Ships Alaskan Spruce to Australia," *American Lumberman* (July 7, 1923): 56.

51. "Big Increase Alaska Spruce for Exporting," *Ketchikan Alaska Chronicle*, March 6, 1923.

52. "Ships Alaskan Spruce to Australia," *American Lumberman* (July 7, 1923): 56.

53. W. J. Frost, incomplete report of timber cruise in southeast Alaska, ca. 1928 (unpublished), 2, Robert A. Kinzie collection, MS 21, Historical Collections, Alaska State Library, Juneau.

54. A. W. Greeley, "Alaska's Acres at Work…at Last," *American Forests* (October 1954): 11.

55. A. W. Greeley, "Alaska Fosters Green Gold Rush," *Western Conservation Journal* (May–June 1956): 52.

56. *Annual Report of the Governor of Alaska, 1957* (Washington: GPO, 1957), 56; George W. Rogers, *Alaska in Transition: The Southeast Region* (Baltimore: Johns Hopkins Press, 1960), 74.

CHAPTER 7

1. Sundry Civil Expenses Appropriations Act of June 4, 1897 (30 Stat. 35).

2. National Forest Management Act of 1976 (90 Stat. 2949).

3. "Sale of Timber," *Wrangell Sentinel*, April 6, 1911.

4. Herman Lahikainen to forest supervisor, October 1912, RG 95, NARA—Pacific Alaska Region, Anchorage.

5. *National Forest Manual: Trespass* (Washington: GPO, 1911), 10–11.

6. Harold Smith (USFS) to Hydaburg Trading Company, October 7, 1919, RG 95, NARA—Pacific Alaska Region, Anchorage.

7. W. G. Weigle, "Tongass National Forest," (manuscript, probably 1913), RG 95, NARA—Pacific Alaska Region, Anchorage.

8. "Salable Timber in Craig District is Surveyed for Use," *Ketchikan Alaska Chronicle*, November 29, 1924; "Busy Season Expected on West Coast," *Ketchikan Alaska Chronicle*, December 9, 1924.

9. Though it was often referred to as a "territory" informally and even in U.S. Government literature, Alaska was officially a "district" until 1912.

10. Act of May 14, 1898 (30 Stat. 414), § 11.

11. "Conditions in Alaska," *Columbia River and Oregon Timberman* (April 1902): 37.

12. Act of February 1, 1905 (33 Stat. 628), § 2.

13. Olmsted, "Inspection Report on the Alexander Archipelago Forest Reserve, 1906," 7.

14. Appropriations Act of March 4, 1907 (34 Stat. 1269).

15. Appropriations Act of March 4, 1913 (37 Stat. 839); reaffirmations began with Act of June 30, 1914 (38 Stat. 425).

16. William B. Greeley to Secretary of Agriculture, January 6, 1928, on file at USFS, Region 10; Ray E. Granvall, Craig L. Amundsen and Larry L. Ismert, *Sitka Spruce: An Economic Perspective* (Wilsonville, Oreg.: Cascade Appraisal Services, 1993), 70.

17. Heintzleman, *Pulp-Timber Resources of Southeastern Alaska*, 24.

18. B. Frank Heintzleman to Charles Flory, February 24, 1931, RG 95, NARA, College Park, Md.

19. "Alaskan Invasion: Japanese Build Pulp Mill, Seek Coal, Oil, Iron in Newest State," *Wall Street Journal*, January 21, 1959.

20. B. Frank Heintzleman to Charles Flory, February 24, 1931.

21. B. Frank Heintzleman to Forester, February 23, 1923.

22. Heintzleman, "Red and Alaska Cedar in Southeastern Alaska," 33; "Shipment Yellow Cedar Sent South," *Petersburg Weekly Report*, March 23, 1923.

23. P. D. Hanson, (Alaska Regional Forester), "Alaska—Last Frontier," *Western Conservation Journal*, July–August 1960, 79; "Initial Log Crib is Towed in From Alaska," *Marine Digest* (May 2, 1959): 13; "Cedar Crib Towed in from Alaskan Port," *Marine Digest* (July 25, 1959): 25; Harris and Farr, *Forest Ecosystem of Southeast Alaska*, 18.

24. C. M. Archbold to Alaska Regional Forester, December 6, 1946, RG 95, NARA—Pacific Alaska Region, Anchorage; Adam Greenwald (former logger), personal communication with author, February 23, 2005; John Daly (former part-owner, Ketchikan Spruce Mills), personal communication with author, September 24, 2004.

25. "Opportunities Seen in Alaska," *Timberman* (November 1945): 98; *Sourdough Notes* (USFS Region 10 newsletter), May 1948; "Logs From Alaska," *Timberman* (January 1948): 226; "Surplus Logs In This

Area Shipped Out," *Daily (Juneau) Alaska Empire*, August 14, 1948; *Sourdough Notes*, January 2, 1949.

26. *Sourdough Notes*, April 1, 1951; House Memorial No. 54, March 21, 1951, *Session Laws of Alaska*, 1951; Senate Memorial No. 6, March 22, 1951, *Session Laws of Alaska*, 1951.

27. "Industrial Alaska," *Timberman* (September 1951) 45.

28. Adam Greenwald, February 23, 2005.

CHAPTER 8

1. *Report of the Governor of Alaska to the Secretary of the Interior, 1919* (Washington: GPO, 1919), 43–44.

2. David C. Smith, *History of Papermaking in the United States (1861–1969)* (New York: Lockwood Publishing Company, 1970), 283, quoted in *Paper Trade Journal*, May 6, 1920.

3. Gifford Pinchot, "Report of the Forester," Annual Reports of the Department of Agriculture for the Year Ended June 30, 1908 (Washington: GPO, 1909): 419.

4. William D. Hagenstein, "The Old Forest Maketh Way for the New," *Environmental Law*, vol. 8, no. 2, (Winter 1978): 482–483; *A National Plan for American Forestry*, 1460. Note: "Ax" is a common alternative spelling for "axe."

5. Henry J. Berger, "Answering Secretary Wallace," *Paper Trade Journal* (October 25, 1923): 38.

6. "Alaska," *Pacific Pulp & Paper Industry* (December 1934): 21.

7. Harold E. Anderson, USFS, "Clearcutting as a Silviculture System in Southeastern Alaska," *Science in Alaska, 1954, Proceedings of the Fifth Alaskan Science Conference* (Washington, D.C.: American Association for the Advancement of Science, 1955), 1; H. E. Andersen, "Clearcutting as a Silvicultural System in Converting Old Forests to New in Southeast Alaska," *Proceedings: Society of American Foresters Meeting, 1955* (Washington: Society of American Foresters, 1956), 59; Sandor and Weisgerber, *Timber Management Plan, Tongass National Forest,* 20.

8. A. E. Helmers, "Alaska Forestry—A Research Frontier," *Journal of Forestry* (June 1960): 465; R. G. Lynch, "Alaska," *American Forests* (January 1960): 54.

9. "National Forest Timber For Sale," *Ketchikan Alaska Chronicle*, July 8, 1922.

10. Taylor, R. F., "The Role of Sitka Spruce in the Development of Second Growth in Southeastern Alaska," *Journal of Forestry*, May 1929, 532.

11. "Tree Seeds Wanted," *Wrangell Sentinel*, September 22, 1910.

12. R. A. Zeller, "Some Points on Timber Removal on the Tongass National Forest, Alaska, and Their Bearing on the Character of the Future Stand," probably 1921 (unpublished), RG 95, NARA—Pacific Alaska Region, Anchorage; R. F. Taylor, *Yield of Second-Growth Western Hemlock-Sitka Spruce Stands in Southeastern Alaska*, USDA Tech. Bull. No. 412 (Washington: GPO, 1934), 4; H. E. Andersen, "Clearcutting as a Silvicultural System in Converting Old Forests to New in Southeast Alaska," *Proceedings: Society of American Foresters Meeting, 1955* (Washington: Society of American Foresters, 1956), 59.

13. "Record Timber Sale Predicted by Forest Head," *Ketchikan Alaska Chronicle*, March 20, 1926.

14. J. H. Thickens and G. C. McNaughton, *Ground-Wood Pulp*, USDA Bull. No. 343, (Washington: GPO, April 26, 1916), 3.

15. Senate Special Committee to Study Problems of American Small Business, *Investigation of Newsprint Shortages and Other Factors Affecting Survival of Smaller Independent Newspapers* (hearings, March 4–7, 1947), 80th Cong., 1st sess., 1947, 179; "Newsprint Paper and Pulp Wood," American Forestry (April 1920): 248; "Fears Foreign Newsprint Control," New York Times, August 3, 1919; A.F. Hawes, "Raw Material for the Paper Industry," *American Forestry* (March 1920): 136.

16. Royal S. Kellogg, "War's Effect on Newsprint Paper," *American Forests* (August 1940): 376; Royal S. Kellogg, *Newsprint Paper in North America* (New York: Newsprint Service Bureau, 1948), 50; Royal S. Kellogg, "The Paper Age," *American Forests and Forest Life* (December 1924): 725.

17. Thomas R. Roach, *Newsprint: Canadian Supply and American Demand* (Durham, N.C.: Forest History Society, 1994), 24.

18. John A. Guthrie, *The Economics of Pulp and Paper* (Pullman, Wash.: State College of Washington Press, 1950), 1; Kellogg, "The Paper Age," 725.

19. *Report of the Governor of Alaska to the Secretary of the Interior* (Washington: GPO, 1885), 11; Bernhard E. Fernow, "Forests of Alaska," *Harriman Alaska Series, Vol. II, History, Geography, Resources* (Washington: Smithsonian Institution, 1910), 255.

20. Kellogg, *Forests of Alaska*, 16–17.

21. Langille, "Tongass National Forest of Southeastern Alaska," 27.

22. *Alaska Sentinel* (Wrangell), March 28, 1907; *Alaska Sentinel* (Wrangell), December 24, 1908; Patricia Roppel, "Alaskan Lumber for Australia," *Alaska Journal* (Winter 1974): 53.

23. *Alaska Sentinel* (Wrangell), February 6, 1908.

24. Walter E. Clark, "Alaska in 1959," *Colliers* (July 19, 1909): 11.

25. William B. Greeley to the Secretary of Agriculture, June 23, 1926, RG 95, NARA, College Park, Md.

26. "Russians Operated First Sawmill in Alaska with a Market in California," 32C.

27. William B. Greeley to the Secretary of Agriculture, May 14, 1921; Weigle, *Tongass National Forest.*

28. Earle H. Clapp and Charles W. Boyce, *How the United States Can Meet Its Present and Future Pulp-Wood Requirements*, USFS Bulletin, partially reprinted in "Tell About Pulp and Paper Possibilities in Territory," *Ketchikan Alaska Chronicle*, September 14, 1924.

29. "Alaska Timber for Pulp Paper," *American Forestry* (July 1913): 493; Sandor and Weisgerber, *Timber Management Plan, Tongass National Forest*, 7–8; S-SALES, D-6, Tongass, Alaska-Pacific Pulp & Paper Company, May 12, 1913, RG 95, NARA—Pacific Alaska Region, Anchorage.
30. "Alaska Timber for Paper Pulp," *American Forestry* (July 1913): 493; Rakestraw, *U.S. Forest Service in Alaska*, 77; Henry S. Graves to Secretary of Agriculture, March 21, 1913, RG 95, NARA—Pacific Alaska Region (Anchorage); C. S. Judd to Lewis Stockley, April 22, 1913, RG 95, NARA—Pacific Alaska Region (Anchorage); C. S. Judd to Mr. Weigle, October 23, 1913, RG 95, NARA—Pacific Alaska Region, Anchorage.
31. *Report of the Governor of Alaska to the Secretary of the Interior* (Washington: GPO, 1914), 13.
32. Graves, "The Forests of Alaska," 33–34, 28.
33. "Think Paper Famine May Be Temporary," *New York Times*, August 11, 1916.
34. "Forest Service has New Advertising Stunt," *American Lumberman* (August 26, 1916): 42; Clinton G. Smith, *Regional Development of Pulpwood Resources of the Tongass National Forest, Alaska*, USDA Bull. No. 950 (Washington: GPO, 1921), 17.
35. "Wrangell Ideal Location for a Large Paper Mill," *Wrangell Sentinel*, September 30, 1916; Sandor and Weisgerber, *Timber Management Plan, Tongass National Forest*, 8; Maurice D. Leehey, Behm Canal Unit, S-SALES, Tongass, September 17, 1917, RG 95, NARA—Pacific Alaska Region, Anchorage; "Many Paper Mills Will Be Running in Southeastern Alaska, Forester Predicts," *Ketchikan Alaska Chronicle*, September 8, 1919.
36. B. T. McBain, "Pulp Possibilities of Alaska and Columbia River," *Timberman* (May 1927): 109.
37. Evangeline Atwood and Robert N. DeArmond (comps.) *Who's Who in Alaskan Politics, 1884–1974* (Portland, Oreg.: Binford & Mort, 1977), 42; "Heintzleman Alaska Forester," *American Forests* (February 1937): 86; *January Field Program, 1919* (Washington: USFS, 1919), 85; B. Frank Heintzleman, "The Forests of Alaska as a Basis for Permanent Development," *Alaska Magazine*, Vol. 1, No. 4 (April 1927): 169; "Alaskan News Print Industry Opening," *Paper Trade Journal* (April 15, 1926): 40; "New Alaska Governor is Mr. Forester," *Ketchikan Alaska Chronicle*, February 24, 1953; "Alaska Pulp Sales," *(U.S. Forest) Service Bulletin*, May 9, 1927; "Bids Are Called For Alaska's First Mill," *Pulp & Paper* (September 1947): 28.
38. Untitled personal recollections manuscript, 37, Melvin L. Merritt Papers, Ax 066, Div. of Spec. Collections and University Archives, University of Oregon, Eugene, Oregon.
39. "Fears Foreign Newsprint Control," *New York Times*, August 3, 1919.
40. Louis Bloch, "Ocean Falls…1912–1917," *Pulp and Paper Magazine of Canada* (September 1947): 66–67; "The Pulp Situation," *Ketchikan Alaska Chronicle*, June 16, 1920; "Capital Only Thing Needed to Produce Paper in Alaska," *Ketchikan Alaska Chronicle*, July 4, 1921.
41. "Alaska Has Great Resource In Timber," *Wall Street Journal*, May 11, 1921; Sherman Rogers, "The Treasure-House of Southeastern Alaska," *Outlook* (December 13, 1922): 659.
42. Joseph Cummings Dort, *Water Powers of Southeastern Alaska*, Federal Power Commission, (Washington: GPO, 1924), 1; Ward T. Bower, *Alaska Fishery and Fur-Seal Industries in 1923*, Bureau of Fisheries Doc. No. 973 (Washington: GPO, 1925), 72.
43. "Alaska Paper Industry Linked with Water Power," *Paper Trade Journal* (October 9, 1924): 41.
44. B. Frank Heintzleman, "Newsprint From The North," *Pacific Pulp & Paper Industry* (February 1927): 12; *Report of the Governor of Alaska to the Secretary of the Interior* (Washington: GPO, 1927), 49; *Annual Report of the Governor of Alaska to the Secretary of the Interior, 1948* (Washington: GPO, 1948), 36.
45. B. Frank Heintzleman, "Water Power in Alaska," *Cordova Daily Times All-Alaska Review for 1929* (Cordova, Alaska: *Cordova Daily Times*, 1929), 10–11; *The Forests of Alaska* (Juneau: USFS, Alaska Region, July 1, 1944), 7.
46. Heintzleman, "Newsprint from the North," 12.
47. *Report of the Forester for FY Ended June 30, 1925* (Washington: GPO, 1925), 2.
48. William B. Greeley, "The Battle of the Secretaries," *American Forests* (May 1951): 11.
49. "Chief Forester Sees Great Opportunities for Development of Lumber and Paper Industries in Alaska," *American Lumberman* (November 13, 1920): 71.
50. "Forester Tells of Alaskan Possibilities," *American Lumberman* (September 11, 1920): 87; "Alaska Our Ace in the Hole," *Paper Trade Journal* (September 6, 1923): 44.
51. The United States Shipping Board was abolished in 1934. "Chief Forester of Alaska Advertising Alaskan Timber," *Ketchikan Alaska Chronicle*, March 11, 1921 (This piece reprints an entire article by Greeley that appeared in the March 1921 issue of *Nation's Business*); "Chief Forester Sees Great Opportunities for Development of Lumber and Paper Industries in Alaska," *American Lumberman* (November 13, 1920): 71–72; William B. Greeley, "What is Wrong with Alaska," *American Forestry* (April 1921): 205; "Alaskan Pulpwood Forests," *American Forestry* (May 1920): 192.
52. "Chief Forester Sees Great Opportunities," 71–72; "Chief Forester of Alaska Advertising Alaskan Timber."
53. *USDA Weekly Newsletter*, May 19, 1920, 1–4.
54. George H. Cecil, "Alaska Pulpwood Resources to be Developed," *Timberman* (June 1920): 36.
55. William B. Greeley to the Secretary of Agriculture, May 4, 1920, RG 95, NARA, College Park, Md.; "Chief Forester of Alaska Advertising Alaskan Timber"; *Report of the Forester for FY Ended June 30, 1921* (Washington: GPO, 1921), 5; *Report of the Forester for FY Ended June 30, 1927* (Washington: GPO, 1927), 23, 24.

56. Heintzleman, "The Forests of Alaska as a Basis for Permanent Development," 171.
57. B. Frank Heintzleman, "Huge Forests to Produce Pulp and Paper," *Cordova Daily Times All-Alaska Review for 1927* (Cordova, Alaska: *Cordova Daily Times*, 1927), 2.
58. William B. Greeley, "Alaska—Uncle Sam's Coming Paper Factory," *World's Work* (November 1928): 85.
59. "Col. Greeley, Good Friend of Alaska, Here With Harding," *Ketchikan Alaska Chronicle*, July 11, 1923; "Chief Forester of Alaska Advertising Alaskan Timber."
60. Sherman Rogers, "The Treasure-House of Southeastern Alaska," *Outlook* (December 13, 1922): 658.
61. "Alaska Forestry Supervisor Hopes To Help Development of Country by Federal Aid," *Ketchikan Alaska Chronicle*, July 24, 1919.
62. James Anderson, "Paper From Alaska," *Scientific American* (January 22, 1921): 75; A.F. Hawes, "Raw Material for the Paper Industry," *American Forestry* (March 1920): 135.
63. "To Open Alaska Forest for Pulp," *Wrangell Sentinel*, December 25, 1919, Leon F. Kneipp to the Secretary of Agriculture, October 23, 1925, RG 95, NARA, College Park, Md.
64. Cecil, "Alaska Pulpwood Resources to be Developed," 37.
65. Clinton G. Smith, *Regional Development of Pulpwood Resources of the Tongass National Forest, Alaska*, USDA Bull. No. 950 (Washington: GPO, 1921), 4.
66. "Pulp and Paper Engineers Coming From East This Month to Look Over Field," *Ketchikan Alaska Chronicle*, August 18, 1920; "Paper Mill on Waterfront is Probable," *Ketchikan Alaska Chronicle*, August 13, 1920; "Hard Work and Not Votes Will Develop Alaska Says Head of Forestry Bureau," *Ketchikan Alaska Chronicle*, August 2, 1920; "U.S. Forester is Pleased by Investigation," *Ketchikan Alaska Chronicle*, August 20, 1920.
67. "Optimism and More Optimism," *Ketchikan Alaska Chronicle*, July 12, 1921.

CHAPTER 9

1. "Speel River Is To Be Developed," *(Juneau) Alaska Daily Empire*, October 5, 1914; "Organize For Speel River Project," *(Juneau) Alaska Daily Empire*, May 5, 1915; "Speel River Is After Ore For Manufacturing," *(Juneau) Alaska Daily Empire*, May 7, 1915.
2. *Seattle Daily Times*, October 14,1914; *(Juneau) Alaska Daily Empire*, October, 5, 1914; May, 5, 1915; May, 7, 1915; Seattle Chamber of Commerce, "Negotiations Being Closed for Billion Feet of Alaska Lumber," *Ketchikan Alaska Chronicle*, January 28, 1921.
3. Handwritten letter on file, Patrick Eugene Kennedy collection, PCA 444, Historical Collections, Alaska State Library, Juneau.
4. "Offers Alaskan Pulp Wood," *American Lumberman* (May 1, 1920): 55.
5. E. E. Carter to the Forester, May 10, 1921, RG 95, NARA (College Park, Md.); "Alaska Making Wood Pulp," *Washington Post*, March 14, 1921.
6. Joseph Cummings Dort, *Water Powers of Southeastern Alaska*, Federal Power Commission, (Washington: GPO, 1924), 5.
7. H. K. Benson, *By-Products of the Lumber Industry*, U.S. Dept. of Commerce, Special Agents Series— No. 110 (Washington: GPO, 1916), 41; "Alaska Pulp Co. Operating to Capacity," *Paper Trade Journal* (January 18, 1923): 52.
8. "Answering Secretary Wallace," *Paper Trade Journal* (October 25, 1923): 38; "Pulp Plant is Operating Now at Full Speed," *Ketchikan Alaska Chronicle*, November 9, 1922; M. B. Summers, report to U.S. Weather Bureau, February 5, 1921, RG 95, NARA, College Park, Md.; House Committee on the Territories, *Supplemental Hearings on H.R. 5694 (and Other Bills) "To Provide for the Administration of National Property and Interests in the Territory of Alaska*, 67th Cong., 1st sess., 1921, 74.
9. "Wrangell Sawmill Cutting Lumber for the First Paper Mill to be Built in Alaska," *Wrangell Sentinel*, September 2, 1920.
10. *Pacific Pulp & Paper Industry* (April 1928): 23.
11. "Alaska Pulp Co. Operating to Capacity," *Paper Trade Journal* (January 18, 1923): 52; "Pulp Plant is Operating Now at Full Speed," *Ketchikan Alaska Chronicle*, November 9, 1922; USFS, *Sale Prospectus for 1,670,000,000 cubic feet National Forest Timber embracing Two Pulpwood Projects in Alaska*, January 15, 1927, Robert A. Kinzie collection, MS 21, Historical Collections, Alaska State Library, Juneau.
12. *Report of the Governor of Alaska to the Secretary of the Interior, 1924*, (Washington: GPO, 1924), 65; Sandor and Weisgerber, *Timber Management Plan, Tongass National Forest*, 7–8; "Producing Lumber in Alaska," *Timberman* (February 1924): 100; John Rustgard, to James Wickersham, March 28, 1921; "Pulp Plant Closed," *Petersburg Weekly Report*, May 27, 1921.
13. "Progress . . .," *Ketchikan Pulp: A $52,500,000 Ultra-Modern Industry* (Ketchikan: *Ketchikan Alaska Chronicle*, 1954), unpaged.
14. "Hundred Million Board Feet Pulp Timber Will Be Sold," *Ketchikan Alaska Chronicle*, April 30, 1920.
15. Cecil, "Alaska Pulpwood Resources to Be Developed," 36–37; Clinton G. Smith, *Regional Development of Pulpwood Resources of the Tongass National Forest, Alaska*, USDA Bull. No. 950 (Washington: GPO, 1921), 1; Clinton G. Smith to Mr. Ballard, June 13, 1921, RG 95, NARA, College Park, Md.
16. Wyckoff, "The Forest Resources of Alaska," 7.
17. "Text of Harding's Speech Given at Seattle," *Ketchikan Alaska Chronicle*, July 31, 1923.
18. Clapp and Boyce, *Present and Future Pulp-Wood Requirements*, 54.
19. "Alaska News Print Industry Opening," *Paper Trade Journal* (April 15, 1926): 40.
20. Sherman Rogers, "The Treasure-House of Southeastern Alaska," *Outlook* (December 13, 1922): 658.

21. Clinton G. Smith, Regional Development of Pulpwood Resources of the Tongass National Forest, Alaska, USDA Bull. No. 950 (Washington: GPO, 1921), 4, 15.
22. "Alaska May Supply World's Pulp," *American Lumberman* (March 14, 1921): 61.
23. William B. Greeley to the Secretary of Agriculture, May 14, 1921, RG 95, NARA, College Park, Md.; R. A. Zeller, "Possibilities for Paper and Power Explained," *Ketchikan Alaska Chronicle*, July 4, 1920; "Forest Service Boosts Alaska," *American Forestry* (July 1920): 443.
24. LeRoy W. Huntington, "Forest Aspects of Alaska," *Timberman* (June 1923): 130.
25. "National Forest Timber Sales," *American Lumberman* (November 13, 1915): 51.
26. Wyckoff, "Alaska Pulp Making," 9.
27. *Report of the Forester for FY Ended June 30, 1921* (Washington: GPO, 1921), 3–4.
28. Smith, *Regional Development of Pulpwood Resources of the Tongass*, 3.
29. "Alaskan Papermakers Have Heavy Freight Handicap," *Wall Street Journal*, October 1, 1921.
30. *Report of the Forester for FY Ended June 30, 1921* (Washington: GPO, 1921), 4.
31. Ken Drushka, *Working in the Woods* (Madiera Park, B.C.: Harbour Publishing, 1992), 59; "Chart Shows British Columbia Timber Titles," *American Lumberman* (November 15, 1924): 71.
32. R. A. Zeller, "Possibilities for Paper and Power Explained," *Ketchikan Alaska Chronicle*, July 4, 1920.
33. Alaska Legislature, House Joint Memorial No. 22, April 28, 1921.
34. City of Ketchikan to Senator Miles Poindexter, March 17, 1921, RG 95, NARA, College Park, Md.
35. USFS, *Sale Prospectus: 335,000,000 Cubic Feet National Forest Pulp Timber West Admiralty Island Unit, Tongass National Forest*, (Washington: GPO, 1921); "100,000,000 Cubic Feet National Forest Timber and Pulpwood for Sale," (advertisement), *Ketchikan Alaska Chronicle*, May 14, 1921; "The Tongass National Forest," *Science* (August 1921): 166.
36. E. E. Carter to the Forester, May 10, 1921, RG 95, NARA (College Park, Md.); Rakestraw, *U.S. Forest Service in Alaska*, 109; David and Brenda Stone, *Hard Rock Gold* (Juneau: Juneau Centennial Committee, 1980), 39, 53.
37. USFS, *Sale Prospectus: 335,000,000 Cubic Feet*, 5.
38. Ibid., 6, 11.
39. House Committee on Agriculture, *Hearings on H. J. Resolution 205: To Authorize the Secretary of Agriculture to Sell Timber Within the Tongass National Forest* (May 26; June 14; July 1, 3, and 9, 1947), 80th Cong., 1st sess., 1947, 27.
40. "Gastineau Co. Property Sold by Auctioneer," *Ketchikan Alaska Chronicle*, December 31, 1924; Stone, *Hard Rock Gold*, 54.
41. "$4,000,000 Paper Mill," *Ketchikan Alaska Chronicle*, July 16, 1921; "Pulp Plant to be Under Construction Coming Winter," *Ketchikan Alaska Chronicle*, October 12, 1923; "Alaska Pulpwood Timber Sold," *American Lumberman* (July 23, 1921): 52; "One Million Cords of Alaska Pulpwood Sold," *American Forestry* (August 1921): 542; "Shrimp Bay Paper Pulp Plant Assured," *Ketchikan Alaska Chronicle*, November 5, 1921.

CHAPTER 10

1. George E. Baldwin, "The Alaskan Question," in *Alaskan Problems*, 62nd Cong., 2d sess., 1912, Sen. Doc. No 572, 5.
2. "Give the Alaskans a Chance," *Washington Post*, January 19, 1915; "Unlocking Alaska," *American Forestry* (July 1914): 470–472; "To Use Alaska's Forests," American Forestry (January 1922): 25.
3. Henry C. Wallace, "President Harding on Alaska," *Timberman* (September 1923): 179–180.
4. William B. Greeley, *Forests and Men* (Garden City, N.Y.: Doubleday & Company, 1951), 96.
5. American Forestry Association news sheet reproduced in "The Fight for Alaska's Forests," *American Forestry* (April 1922): 202.
6. Henry C. Wallace, "President Harding on Alaska," *Timberman* (September 1923): 179–180.
7. "Chief Forester Made No Criticism," *American Lumberman* (March 11, 1922): 70.
8. "Pulp and Paper," *Ketchikan Alaska Chronicle*, July 8, 1923.
9. "The Fight for Alaska's Forests," 202.
10. Charles G. Ross, "With Harding in Alaska," *American Forestry* (October 1923): 582.
11. Wallace, "President Harding on Alaska," 179–180.
12. "Forest Service Should Retain Control of Alaska's Forest Resources," *Timberman* (July 1923): 23.

CHAPTER 11

1. "334,000,000 Cubic Feet National Forest Timber and Pulpwood for Sale," (advertisement) *Ketchikan Alaska Chronicle*, March 2, 1923.
2. B. Frank Heintzleman, "Pulp-Timber Resources of Southeastern Alaska," *Timberman* (March 1929): 166.
3. "Ten Million to Be Spent at Thomas Bay," *Petersburg Weekly Report*, July 14, 1922; "Paper Mill Magnate Secures Many Pictures From This Region," *Wrangell Sentinel*, August 9, 1923.
4. USFS timber sale notice, "334,000,000 Cubic Feet National Forest Timber and Pulpwood," *American Lumberman* (April 28, 1923): 51; USFS, *Sale Prospectus: 334,000,000 Cubic Feet National Forest Pulp Timber, Cascade Creek Unit, Tongass National Forest*, (May 1, 1923), 5; William B. Greeley to the Secretary (Agriculture), August 22, 1923, RG 95, NARA (College Park, Md.); "Power License in Alaska Revoked," *Paper Trade Journal* (July 8, 1926): 53; "Dougherty Tells Ketchikan People of Development,"

Petersburg Weekly Report, July 21, 1922; "Big Pulp Plant Planned Cascade Creek," *Ketchikan Alaska Chronicle*, August 2, 1924.

5. "For Print Paper Industry in Alaska," *Paper Trade Journal* (February 22, 1923): 56.
6. "Wallace Foresees Great Alaskan Paper Company," *Ketchikan Alaska Chronicle*, November 7, 1923.
7. Henry J. Berger, "Answering Secretary Wallace," *Paper Trade Journal* (October 25, 1923): 38.
8. "Government Sells Alaska Pulp Wood," *New York Times*, August 24, 1923; "Power License in Alaska Revoked," *Paper Trade Journal* (July 8, 1926): 53; "Power License Revoked," *Petersburg Herald*, June 5, 1926.
9. Henry J. Berger, "Alaska's Problem," *Paper Trade Journal* (September 27, 1923): 44.
10. Berger, "Answering Secretary Wallace," 38; Berger, "Alaska's Problem," 44.
11. "Mr. Wallace Protests Alaska Editorial," *Paper Trade Journal* (October 25, 1923): 39.
12. Heintzleman, "Newsprint From the North," 11.
13. "Transportation Problem is Holding Back Paper Mills," *Ketchikan Alaska Chronicle*, May 24, 1920.
14. "Alaska Pulp Men Seek Special Rate," *Ketchikan Alaska Chronicle*, October 12, 1921.
15. "Forest Service Offers Alaska Pulp Timber," *Timberman* (February 1927): 104; "Opportunities for Development of the Newsprint Paper Manufacturing Industry on the Tongass National Forest, Southeastern Alaska," February 1926, unknown author, but almost certainly a USFS document, RG 95, NARA, College Park, Md.
16. "Discrimination Against Alaskan Sawmills," *Timberman* (January 1929): 35; L. Ethan Ellis, *Print Paper Pendulum: Group Pressures and the Price of Newsprint* (New Brunswick, N.H.: Rutgers University Press, 1948), 132.
17. "Crown-Zellerbach Merger," *Timberman*, (June 1928): 106.
18. Warren Bullock, writing in *American Forestry*, reported in 1923 that some two million tons of waste paper, worth about $50 million, was recovered annually in the U.S. According to Bullock, six tons of recovered paper produced the equivalent in pulp of an acre of virgin spruce timber. The recovered paper was utilized not for newsprint paper, but for book paper and paper board. As early as 1922 the Forest Service's Forest Products Laboratory at Madison, Wisconsin was researching the de-inking of pulp manufactured from old newspapers. The de-inked pulp was to be utilized to manufacture new newsprint paper. "Forest Service Issues Alarmist Propaganda," *American Lumberman* (January 24, 1920): 37; Warren B. Bullock, "Saving Forests by Saving Paper," *American Forestry* (November 1923): 654; "Forest Laboratory Expert Is In Territory to Give Service," *Ketchikan Alaska Chronicle*, September 20, 1922.
19. "National Forester Tells of Alaskan Forest Possibilities," *Petersburg Weekly Report*, January 5, 1923.
20. Heintzleman, *Pulp-Timber Resources of Southeastern Alaska*, 22–23.

CHAPTER 12

1. "Change Made for Disposing Alaska Timber," *Ketchikan Alaska Chronicle*, December 28, 1925.
2. "Forestry Will Seek Pulp Men," *Ketchikan Alaska Chronicle*, November 30, 1925. (Indirect quote of Charles Flory, district forester)
3. Heintzleman, "Newsprint from the North," 11; "C. H. Flory is Enthusiastic Over Flights," *Ketchikan Alaska Chronicle*, July 1, 1926.
4. "Aerial Survey of Alaskan Forests," *American Lumberman* (August 14, 1926): 73; "Aerial Survey of Alaska," *American Lumberman* (October 23, 1926): 74; Henry J. Berger, "Prospects for Making Paper in Alaska," *Paper Trade Journal* (October 21, 1926): 38.
5. In terms of forest management, the expected growth and yield of trees and forests is commonly integrated into a concept called a rotation age. A rotation age, in even-aged systems, is simply the period (in years) between regeneration establishment and the final cutting. Clapp and Boyce, *Present and Future Pulp-Wood Requirements*, 54; Heintzleman, "Newsprint From The North," 11; Bill Keil, "Decay Takes Toll on Alaska's Mature Timber," *Timberman* (September 1960): 39; Heintzleman, *Pulp-Timber Resources of Southeastern Alaska*, 18; Paul Hennon (forest pathologist, USFS), e-mail message to author, August 25, 2005; James W. Kimmey, *Cull Factors for Sitka Spruce, Western Hemlock and Western Redcedar in Southeast Alaska*, Station Paper No. 6 (Juneau: USFS, Alaska Forest Research Center, August 1956): 27.
6. Frost, incomplete report of timber cruise in southeast Alaska, 1.
7. Taylor, "Early Forest Research in Alaska," 31–41; R. F. Taylor, *Yield of Second-Growth Western Hemlock-Sitka Spruce Stands in Southeastern Alaska*, USDA Tech. Bull. No. 412 (Washington: GPO, 1934), 6; Alaska Resources Committee, *Alaska*, 98; "U.S. Proposes to Sell Big Timber Stands For Private Pulp Industry in Alaska," *Pacific Pulp & Paper Industry* (May 1944) 27; B. Frank Heintzleman, "Forests of Alaska," *Trees, The Yearbook of Agriculture, 1949* (Washington: GPO, 1949), 360–372.
8. *Report of the Chief of the Forest Service*, 1946, 34.
9. Heintzleman, "The Forests of Alaska as a Basis for Permanent Development," 174.
10. Heintzleman, "Forests of Alaska," *Trees*, 360–372.
11. Heintzleman, "Newsprint From the North," 12.
12. "Milestones in Alaska's History," *Pacific Pulp & Paper Industry* (May 1927): 38; B. Frank Heintzleman, "Huge Forests to Produce Pulp and Paper," *Cordova Daily Times All-Alaska Review for 1927* (Cordova, Alaska: *Cordova Daily Times*, 1927), 2.

13. Charles R. Berry, "Economic Timber Policy Urged," *Timberman* (April 1924): 180; B. T. McBain, "Pulp Possibilities of Alaska and Columbia River," *Timberman* (May 1927): 109.
14. "A Questionable Policy," *West Coast Lumberman* (May 1, 1927): 31.
15. "Pulp Production in Washington," *American Lumberman* (January, 12 1929): 56.
16. "Federal Policy on Timber Sales," *American Lumberman* (April 9, 1927): 73.
17. M. R. Higgins, "Scientific Economy Vital in Pulp Production," *Pacific Pulp & Paper Industry* (February 1927): 8.
18. Allen H. Hodgson, "Pulpwood from Logging Waste," *Timberman* (December 1928): 182; Kellogg, "Making News Print Paper," 35.
19. "How About the Forest Reserves?" *Pacific Pulp and Paper Industry* (September 1928): 27; House Committee on Agriculture, *Hearings on H. J. Resolution 205*, 27.
20. B. T. McBain, "Pulp Possibilities of Alaska and Columbia River," *Timberman* (May 1927): 109.
21. "Paper Famine if Forests are Wasted," *American Forestry* (February 1920): 94.
22. *Report of the Forester for FY Ended June 30, 1927* (Washington: GPO, 1927), 23, 24.
23. E. F. [Emmanuel Fritz], "Review of 'The Distribution and the Mechanical Properties of Alaska Woods,'" *Journal of Forestry* (May 1931): 849.

CHAPTER 13

1. "Will Examine Resources of Pulp Timber," *Ketchikan Alaska Chronicle*, November 23, 1925; "New Forestry Policy," *Ketchikan Alaska Chronicle*, December 2, 1925; "Heintzleman Will Address Paper Makers," *Ketchikan Alaska Chronicle*, February 19, 1926; "Pulp Makers to Inspect Local Field," *Ketchikan Alaska Chronicle*, May 26, 1926; "Heintzleman is Optimistic Over Forest Development," *Ketchikan Alaska Chronicle*, October 22, 1926.
2. "Hearst to Build Paper Mill," *Ketchikan Alaska Chronicle*, November 5, 1926; "Zellerbachs Invade Field," *Ketchikan Alaska Chronicle*, November 24, 1926; "Zellerbachs Announce Plans for Huge Developments Here," *Ketchikan Alaska Chronicle*, December 13, 1926.
3. "Crown and Zellerbach Join Forces," *Pacific Pulp & Paper Industry* (April 1928): 23; William B. Greeley to files, September 1, 1923, RG 95, NARA, College Park, Md.; "S.F. Company Plans for Greatest Paper Mill Plant on U.S. Territory in Alaska," *Ketchikan Alaska Chronicle*, February 1, 1923.
4. "Large Sales of National Forest Timber to Establish Paper Industry in Alaska," *West Coast Lumberman* (February 1, 1927): 32.
5. "Print Paper From Alaska," *Washington Post*, January 17, 1927.
6. "Forest Service Offers Alaska Pulp Timber," *Timberman* (February 1927): 103; "Large Sales of National Forest Timber to Establish Paper Industry in Alaska," *West Coast Lumberman* (February 1, 1927): 32–33; Arthur M. Hyde to President of U.S. Senate, March 19, 1930, in *Pulpwood Supply in Alaska*, 71st Cong., 2d sess., Sen. Doc. No. 120 (Washington: GPO, 1930), 151; USFS, *Sale Prospectus: 1,670,000,000 Cubic Feet*; Ross, "With Harding in Alaska," 582.
7. William B. Greeley to Secretary (Agriculture), June 23, 1926, RG 95, NARA, College Park, Md.
8. Federal Water Power Act (41 Stat. 1063); "Alaska Pulp Making," *Pathfinder* (Valdez, Alaska) (November 1920): 7–10; "Liberal Terms Offered Pulp Manufacturers in Southeastern Alaska," *Wrangell Sentinel*, March 10, 1921.
9. "Forest Service Offers Alaska Pulp Timber," *Timberman* (February 1927): 103, 105; B. Frank Heintzleman, "Alaska and the Paper Industry," *Pacific Pulp & Paper* (January 1928): 72; "Large Sales of National Forest Timber to Establish Paper Industry in Alaska," *West Coast Lumberman* (February 1, 1927): 33.
10. "Cameron to Study Alaska Project," *Pacific Pulp & Paper Industry* (June 1927): 23.
11. "National Forest Timber Sales," *American Lumberman* (April 23, 1927): 74; "Zellerbach & Cameron Awarded Alaska Pulp Timber Tracts," *Pacific Pulp & Paper Industry* (May 1927): 10.
12. "Engineers Go North for Alaska Survey," *Pacific Pulp & Paper Industry* (March 1928): 29; Heintzleman, "Alaska and the Paper Industry," 74.
13. "Alaska Foresters Praised," *Pacific Pulp & Paper Industry* (July 1927): 44; "C. H. Flory and His Assistants are Commended," *(Juneau) Daily Alaska Empire*, May 6, 1927.
14. *(U.S. Forest) Service Bulletin*, May 9, 1927.
15. Roach, *Newsprint*, 6; Kellogg, *Newsprint Paper in North America*, 50; "News Print Manufacturers 'In the Dark,'" *Pacific Pulp & Paper Industry* (July 1928): 36.
16. "Kellogg Urges Sound Development," *Pacific Pulp & Paper Industry* (July 1927): 27.
17. "Paper Making Sure to Come to Territory," *(Juneau) Daily Alaska Empire*, August 20, 1927.
18. "Alaska's Forests Promise Steady Industry," *Pacific Pulp & Paper Industry* (June 1928): 62.
19. "News Print Manufacturers 'In the Dark,'" *Pacific Pulp & Paper Industry* (July 1928): 36.
20. "News Print in Canada," *Paper Trade Journal* (March 29, 1928): 50.
21. "Crown-Zellerbach Merger," 106; "Crown Zellerbach Offers New Preferred Issue," *Pacific Pulp & Paper Industry* (January 1929): 28; "Crown-Zellerbach Merger," *Pacific Pulp & Paper Industry* (1929 Review Number): 38.
22. "Federal Trade Commission Sends Newsprint Paper Report to Senate," Federal Trade Commission news release, July 3, 1930.
23. "Zellerbach Paper Co. Denies Charges," *Paper Trade Journal* (March 27, 1924): 34; "Coast Concern Charged With Anti-Trust Act Violation," *Pacific Pulp & Paper Industry* (February 1934): 16; Crown

Zellerbach Corporation and Crown Willamette Paper Company, *Official Supplementary Statement of Information Relating to Plan of Reorganization of Crown Zellerbach Corporation and Crown Willamette Paper Company*, 1937.

24. B. Frank Heintzleman to Forester, February 24, 1928, RG 95, NARA, College Park, Md.; "Engineers Go North for Alaska Survey," *Pacific Pulp & Paper Industry* (March 1928): 29; "Meldrum Heads Alaska Timber Cruisers," *Pacific Pulp & Paper Industry* (April 1929): 43.

25. "Urges High Tariff to Bar Alien Food," *New York Times*, July 31, 1928; "Alaska's Paper Resources," *Paper Trade Journal* (November 15, 1928): 38.

26. "Alaska's Paper Resources," *Paper Trade Journal* (November 15, 1928): 38.

27. While some considered the grizzly bear a dangerous menace that should be exterminated, a number of influential conservationists advocated—unsuccessfully—for the designation of Admiralty Island as a bear "sanctuary" in which hunting for the bruins would be prohibited. Frank Heintzleman wanted to ensure industrial access to the island's timber resources, and opposed its designation as a bear sanctuary. "Withdrawing…highly valuable lands and holding them indefinitely as wild areas solely to perpetuate the bear is wholly unnecessary and unjustified," he wrote. Heintzleman added that cutting all of Admiralty Island's merchantable timber over a rotation period of at least 75 years would "not materially affect the bear." B. Frank Heintzleman, "Managing the Alaska Brown Bear," *American Forests* (June 1932): 329; "Alaska Aerial Survey Discovers Power Site," *Paper Trade Journal* (September 5, 1929): 40; "Alaska Surveys Completed," *Pacific Pulp & Paper Industry* (November 1929): 56; Harold E. Smith, "Jack Thayer Was Killed By a Bear," *Alaska* (August 1971): 23.

28. "Those Alaska Paper Mill Projects," *Pacific Pulp & Paper Industry* (October 1929): 40; Wyckoff, "History of Lumbering in Southeast Alaska," 40; *Report of the Forester for FY Ended June 30, 1927* (Washington: GPO, 1927), 23; "Zellerbach & Cameron Awarded Alaska Pulp Timber Tracts," *Pacific Pulp & Paper Industry* (May 1927): 10; W. L. Cook, "Duty to Be Cautious," *Paper Trade Journal* (April 3, 1947): 45; B. Frank Heintzleman to Forester, August 24, 1928, RG 95, NARA (College Park, Md.); "Alaska Paper Development Project," *Timberman* (January 1930): 112.

29. *Report of the Forester for FY Ended June 30, 1927* (Washington: GPO, 1927), 23; "Alaska Paper Development Project," *Timberman* (January 1930): 112.

30. Don Meldrum, "Brief and Supplemental Reports on Pulptimber Allotment "A", Tongass National Forest, Southeastern Alaska," 1929 (unpublished), 23, 24, 45, Robert A. Kinzie collection, MS 21, Historical Collections, Alaska State Library, Juneau.

31. "Paper Mills in Alaska," *Pacific Pulp & Paper Industry* (July 1930): 27; "Report on Alaska Paper Projects Finished," *Pacific Pulp & Paper Industry* (August 1930): 34; "Alaska Paper Mills," *Pacific Pulp & Paper Industry* (December 1930): 30.

32. "Alaska Projects Seem Nearer," *Pacific Pulp & Paper Industry* (April 1930) 42; Stone, *Hard Rock Gold*, 39, 53; "Alaska Road Hints Paper Development," *Pacific Pulp & Paper Industry* (June 1931): 35; B. Frank Heintzleman, "Managing the Alaska Brown Bear," *American Forests* (June 1932): 329.

33. "The World Wags On," *West Coast Lumberman* (November 15, 1928): 18.

34. "1928 Ends With Firm Basis for Prosperous 1929," *American Lumberman* (December 29, 1928): 31.

35. "The American Paper & Pulp Association Opposes Alaska Development," *Pacific Pulp & Paper Industry* (August 1931): 13.

36. Kellogg, *Newsprint Paper in North America*, 50; Roach, *Newsprint*, 11, 17; *Lockwood's Directory*, 55th ed. (New York: Lockwood Trade Journal, Inc., 1930), preface.

37. USFS, *A National Plan for American Forestry*, 581; Senate Special Committee to Study Problems of American Small Business, *Investigation of Newsprint Shortages and Other Factors Affecting Survival of Smaller Independent Newspapers* (hearings, March 4–7, 1947), 80th Cong., 1st sess., 1947, 138.

38. "The American Paper & Pulp Association Opposes Alaska Development," 13.

39. Leon F. Kneipp to Secretary of Agriculture, August 17, 1931, RG 95, NARA, College Park, Md.

40. Letter to editor, *Pacific Pulp & Paper Industry* (September 1931): 18.

41. "Alaska," *Pacific Pulp & Paper Industry* (December 1934): 21; "Southeastern Alaska Logging Practice," *Timberman* (March 1935): 53.

42. USFS, *A National Plan for American Forestry*, 1380; "Dill Wants Bars for Wood Pulp," *Spokane Chronicle*, February 14, 1934.

43. Alaska Territorial Legislature, House Joint Memorial No. 16, 1935. Alaska State Archives.

44. Concurrent Resolution No. 24, 75th Cong., 1st sess., 2 F.R. 9592, August 21, 1937; National Resources Committee, *National Resources Planning Facts* (Washington: GPO, 1939), 6.

45. Alaska Resources Committee, *Alaska*, 104.

46. "Heintzleman Sees Possibilities for Alaska," *Pacific Pulp & Paper Industry*, 33; Heintzleman, *Pulp Timber Resources of Southeastern Alaska*, 28; "U.S. Newsprint Production Lowest in 35 years," *Pacific Pulp & Paper Industry* (May 1939): 34.

47. "Chief Forester Views Alaska Timber," *Timberman* (July 1939): 74.

48. "Silcox Looks at Alaska's Forests," *West Coast Lumberman* (September 1939): 18.

CHAPTER 14

1. *Forestry in Wartime: Report of the Chief of the Forest Service*, 1942, 10; E. E. Carter to Milton S. Briggs (Commodity Credit Corporation, USDA), March 6, 1945, RG 95, NARA—Pacific Alaska Region, Anchorage.

2. "Uncle Sam to Build Vast Air Fleet," *American Lumberman* (June 23, 1917): 38; Disque, *History of Spruce Production Division*, introduction page VII; Burt P. Kirkland, "The Iron Horse of the West," *American Forestry* (April 1923): 205.
3. Disque, *History of Spruce Production Division*, 55; *Petersburg Weekly Report*, January 18, 1918; Weigle, "The Production of Airplane Spruce in Alaska."
4. Erle Kauffman, "Spruce Goes Back to War," *American Forests* (August 1940): 364.
5. "Chief Forester of U.S. Tell of High Quality of Alaska Spruce—Passed Every Test for Airplane Construction During World War," *Wrangell Sentinel*, August 25, 1921; William B. Greeley to Major W. H. Frank, October 18, 1922, RG 95, NARA (College Park, Md).
6. Arthur W. Priaulx, "Aircraft Lumber," *American Forests* (July 1943): 342; James Montanges, "This is the Mosquito!" *American Forests* (January 1944): 14; "XP-77—New All-Wood Fighter Plane," *American Forests* (June 1945): 274; Frank H. House, *Timber at War* (London: Ernst Benn Limited, 1965), 47; Virtual Aircraft Museum, http://www.aviastar.org/air/usa/bell_xp-77.php (accessed December 20, 2007); Empire Forestry Association, London, *The Forest, Forestry, and Man* (London: Empire Forestry Association, 1947), 13.
7. "Logging on the Queen Charlotte Islands," *Timberman* (August 1945): 134.
8. Kauffman, "Spruce Goes Back to War," 364.
9. H. H. Chapman, "The War and the Parks," *Timberman* (June 1943): 16.
10. C. H. Forward, "Alaska Airplane Spruce," *West Coast Lumberman* (January 1940): 19; "Logging Airplane Spruce," *West Coast Lumberman* (September 1940): 50.
11. "Confer on Airplane Spruce," *American Lumberman* (December 28, 1940): 39; "Spruce Ceiling Established," *West Coast Lumberman* (May 1942): 52.
12. "Sitka Spruce Logs Frozen," *West Coast Lumberman* (August 1942): 65; "Aircraft Spruce Workers 'Frozen'," *West Coast Lumberman* (November 1942): 84.
13. "Lumber 'Very Critical' Material," *West Coast Lumberman* (November 1942): 72d; "Lumber in the War Theatres," *West Coast Lumberman* (February 1945): 45.
14. James W. Girard, *The Man Who Knew Trees: The Autobiography of James W. Girard*, Series Pub. No. 4 (St. Paul, Minn.: Forest Products History Foundation, Minnesota Historical Society, 1949), 30.
15. *Forestry in Wartime: Report of the Chief of the Forest Service*, 1942, 10; "U.S. Will Draw On Alaskan Spruce," *American Forests* (July 1942): 328; "First Raft of Alaska Spruce Arrives in Puget Sound," *West Coast Lumberman* (February 1943): 32b; "J. M. Wyckoff to Handle Spruce Sales," *Ketchikan Alaska Chronicle*, August 30, 1943; "Spruce Log Production in Full Swing," *Ketchikan Alaska Chronicle*, May 5, 1944; "ASLP Output May Total 85 Millions," *Ketchikan Alaska Chronicle*, June 3, 1944.
16. "First Raft of Alaska Spruce Arrives in Puget Sound," *West Coast Lumberman* (February 1943): 32d.
17. E. P. Stamm, A. J. F. Brandstrom, and P. S. Bonney, "Confidential Report to U.S. War Production Board on Plans and Feasibility of Logging and Transporting High Grade Sitka Spruce From Alaska to Puget Sound for Airplane Stock," February–March 1942, RG 95, NARA—Pacific Alaska Region, Anchorage; Rakestraw, *U.S. Forest Service in Alaska*, 120–124; "Plane Spruce Program Progressing," *Ketchikan Alaska Chronicle*, July 11, 1942.
18. Brixie Crabtree (former ASLP logger), personal communication with author, December 8, 2004; advertisement, *Ketchikan Alaska Chronicle*, August 3, 1944.
19. "Alaska Spruce Log Program Progress," *Timberman* (June 1943): 93; "Logging Jobs Now 'Critical,'" *Amerian Forests* (March 1945): 130.
20. "Alaska Spruce Log Program," *Timberman* (November 1942): 24; Rakestraw, *U.S. Forest Service in Alaska*, 120–124.
21. "Alaska Spruce," *Timberman* (February 1943): 10–12; "3 ASLP Camps to Log This Winter," *Ketchikan Alaska Chronicle*, December 24, 1943; "Logging Camps Close Temporarily," *Ketchikan Alaska Chronicle*, December 15, 1942; Brixie Crabtree (former ASLP logger), personal communication with author, December 8, 2004.
22. "First Raft of Alaska Spruce Arrives in Puget Sound," *West Coast Lumberman* (February 1943): 32d; Leonard Campbell (Wrangell, Alaska, tugboat business owner), personal communication with author, March 11, 2006.
23. "Alaska Spruce Log Program Progress," 93; Commodity Credit Corporation Logging Contract No. ASLP-089, Bowen and Dillinger Logging Company, RG 95, NARA—Pacific Alaska Region, Anchorage.
24. "First Raft of Alaska Spruce Arrives in Puget Sound," 32d; "Complete Description of the Davis Ocean Going Log Raft," *West Coast Lumberman* (December 15, 1917): 28–29; F. Wood, "Davis Raft Building," *Timberman* (November 1941): 14; Drushka, *Working in the Woods*, 87; Walter G. Hardwick, *Geography of the Forest Industry of Coastal British Columbia* (Vancouver, B.C.: University of British Columbia, 1963), 35.
25. The Morrison Mill Company had specialized in the cutting of spruce for more than 20 years, and in the last 7 years had cut only this species. "First Raft of Alaska Spruce Arrives in Puget Sound," 32d; "Specialists in Sitka Spruce," *Timberman* (December 1944): 37; "Alaska Spruce," 10–12; "Spruce Flowing From Alaska," *West Coast Lumberman* (October 1943): 74; "Tugs Unable to Keep Up With Loggers," *Ketchikan Alaska Chronicle*, August 19, 1943; G. K. Clark to files, June 17, 1944, RG 95, NARA—Pacific Alaska Region, Anchorage.
26. "Spruce Raft Breaks Apart," *Ketchikan Alaska Chronicle*, January 19, 1944.

27. "Alaska Spruce Project Gets Under Way," *West Coast Lumberman* (October 1942): 11–12; Commodity Credit Corporation Logging Contract No. ASLP-089, Bowen and Dillinger Logging Company, RG 95, NARA—Pacific Alaska Region, Anchorage.

28. "Alaska Spruce Log Program Progress," *Timberman* (June 1943): 93; "Alaska's Forests Yield Airplane Spruce," West Coast Lumberman (July 1943): 12; "Spruce Flowing From Alaska," *West Coast Lumberman* (October 1943): 74.

29. "Alaska's Forest Yield Airplane Spruce," *West Coast Lumberman* (August 1943): 55; Buol Logging Company, Edna Bay Rate Schedule, RG 95, NARA—Pacific Alaska Region, Anchorage.

30. "ASLP May Be Ended Next May," *Ketchikan Alaska Chronicle*, February 9, 1944.

31. Rakestraw, *U.S. Forest Service in Alaska*, 120–124.

32. E. E. Carter to Milton S. Briggs (Commodity Credit Corporation, USDA), March 6, 1945, RG 95, NARA—Pacific Alaska Region, Anchorage.

33. Ibid.

34. Ibid.

35. "Drastic Curtailment in Aero Spruce," *Timberman* (June 1944): 112; "Heintzleman Visits," *West Coast Lumberman* (February 1945): 102.

36. *Annual Report of the Governor of Alaska to the Secretary of the Interior for the Fiscal Year Ended June 30, 1943* (Washington: GPO, 1943), 16; Theodore Catton, *Land Reborn: A History of the Administration and Visitor Use in Glacier Bay National Park and Preserve* (Anchorage: National Park Service, 1995), 92–93; Frank T. Been, "Field Notes and Interviews, Inspection of Glacier Bay and Sitka National Monuments, August 12 to September 4, 1942, 24–25." Report on file at Glacier Bay National Park and Preserve, Gustavus, Alaska; "An International Development Program For North Pacific Region," *Pacific Pulp & Paper Industry* (May 1943): 23–24.

37. William B. Brown (historian), personal communication with author, December 4, 2004.

38. Senate Committee on Interior and Insular Affairs, *Providing for the Admission of Alaska Into the Union*, 81st Cong., 2d. sess., Sen. Rept. No. 1929, 1950, 3; National Resources Planning Board, "Post War Economic Development of Alaska," *Regional Development Plan Report for 1942* (Washington: GPO, 1942), IV.

39. *Resolution 205*, 47.

40. "Heintzleman Says He Wants Statehood," *Ketchikan Alaska Chronicle*, March 13, 1953.

CHAPTER 15

1. "'Catastrophic Shortages in Paper Threaten'—APPA," *Pacific Pulp & Paper Industry* (February 1944): 6; "The No. 1 Problem," *Pulp & Paper Industry* (May 1945): 23.

2. Martin W. Grefnes, "Alaska Pulp and Paper Possibilities," *Timberman* (October 1941): 32; "An International Development Program For North Pacific Region," *Pacific Pulp & Paper Industry* (May 1943): 24.

3. "Proposal for the Development of the Pulp and Paper-Making Industry in Alaska," USFS, April 10, 1944.

4. "Postwar Alaska Pulp and Paper Project," *Journal of Forestry* (June 1944): 415; *The Forests of Alaska* (Juneau: USFS, Alaska Region, July 1, 1944), 5.

5. "U.S. Proposes to Sell Big Timber Stands For Private Pulp Industry in Alaska," *Pacific Pulp & Paper Industry* (May 1944): 29.

6. Irving Brant to Harold L. Ickes, January 24, 1939, reprinted in Edgar B. Nixon (comp. and ed.), *Franklin D. Roosevelt and Conservation, 1911–1945, Vol. 1* (Hyde Park, N.Y.: National Archives and Records Service, F.D.R. Library, 1957), 295; "Alaska Pulptimber Offerings Will Be Discussed; Secretary Ickes States His Views on Indian Claims," *Pacific Pulp & Paper Industry* (December 1944): 10; "Alaskans Deplore Loss of Pulp Industry," *Pulp & Paper Industry* (September 1945): 29; "Tongass Topic Tells Turns," *Ketchikan Pulp: A $52,500,000 Ultra-Modern Industry* (Ketchikan: Ketchikan Alaska Chronicle, 1954), unpaged.

7. "War Trend Postpones Alaska Mill Negotiations," *Pulp & Paper Industry* (February 1945): 44; "Chief Forester Reports," *Timberman* (February 1945): 33; A. E. Helmers, "Alaska Forestry—A Research Frontier," *Journal of Forestry* (June 1960): 466.

8. "Pulp and Paper Developments in Alaska," *Industry Report, Pulp and Paper Semiannual Review*, (Washington: U.S. Dept. of Commerce, 1948), 7–8; "Heintzleman Sees More Pulp Plants in Southeast Alaska," *Wrangell Sentinel*, September 5, 1952. (reprinted from the *Daily Alaska Empire*).

9. "Opportunities Seen in Alaska," *Timberman* (November 1945): 98.

10. "Postwar Plans & Prospects: A Preview of the North American Industry's Future," *Pacific Pulp & Paper Industry* (May 1944): 18; "Report of the Consumer Goods Subcommittee," *Industry Report, Pulp and Paper, Third Quarter Review* (Washington: U.S. Dept. of Commerce, 1947), 7–8; *Resolution 205*, 16.

11. "Canada and Paper," *Paper Trade Journal* (August 11, 1946): 34.

12. Roach, *Newsprint*, 29.

13. "Sulphite Pulp Mill for Prince Rupert," *Paper Trade Journal* (January 24, 1924).

14. "Pulp Mill Outlook Worries Alaskans," *New York Times*, June 26, 1951.

15. Bob Callan, "Pulp From Alaska," *Alaska Life* (January 1947): 4; William J. Stanton, *Alaska Recreation Survey* (Part 1, Vol. 1): *Economic Aspects of Recreation in Alaska* (Washington: GPO, 1953), 63.

CHAPTER 16

1. USFS, "Plan of Development, Region 8, January 20, 1932," 30, RG 95, NARA—Pacific Alaska Region, Anchorage.
2. An Act Providing a Civil Government for Alaska, May 17, 1884 (23 Stat. 26).
3. Morris Act (Chippewa Indians of Minnesota), June 27, 1902 (32 Stat. 400).
4. Tlingit and Haida Jurisdictional Act of June 19, 1935 (49 Stat. 388).
5. Alaska Amendment of May 1, 1936 (49 Stat. 1250).
6. Part 201.21b, Native fishing rights, 7 F.R. 2480, March 31, 1942.
7. "Plan to Give Alaska Resources to Indians Resisted," *West Coast Lumberman* (January 1945): 58.
8. "Alaska Pulptimber Offerings Will Be Discussed; Secretary Ickes States His Views on Indian Claims," *Pacific Pulp & Paper Industry* (December 1944): 10; "Mr. Krug's Great Opportunity," *West Coast Lumberman* (April 1946): 106.
9. H. L. Faulkner to Alaska Packer's Association, July 31, 1944, APA Collection, Center For Pacific Northwest Studies, Bellingham, Wash.
10. *Resolution 205*, 194; Robert E. Price, *The Great Father in Alaska* (Douglas, Alaska: First Street Press, 1990), 113.
11. "Alaska Indians Lose First Round," *West Coast Lumberman* (April 1945): 103.
12. "Ickes Grants Pulpwood Areas To Alaska Indians—His Wards," *Pulp & Paper Industry* (August 1945): 34.
13. "Distribution Seen Newsprint Snag," *New York Times*, March 7, 1949; "Alaska is Boomed as Paper Source," *New York Times*, December 17, 1946; "Krug Denies Charge He Blocks Mills," *Paper Trade Journal* (March 11, 1948): 7.
14. *Resolution 205*, 3; "Alaska is Boomed as Paper Source," *New York Times*, December 17, 1946.
15. *Resolution 205*, 3.
16. "Alaska Newsprint Promoted By U.S.," *New York Times*, January 12, 1947; "Newsprint Industry Krug Aim in Alaska," *New York Times*, March 23, 1947; "Woodpulp Industry Is Urged For Alaska," *New York Times*, April 11, 1947.
17. *Resolution 205*, 2.
18. 93 C.R. 4118 (1947).
19. *Resolution 205*, 180.
20. Ibid., 45; George W. Rogers, "The Human Factor in Alaska's Economic Development," *Science in Alaska, 1963*, Proceedings of the Fourteenth Alaskan Science Conference (Washington, D.C.: American Association for the Advancement of Science, 1964), 87.
21. "Bids Are Called For Alaska's First Mill," *Pulp & Paper* (September 1947): 28.
22. Tongass Timber Act, August 8, 1947, (61 Stat. 920).
23. USFS statement, Tongass Timber Act, 1947, RG 95, NARA, College Park, Md.
24. "Bids Are Called For Alaska's First Mill," *Pulp & Paper* (September 1947): 28.
25. Harold L. Ickes, "Man to Man," *New York Post*, August 21, 1947. (reprinted by Alaska Native Brotherhood in a pamphlet titled "Alaska's Teapot Dome.")
26. Harold L. Ickes, *The Autobiography of a Curmudgeon* (New York: Reynal & Hitchcock, 1943), x.
27. "Indians In Alaska Win U.S. Land Suit," *New York Times*, October 8, 1959: 18; Claus-M Naske and Herman E. Slotnick, *Alaska, A History of the 49th State* (Grand Rapids, Mich.: William B. Eerdmans Publishing Company, 1979), 203.
28. "Pulp Mills for Alaska," *American Forests* (October 1947): 447; "Alaska Newsprint Promoted By U.S.," *New York Times*, June 12, 1947; "Alaska Is Opened To Paper Industry," *New York Times*, August 9, 1947; "Alaska Pulpwood To Be Sold By U.S.," *New York Times*, August 26, 1947; "Capehart Cites Problems in Alaskan Pulp Development," *Paper Trade Journal* (February 12, 1948): 18.
29. "Brown and Capehart See More Newsprint," *Paper Trade Journal* (October 23, 1947): 9.
30. "Capehart's Group Tours In Alaska's Tongass Forest," *Paper Trade Journal* (September 4, 1947): 7; "Alaskan Dreamland," *Paper Trade Journal* (December 26, 1946): 42.
31. *Report of the Chief of the Forest Service* (Washington: GPO, 1947), 18; Erle Kauffman, "Timber Homestead in Alaska," *American Forests* (November 1947): 488.
32. H. D. Crawford, "Alaska Can Ease Our Newsprint Shortage," *American Forests* (November 1951): 46.
33. Senate Special Committee to Study Problems of American Small Business, *Investigation of Newsprint Shortages and Other Factors Affecting Survival of Smaller Independent Newspapers* (hearings, March 4–7, 1947), 80th Cong., 1st sess., 1947, 139, 145, 213 214.
34. Senate Special Committee to Study Problems of American Small Business, *Investigation of Newsprint Shortages and Other Factors Affecting Survival of Smaller Independent Newspapers* (hearings, March 4–7, 1947), 80th Cong., 1st sess., 1947, 215, 219.
35. Senate Special Committee to Study Problems of American Small Business, *Investigation of Newsprint Shortages and Other Factors Affecting Survival of Smaller Independent Newspapers* (hearings, March 4–7, 1947), 80th Cong., 1st sess., 1947, 218.
36. "Alaska Beckons," *Business Week* (September 6, 1947): 18.
37. "Government Extends Alaskan Bid Date," *Paper Trade Journal* (December 11, 1947): 10.
38. "Alaska Beckons," *Business Week* (September 6, 1947): 18.

1. "Parent Company Leads Field," *Ketchikan Pulp: A $52,500,000 Ultra-Modern Industry* (Ketchikan: *Ketchikan Alaska Chronicle*, 1954), unpaged; Gordon D. Marckworth, Western Conservation Journal (November–December 1959): 20; Ketchikan Pulp Company, *Wood Pulp Manufacture in Southeastern Alaska*, 1950–, unpaged.

2. "'Tongass Topic Tells Turns," *Ketchikan Pulp: A $52,500,000 Ultra-Modern Industry*, unpaged.

3. "Samples Show Alaska Pulp Equals Washington's," *Wrangell Sentinel*, October 31, 1947; "Tongass Topic Tells Turns," *Ketchikan Pulp: A $52,500,000 Ultra-Modern Industry*, unpaged; Nard Jones, "Ketchikan's 'Catch,'" *American Forests* (October 1954): 38; "Ketchikan Partner Leads the World in the Consumption of Dissolving Woodpulp," *Pulp & Paper* (October 1954): 124.

4. "Government Extends Alaskan Bid Date," *Paper Trade Journal*, (December 11, 1947): 10; Rakestraw, *U.S. Forest Service in Alaska*, 127; "Birth of an Industry: One Bid For a Stand of South Alaska Timber," *Wall Street Journal*, August 2, 1948; "Pulp Contract Let!" *Ketchikan Alaska Chronicle*, August 2, 1948, special edition; "500-Ton Pulp Mill Assured to Employ About 1200 Men," *Ketchikan Alaska Chronicle*, August 2, 1948; Charles Lockwood, "The Pulp Mill Boom Is On!" *Alaska Life* (November 1948): 4.

5. "Ketchikan Co. Buys Timber in Alaska," *New York Times*, July 28, 1951.

6. "FS Sells Largest Amount of Timber to Ketchikan Pulp," *Wrangell Sentinel*, July 27, 1951; Puget Sound Pulp and Timber, Annual Report, 1953; USFS, Timber Sale Agreement No. A10FS-1042, July 26, 1951, § 1(h), § 3(b); *Reid Brothers Logging Company v. Ketchikan Pulp Company and Alaska Lumber and Pulp Company*, 699 F.2d 1295 (9th Cir. 1983); "Forest Industries News," *Lumberman* (October 1951): 12.

7. *Report of the Chief of the Forest Service, 1948*, 25; "USFS Calls 1952 Season Its Greatest," *Ketchikan Alaska Chronicle*, January 5, 1953; "Green Gold Brings to Alaska a New Era That Outshines Even Its Glamorous Past," *Pulp & Paper* (October 1954): 74.

8. "Pulp and Paper Developments in Alaska," *Industry Report, Pulp and Paper Semiannual Review* (Washington: U.S. Dept. of Commerce, 1948), 8; James LaBau (retired USFS inventory project leader, PNW), personal communication with author, August 30, 2009.

9. The pulp manufactured at the Ketchikan Pulp Company from Southeast Alaska's western hemlock and Sitka spruce timber was about 91 percent cellulose. Pulp manufactures divide cellulose into three types—alpha, beta, and gamma—based on its reaction to a caustic soda solution. The pulp mill at Ketchikan produced pulp rich in alpha cellulose—a long polymer that is resistant to common chemical agents and to atmospheric conditions. So long as it is kept dry, it also resists bacteria and fungi. Ketchikan Pulp Company's production had an alpha cellulose component in the 93 percent range, and is commonly referred to as high-alpha pulp, and sometimes as special-alpha pulp. H. N. Lee, et al., *The Manufacture of Pulp and Paper* (New York: McGraw-Hill Book Company, 1937), 48; Irene Durbak, *Dissolving Pulp Industry: Market Trends*, Gen. Tech. Rept. FPL-GTR-77 (Madison, Wisc.: Forest Products Laboratory, 1993), 1; *Alaska Lumber & Pulp Company, Inc., Alaska Pulp Company, Ltd.*, corporate publication, TS1173.A46, Historical Collections, Alaska State Library, Juneau; National Resources Planning Board, "Post War Economic Development of Alaska," *Regional Development Plan Report for 1942* (Washington: GPO, 1942), 8.

10. Louis E. Wise, "Spinners of Wood," *American Forests and Forest Life* (March 1924): 137; John A. Guthrie, *The Economics of Pulp and Paper* (Pullman, Wash.: State College of Washington Press, 1950), 66.

11. "Alaska Is New Frontier for Pulp," *Western Conservation Journal* (July–August 1955): 16; *The Story of Wood Pulp Manufacture in Alaska*, Ketchikan Pulp Company, no date, unpaged, Tongass Historical Museum, Ketchikan; Jesse W. Markham, *Competition in the Rayon Industry* (Cambridge, Mass.: Harvard University Press, 1952), 217.

12. Robert S. Aries and William Copulsky, "The U.S. Market for Dissolving Wood Pulp," *Pulp and Paper Magazine of Canada* (February 1950): 47–50.

13. "Every Ketchikan Citizen Must Help Make Ketchikan Pulp Company Mill a Success," *Ketchikan Alaska Chronicle*, March 15, 1954, 2.

14. The financing of the enterprise was handled by Morgan, Stanley & Company, and Dillon, Read & Company, of New York. Investors were the Equitable Life Assurance Society and the Metropolitan Life Insurance Company, also of New York. Lawrence E. Davies, "Alaska Dedicates Great Pulp Plant," *New York Times*, July 15, 1954.

15. "Magnitude of Alaska Mill is Impressive," *Pulp & Paper* (October 1954): 82; *Ketchikan Pulp: A $52,500,000 Ultra-Modern Industry*, unpaged; H. L. Clark, "Job Prospects in Alaska," *The Labor Market and Employment Security* (U.S. Dept. of Labor, Bureau of Employment Security) (March–April 1957): 10.

16. Ralph Dale (former plant engineer, Ketchikan Pulp Company), personal communication with author, March 22, 2005.

17. *Ketchikan Pulp Company: Our First 20 Years* (Ketchikan: Ketchikan Pulp Company, 1974), 21; "Log to Finished Pulp," *Ketchikan Pulp: A $52,500,000 Ultra-Modern Industry*, unpaged; Ralph Dale, March 22, 2005.

18. *Ketchikan Pulp Company*, 8; Ralph Dale, March 10, 2005; Ketchikan Pulp Company Annual Report, 1957.

19. In comparison to KPC's MgO process, the sulfite pulping process employed at the Columbia Cellulose mill at Prince Rupert, British Columbia, consumed (and presumably discharged) about 26,000 tons of sulfur annually, or about 400 pounds for each ton of pulp manufactured. USFS, Timber Sale Agreement No. A10FS-1042, July 26, 1951, § 1(i); "Pollution Studies Yield Results," *Pacific Fisherman* (March 1954): 39; advertisement, Babcock & Wilcox Company, *Pulp & Paper* (November 12, 1962): 49 (Babcock &

Wilcox supplied the MgO process equipment at KPC.); John Guthrie, mill manager, Columbia Cellulose Company, remarks as part of a Panel Discussion on "The Future of Industry in Central British Columbia," at the Convention of the Associated Boards of Trade in Prince George, BC, September 15, 1961.

20. Harris and Farr, *Forest Ecosystem of Southeast Alaska*, 8, quoting Jones, 1954; "Mayor's Message Praises Pulp," *Ketchikan Pulp: A $52,500,000 Ultra-Modern Industry*, unpaged.

21. B. Frank Heintzleman, "Heintzleman Sees New Era," *Ketchikan Pulp: A $52,500,000 Ultra-Modern Industry*, unpaged.

22. "Frank Heintzleman Receives Honor Award in Washington, D.C.," *Wrangell Sentinel*, May 14, 1952.

23. Ketchikan Pulp Company Annual Report, 1956, 13.

24. *The Story of Wood Pulp Manufacture in Alaska*, Ketchikan Pulp Company, no date, unpaged, Tongass Historical Museum, Ketchikan.

25. *Ketchikan Pulp Company: Our First 20 Years*, 20–22; *Ketchikan Pulp: A $52,500,000 Ultra-Modern Industry*, unpaged.

26. Louis Blackerby, "New Mills Are Under Way, Wood Trends Changing," *Pulp & Paper* (1958 Review Number): 153; *Ketchikan Pulp Company: Our First 20 Years*, 9; Ralph Dale, March 22, 2005; USFS, Tongass National Forest, 1982 Timber Supply and Demand Report.

27. *Ketchikan Pulp Company: Our First 20 Years*, 8–9; "From Ketchikan to Barrow," *Alaska Sportsman* (October 1954): 23; "Ketchikan Pulp Blazes Trail," *1955 Yearbook of Alaska Timber Industries* (Ketchikan: *Ketchikan Alaska Chronicle*, 1955), unpaged; "Rupert Ferry the Berries," *Ketchikan Pulp: A $52,500,000 Ultra-Modern Industry*, unpaged.

28. Ketchikan Pulp Company Annual Report, 1955, 2, 4.

29. B. Frank Heintzleman, "Public Incentives to the Pulp Industry," *Proceedings of the Alaskan Science Conference*, National Academy of Sciences (Washington, D.C.: National Academy of Sciences, 1951), 131.

30. B. Frank Heintzleman, "What This Industry Means to Alaska," *Pulp & Paper* (October 1954): 77.

31. B. Frank Heintzleman, "Great Forest Industry Future Envisaged in State of Alaska," *Pulp & Paper* (February 1960): 92.

32. Chapter 33, Laws of Alaska, 1953.

33. "Ketchikan Co. Buys Timber in Alaska," *New York Times*, July 28, 1951.

34. USFS, Timber Sale Agreement No. A10FS-1042, July 26, 1951, § 1(d); USFS notes on meeting with Japanese, 1953, RG 95, NARA—Pacific Alaska Region, Anchorage; Arthur M. Brooks, "Major Problems of the Private Timber Operators in Southeastern Alaska," *Science in Alaska*, 1954, Proceedings of the Fifth Alaskan Science Conference (Washington, D.C.: American Association for the Advancement of Science, 1955), 5.

35. "Ketchikan Pulp Affects Timber," *Ketchikan Pulp: A $52,500,000 Ultra-Modern Industry*, unpaged; "Significant Industry News," *Lumberman* (March 1959): 8; "Logging in Alaska is Tougher," *Lumberman* (November 1955): 65; "Logging in Alaska is Much Like Capturing a Beachhead—But without the Shooting," *Pulp & Paper* (October 1954): 138; A. M. Brooks, "Opportunities for Log Operators Described by Pulp Mill Official," *1955 Yearbook of Alaska Timber Industries*, unpaged.

36. "Sustained Yield, From the Fisheries…From the Forests," *Pacific Fisherman* (July 1947): 24.

37. C. Howard Baltzo, in House Special Subcommittee on Alaskan Problems, Committee on Merchant Marine and Fisheries, *Hearing on H.R. 1515, A Bill to Provide for the Gradual Elimination of Salmon Traps in the Waters of Alaska*, Part 2, 81st Cong., 1st sess., 1949, 633–634; "Liberate Alaska From the Fish Trap," published in January 1949 by Cordova and Ketchikan Fishermen, 87–88. Whether Baltzo was aware of it or not, there was an existing federal statute that prohibited polluting navigable waters. It was the Refuse Act of 1899, the same legislation that was ignored several decades earlier when sawmill waste fouled the waters around some Southeast Alaska communities. Beginning in 1936, federal courts began interpreting "refuse" to include industrial pollutants, such as those discharged by pulp mills. Government regulators, however, would continue to virtually ignore this legislation until the late 1960s. See: Diane D. Eames, "The Refuse Act of 1899: Its Scope and Role in Control of Water Pollution," *California Law Review* (November 1970): 1444–1473.

38. "One Hundred Alaskans," *Pacific Fisherman* (August 1949): 45.

39. Corey Ford and Frank Dufresne, "Lost Paradise," *Field & Stream* (September 1956): 63.

40. Richard Starnes, "Night Comes to Admiralty!" *Field & Stream* (August 1965): 18.

41. USFS, Timber Sale Agreement No. A10FS-1042, July 26, 1951, §21; W. R. Meehan, W. A. Farr, D. M. Bishop, and J. H. Patric, "Some Effects of Clearcutting on Salmon Habitat of Two Southeast Alaska Streams," USDA Forest Service Research Paper PNW-82, (Juneau: Institute of Northern Forestry, 1969), 36, 41; "The Effect of Logging on Twelve Salmon Steams in Southeast Alaska," (Juneau: USDA Forest Service, Alaska Region, April 1965), 2.

42. "Logging in Alaska is Tougher," *Lumberman* (November 1955): 68.

43. Alaska Resources Committee, *Alaska*, 98; Don Meldrum, "Brief and Supplemental Reports on Pulptimber Allotment 'A', Tongass National Forest, Southeastern Alaska," 1929 (unpublished), 22, Robert A. Kinzie collection, MS 21, Historical Collections, Alaska State Library, Juneau.

44. Frost, incomplete report of timber cruise in southeast Alaska, 2.

45. "Logging in Alaska is Much Like Capturing a Beachhead—But without the Shooting," *Pulp & Paper* (October 1954): 137, 140; A. S. Harris, "Tree Reproduction Development on a Mile-Square Clear Cutting," *Science in Alaska, 1962*, Proceedings of the Thirteenth Alaskan Science Conference

(Washington, D.C.: American Association for the Advancement of Science, 1963), 87; Anonymous, personal communication with author.

46. *Maybeso Experimental Forest* (pamphlet), USFS, Pacific Northwest Research Station (September 2004); "Late Forest Industries News," *Lumberman* (June 1954): 12; A. E. Helmers, "Alaska Forestry—A Research Frontier," *Journal of Forestry* (June 1960): 466; Arland S. Harris, *Natural Reforestation on a Mile-Square Clearcut in Southeast Alaska*, PNW-52 (Juneau: Pacific Northwest Forest and Range Experiment Station, Institute of Northern Forestry, 1967), 1; Andersen, "Clearcutting as a Silvicultural System," 61; Henry E. Hays, "Status of Regeneration on Areas Logged on the South Tongass Since 1954," *Science in Alaska, 1960*, Proceedings of the Eleventh Alaskan Science Conference (Washington, D.C.: American Association for the Advancement of Science, 1961), 146.

47. USFS, Timber Sale Agreement No. A10FS-1042, July 26, 1951, § 23; George Sundberg, "Statistics Show Rapid Growth of Southeastern Alaska," *Ketchikan Alaska Chronicle*, November 19, 1954.

48. Alaska Resource Development Board, *Biennial Report of the Alaska Resource Development Board, 1955–57*, (Juneau: Alaska Resources Development Board, 1957), 13–14.

49. "Definite Improvement is Noted in Ketchikan Pulp Operations," *Daily Alaska Empire*, September 15, 1957; "Green Gold Lures Gyppo Logger to Alaska Forests," *Timberman* (January 1961): 35.

50. Mason B. Bruce, "National Forests in Alaska," *Journal of Forestry* (June 1960): 441.

CHAPTER 18

1. "On MacArthur Staff," *Timberman* (June 1946): 122.

2. "Chamber Told of Japanese Pulp Plans," *Ketchikan Alaska Chronicle*, March 11, 1954.

3. "Japanese Buy Big Block of B.C. Timber," *West Coast Lumberman* (August 1937): 62; "Japan Now Taking B.C. Pulp," *Timberman* (May 1949): 145.

4. Boris V. Ivanov, "Closer Relations In Japan Soviet Timber Trade," *Japan Lumber Journal* (October 5, 1964): 4; "Industry News From Japan," *Lumberman* (September 1955): 122.

5. *Alaska Pulp Corporation v. United States*, Fed. Cl. 400 (2004), at 9.

6. William C. McIndoe, "Economic Possibilities for Rayon Production in the Pacific Northwest," June 1, 1937: 28, U.S. Army Corps of Engineers, Office of the Division Engineer, North Pacific Division, Portland, Oreg.; Textile Mission to Japan, *Report of the Textile Mission to Japan* (London: His Majesty's Stationery Office, 1946), 10; Jesse W. Markham, *Competition in the Rayon Industry* (Cambridge, Mass.: Harvard University Press, 1952), 210; Institute of Trade Research & Statistics, *Textile Exports of Japan* (Osaka: Institute of Textile Trade Research & Statistics, 1960), 96; Japan Textile Council, *Textile Japan* (Tokyo: Japan Textile Council, 1957), 70.

7. Donald C. Warner (U.S. Dept. of State) to O. F. Benecke (Juneau Chamber of Commerce), March 13, 1953, RG 95, NARA—Pacific Alaska Region, Anchorage.

8. Fred Niendorff, "Pulp Industry Fears Secret Japan Deal," *Seattle Post-Intelligencer*, December 30, 1952.

9. A. W. Greeley to Matt O'Conner (*Seattle Times*), December 11, 1953, RG 95, NARA—Pacific Alaska Region, Anchorage.

10. Ibid.

11. Lawrence E. Davies, "Alaska Dedicates Great Pulp Plant," New York Times, July 15, 1954; "Japanese Commission Visits Alaska," *Lumberman* (February 1953): 69.

12. Thomas Morgan to Charles E. Burdick. January 10, 1953, RG 95, NARA—Pacific Alaska Region, Anchorage.

13. "Norblad Hits Japanese Ventures in Alaska," *Ketchikan Alaska Chronicle*, January 26, 1953; Willis C. Armstrong (acting director, Office of International Materials Policy) to Editor, *Pulp & Paper* (October 1953): 54–56.

14. USFS, Validation Report on Alaska Pulp Company, Ltd, (1982), 3; "Alaska Gets 2nd Mill," *Western Conservation Journal* (November–December 1958): 26–28; "Alaskan Invasion: Japanese Build Pulp Mill, Seek Coal, Oil, Iron in Newest State," *Wall Street Journal*, January 21, 1959; "Japanese Build Alaska Pulp Mill," *New York Times*, July 13, 1958; Louis H. Blackerby, "First Japanese Mill in North America Pioneers New Methods," *Pulp & Paper* (February 1960): 81.

15. "Alaska Gets 2nd Mill," *Western Conservation Journal* (November–December 1958): 26–28; Louis H. Blackerby, "First Japanese Mill in North America Pioneers New Methods," *Pulp & Paper* (February 1960): 80. Robertson, a former mayor of Juneau, had a long history with pulp mill interests. In 1920, he was the Alaska agent for the never-got-off-the-ground Alaskan-American Paper Corporation (New York City).

16. "Tapping of Rich Forest Breeds a Batch of New Wood–Product Plants," *Wall Street Journal*, June 20, 1953; Everett Central Labor Council, International Brotherhood of Pulp, Sulphite & Paper Mill Workers (locals 153 and 155), and West Coast Lumbermen's Association to Senator Warren Magnuson, February–March 1953, RG 95, NARA—Pacific Alaska Region (Anchorage); Hubert J. Gellert, *Japanese Companies in Alaska* (Fairbanks: Institute of Social, Economic and Governmental Research, University of Alaska, 1967), 8–9.

17. "Japanese Seek American Funds," *Lumberman* (August 1953): 60.

18. "Late Forest Industries News," *Lumberman* (February 1956): 12; Arthur W. Greeley, "Sitka Site of New Pulp Mill?" *1955 Yearbook of Alaska Timber Industries* (Ketchikan: *Ketchikan Alaska Chronicle*, July, 1955), unpaged.

19. "Plan to Ship Waste From Alaska to Japan," *Lumberman* (September 1953): 98; "Chips For Japan," *Lumberman* (February 1954): 108.
20. USFS Timber Sale Contract No. 12-11-010-1545, January 25, 1956; "Late Forest Industries News," *Timberman* (March 1956): 21; "Japanese Join Project For Pulp Mill in Alaska," *New York Times*, August 17, 1957.
21. "Alaska Gets 2nd Mill," 26–28; USFS, Timber Sale Contract No. 12-11-010-1545, October 15, 1957; *Reid Brothers Logging Company v. Ketchikan Pulp Company*; Louis H. Blackerby, "U.S. Far West," *Pulp & Paper* (1960 Review Number): 160).
22. Calculations by Clarence Adams, USFS, Umpqua National Forest, October 15, 2004.
23. USFS, Timber Sale Agreement No. A10FS-1042, July 26, 1951, § 8; USFS, Timber Sale Contract No. 12-11-010-1545, January 25, 1956, § 7b.
24. "Japanese Aid Alaska Project with Finances and Purchases," *New York Times*, August 5, 1960; "Japanese Return to Private U.S. Money Market," *New York Times*, October 17, 1957; *Alaska Pulp Corporation v. United States*, Fed. Cl. 400 (2004), at 9. (quoting Dr. Ulrike Schaede).
25. *Alaska Pulp Corporation v. United States*, Fed. Cl. 400 (2004), at 10.
26. *Alaska Pulp Corporation v. United States*, Fed. Cl. 400 (2004), at 9 (quoting Dr. Ulrike Schaede).
27. "Japanese Return to Private U.S. Money Market," *New York Times*, October 17, 1957; "Alaskan Invasion"; Louis H. Blackerby, "First Japanese Mill in North America Pioneers New Methods," *Pulp & Paper* (February 1960): 80; *Alaska Pulp Corporation v. United States*, Fed. Cl. 400 (2004), at 10. The eight firms were: General Electric Company; Chicago Bridge & Iron Company; Improved Machinery, Inc.; Rice, Barton Corp.; Ingersoll-Rand Company; Sumner Iron Works; Bingham Pump Company; and Isaacson Iron Works.
28. Ralph Dale, March 22, 2005; Hubert J. Gellert, *Japanese Companies in Alaska* (Fairbanks: Institute of Social, Economic and Governmental Research, University of Alaska, 1967), 2, "Japanese Build Alaska Pulp Mill," *New York Times*, July 13, 1958.
29. To rip particularly large logs into manageable pieces, a "log splitter"—basically an electrically-powered chain saw with a 12-foot bar—was installed in the mill's log pond in early 1961.
30. Blackerby, "First Japanese Mill in North America Pioneers New Methods," 86; *Pulpzette* (APC's in-house publication), April 1961; Session Laws of Alaska, 1957, Chapter 129; Hubert J. Gellert, *Japanese Companies in Alaska* (Fairbanks: Institute of Social, Economic and Governmental Research, University of Alaska, 1967), 15.
31. Ernestine C. Veatch, "Pulp to be on Machine Within a Few Days Now," *Daily Sitka, Alaska Sentinel*, November 19, 1959; "ALP Co. to Ship it First Pulp to Japan Soon," *Daily Sitka, Alaska Sentinel*, December 29, 1959.
32. "Alaska Lumber & Pulp Co. Plant Dedicated Today," *Daily Sentinel* (Sitka), June 29, 1960; "Japanese Aid Alaska Project With Finances and Purchases," *New York Times*, August 5, 1960; "Japanese Build Alaska Pulp Mill," *New York Times*, July 13, 1958; "Alaska Lumber & Pulp Co. Plant Dedicated Today," *Daily Sentinel* (Sitka), June 29, 1960.
33. "Alaska Logging Methods," *Pulp & Paper* (February 1960): 106.
34. "Pulp Mill Opens," *Marine Digest* (July 16, 1960): 9; B. Frank Heintzleman, "Great Forest Industry Future Envisioned in State of Alaska," *Pulp & Paper* (February 1960): 93.
35. "Ceremony at ALP Co. Marks First Shipment of Pulp," *Daily Sitka, Alaska Sentinel*, January 4, 1960.

CHAPTER 19

1. Employment/payroll data is for the entire state, though virtually all of the production was from Southeast Alaska. George W. Rogers, "The Human Factor in Alaska's Economic Development," Science in Alaska, 1963, *Proceedings of the Fourteenth Alaskan Science Conference* (Washington, D.C.: American Association for the Advancement of Science, 1964), 87; Federal Field Committee for Development Planning in Alaska, *Economic Development in Alaska* (U.S. Dept. of Commerce, August, 1966), 21; Federal Field Committee for Development Planning in Alaska, *A Subregional Economic Analysis of Alaska* (U.S. Dept. of Commerce, 1968), 26; O. Keith Hutchison and Vernon J. LaBau, *Forest Ecosystem of Southeast Alaska: No. 9, Timber Inventory, Harvesting, Marketing, and Trends*, Gen. Tech. Rept. PNW-34 (Portland, Oreg.: Pacific Northwest Forest and Range Experiment Station, 1975), abstract.
2. Arthur W. Greeley, "Alaska's Acres at Work…at Last," *American Forests* (October 1954): 52.
3. "Alaska Timber: Tapping of Rich Forest Breeds a Batch of New Wood-Product Plants," *Wall Street Journal*, June 20, 1953; "1953 Major Trends and Developments," *Lumberman* (March 25, 1954): 64; "U.S. Sells Big Stand of Alaskan Timber," *New York Times*, August 19, 1955; *1954 Annual Report for the Fiscal Year Ended June 30*, Governor of Alaska to the Secretary of the Interior (Washington: GPO, 1954), 42.
4. Greeley, "Alaska Fosters Green Gold Rush," 52.
5. B. Frank Heintzleman, "Great Forest Industry Future Envisioned in State of Alaska," *Pulp & Paper* (February 1960): 92; "Alaska Will Select Its Timber Holdings," *Timberman* (September 1960): 40.
6. USFS, Draft Environmental Impact Statement, Pacific Northern Timber Company Timber Sale, 1977–1981 Operating Period (1976), vii; "Pulp Concern Asks Delay in Mill Plans," *New York Times*, May 22, 1958; "Next Move is up to Georgia-Pacific," *Pulp & Paper* (April 17, 1961): 9; "Alaska Pulp Project Called Off," *San Francisco Chronicle*, June 20, 1961; "Puget becomes Georgia-Pacific Division," *Pulp &*

Paper (August 5, 1963): 23; *International Directory of Company Histories*, Vol. IV (Chicago: St. James Press, 1991), 304.

7. "U.S. Plywood Buys Timber in Alaska," *New York Times*, February 8, 1968; Rakestraw, *U.S. Forest Service in Alaska*, 164.

CHAPTER 20

1. *1950 Report of the Chief of the Forest Service*, 38; *Report of the Chief of the Forest Service, 1959*, 2, 10; Charles F. Wilkinson and H. Michael Anderson, "Land and Resource Planning in the National Forests," *Oregon Law Review*, Vol. 64, No. 1 & 2 (1985): 29.

2. Albert W. Wilson, "Warning to Pulp and Paper by U.S. Chief Forester," *Pulp & Paper* (April 1960): 69.

3. Multiple-Use Sustained-Yield Act, Public Law 86-517 (June 12, 1960); *Report of the Chief of the Forest Service, 1960*, 32; 16 U.S.C. § 529 (2000); Wilkinson and Anderson, "Land and Resource Planning in the National Forests," 30; *Sourdough Notes*, November 1960; *Sourdough Notes*, October 1960.

4. Alaska Native Claims Settlement Act, Public Law 92-203 (December 18, 1971).

5. *West Virginia. Div. of the Izaak Walton League of America, Inc. v. Butz*, 522 F.2d 945 (4th Cir.1975).

6. *Zieske v. Butz*, 406 F.Supp. 258 (D. Alaska 1975).

7. National Forest Management Act of 1976, Public Law 94-588 (October 22, 1976), § 6(e)(1).

8. Public Law 94-588, § 15(a), 14(c).

9. 36 C.F.R. § 219.27(d) (1985); Public Law 94-588 § 15(b).

10. Clean Air Act Amendments of 1970, Public Law 91-604 (December 31, 1970); Clean Water Act, Public Law 92-500 (October 18, 1972).

11. Don Muller (former chemist, Alaska Pulp Corp.), personal communication with author, February 15, 2006.

12. John Lindback, "Lawmakers Likely to OK Loan Guarantee for Mill," *Anchorage Daily News*, May 3, 1986.

13. Douglas Mertz (former assistant attorney general, State of Alaska), e-mail message to author, February 21, 2006.

14. Alaska National Interest Lands Conservation Act, Public Law 96-487 (December 2, 1980).

15. Public Law 96-487 § 705(a).

16. Joseph Mehrkens (former economist, USFS), personal communication with author, November 17, 2005

17. Tongass Timber Reform Act, Public Law 101-626 (November 28, 1990); Public Law 101-626, § 301(c)(1).

18. Public Law 101-626, § 101.

19. *Use of the National Forests* (Washington: GPO, 1905), 11.

20. "'Greeleyism' in U.S. Forest Service," *International Woodworker* (December 17, 1947): 2.

21. *Reid Brothers Logging Company v. Ketchikan Pulp Company and Alaska Lumber and Pulp Company*, No. C75-165S (W.D. Wash., March 13, 1975).

22. House Subcommittee on Mining, Forest Management, and Bonneville Power Administration of the Committee on Interior and Insular Affairs, *Oversight Hearing on Timber Industry Practices in the Tongass National Forest, Alaska*, Serial No. 98-19, 98th Cong., 1st sess., 1983, 367.

23. *Reid Brothers Logging Company v. Ketchikan Pulp Company* (June 8, 1981).

24. House Subcommittee on Mining, Forest Management, and Bonneville Power Administration of the Committee on Interior and Insular Affairs, *Oversight Hearing on Timber Industry Practices in the Tongass National Forest, Alaska*, 98th Cong., 1st sess., Serial No. 98-19, 1983, 111–112.

25. *Reid Brothers Logging Company v. Ketchikan Pulp Company* (June 8, 1981).

26. Kathie Durbin, *Tongass: Pulp Politics and the Fight for the Alaska Rain Forest* (Corvallis: Oregon State University Press, 1999), 131; *Reid Brothers Logging Company v. Ketchikan Pulp Company and Alaska Lumber and Pulp Company*, 699 F.2d 1295 (9th Cir. 1983).

27. House Subcommittee on Mining, Forest Management, and Bonneville Power Administration of the Committee on Interior and Insular Affairs, *Oversight Hearing on Timber Industry Practices in the Tongass National Forest, Alaska*, 98th Cong., 1st sess., Serial No. 98-19, 1983, 99, 101.

28. Alfred A. Wiener, "Review of Potetential Damages Report," May 20, 1982. Specific areas of consideration in Wiener's reported included with letter by R. Max Peterson, Chief, USFS, to Rep. James Weaver, Chairman, Subcommittee on Mining, Forest Management and Bonneville Power Administration, August 26, 1983. Reprinted in: House Subcommittee on Mining, Forest Management, and Bonneville Power Administration of the Committee on Interior and Insular Affairs, *Oversight Hearing on Timber Industry Practices in the Tongass National Forest, Alaska*, 98th Cong., 1st sess., Serial No. 98-19, 1983, 325.

29. Helmut F. Furth, Deputy Assistant Attorney General, Antitrust Divison, Department of Justice, to A. James Barnes, General Counsel, Department of Agriculture, January 4, 1983. Reprinted in: House Subcommittee on Mining, Forest Management, and Bonneville Power Administration of the Committee on Interior and Insular Affairs, *Oversight Hearing on Timber Industry Practices in the Tongass National Forest, Alaska*, 98th Cong., 1st sess., Serial No. 98-19, 1983, 150.

30. National Forest Roads and Trails Act of 1964, Public Law 88-657 (October 13, 1964).

31. A bill relating to certain Forest Service timber sale contracts involving road construction, Public Law 94-154 (December 16, 1975).

32. GAO/RCED-95-2 at 15 (January 1995).

33. *Alaska Pulp Corporation v. United States*, Fed. Cl. 400 (2004), at 34; *Pulp & Paper 1997 North American Factbook* (San Francisco: Miller Freeman Publications, 1997), 370.
34. "NLRB Reaches $11.75 Million Backpay Settlement with Alaska Pulp Corporation," *PR Newswire*, May 13, 2003.
35. "L-P Withdraws Sale Offer," *Ketchikan Daily News*, June 26, 1984; "Closure Ordered; no Date to Open," *Ketchikan Daily News*, June 22, 1984.
36. Wayne Weihing (former employee, Ketchikan Pulp Co.), personal communication with author, September 5, 2005.
37. Alaska Pulp Corporation v. United States, Fed. Cl. 400 (2004), at 40.
38. "Ketchikan Pulp Company to Close Pulp Mill; Will Negotiate with Government on Continued Operation of Sawmills," *Business Wire* (October 7, 1996); David Whitney, "Pulp Mill Seeks Extension Company Turns To Congress For Longer Tongass Contract," *Anchorage Daily News*, March 19, 1996.
39. "Sitka to Cease Operations," *Pulp & Paper* (August 1993): 29.
40. *Alaska Pulp Corporation v. United States*, Fed. Cl. 400 (2004), at 2, 56.
41. No. 95-463C, Nos. 94-96C & 95-775C (Consolidated), and No. 95-816C.
42. "Louisiana-Pacific's Ketchikan Pulp Company Subsidiary Confirms Report of Compromise Proposal to Clinton Administration," *Business Wire* (September 23, 1996).
43. "Ketchikan Pulp Co. Reaches Agreement on Timber Supply Allowing Sawmills to Operate for Three More Years," *Business Wire* (February 21, 1997).
44. *Pulp & Paper 1997 North American Factbook* (San Francisco: Miller Freeman Publications, 1997), 372; "B.C. Government Did Not Consult Natives: Court," *Globe and Mail*, December 31, 2004 (internet); "Projects approved for Prince Rupert, Castlegar," *Pulp & Paper Canada* (January 1976): 9.
45. "Chief Forester of Alaska Advertising Alaskan Timber," *Ketchikan Alaska Chronicle*, March 11, 1921.
46. USFS, *Tongass Land and Resource Management Plan Amendment*, Draft Environmental Impact Statement, R10-MB-602a (January 2007), 3-410 through 3-414.
47. *Tongass Timber Reform Act*, 100th Cong. 2d. sess, Ho. Rept. No. 100-600, 1988, 4.

AFTERWORD

1. Clive S. Thomas, editor, *Alaska Public Policy Issues: Background and Perspectives* (Juneau: Denali Press, 1999), 29, 41–42.
2. Rakestraw, *U.S. Forest Service in Alaska*, 176.
3. David C. Smith, "Pulp, Paper, and Alaska," *Pacific Northwest Quarterly* (April 1975): 61.
4. USFS, *Tongass Land Management Plan Revision*, Final Supplemental Environmental Impact Statement, R10-MB-481f (February 2003), 3–107.
5. McDowell Group, *Timber Markets Update and Analysis of Integrated Southeast Alaska Forest Products Industry*, (Juneau: McDowell Group, September 2004); Allen M. Brakley, Thomas D. Rojas, and Richard W. Haynes, *Timber Products Output and Timber Harvests in Alaska: Projections for 2005–2025*, Gen. Tech. Rept. PNW-GTR-677 (Portland: USFS, Pacific Northwest Research Station, July 2006).
6. McDowell Group, *Timber Markets Update*, 1, 3, 27, 32, 39; Jim Calvin (McDowell Group) to Murray Walsh (Southeast Conference), December 30, 2006, available at http://www.fs.fed.us/r10/ro/policy-reports/for_mgmt/index.shtml (accessed March 12, 2007).
7. R10 Summary_current_exports-1999–2005, available at http://www.fs.fed.us/r10/ro/policy-reports/for_mgmt/index.shtml, (accessed March 12, 2007).
8. Public Law 108-7, §318.
9. Dennis E. Bschor (Regional Forester, Region 10, Juneau Alaska) to Forest Supervisor (Tongass National Forest), March 14, 2007.
10. Southeast Conference, young-growth inventory project, scope of work, 2010; USDA News Release No. 0288.10, May 26, 2010, "USDA Pursues Jobs, Community Stability While Developing New Approach to Forest Management," in *Southeast Alaska Letter to Tongass Future Roundtable Highlights "Transition Framework" for Economic Development and Timber Harvesting Outside of Roadless Areas*.
11. Susan J. Alexander, Eric B. Henderson and Randy Coleman, USFS, Alaska Region, Economic Assessment of Southeast Alaska: Transitioning to a Young Growth Timber Industry and Stimulating Job Creation (unpublished manuscript, 2010), 1.
12. USDA News Release No. 0288.10, May 26, 2010, USDA Pursues Jobs, Community Stability While Developing New Approach to Forest Management in Southeast Alaska *Letter to Tongass Future Roundtable Highlights "Transition Framework" for Economic Development and Timber Harvesting Outside of Roadless Areas*.
13. Susan J. Alexander, Eric B. Henderson and Randy Coleman, USFS, Alaska Region, Economic Assessment of Southeast Alaska: Transitioning to a Young Growth Timber Industry and Stimulating Job Creation (unpublished manuscript, 2010), 31; USFS, Land and Resource Management Plan – Tongass National Forest, 1997, R10-MB 338-CD, Final Environmental Impact Statement, 3-303.
14. USFS, Tongass National Forest Land and Resource Management Plan Amendment, 2008 Environmental Impact Statement, Record of Decision, January 2008, 6.
15. "Proposed Land and Resource Plan, Tongass National Forest" (January 2007), R10-MB-602b, Draft Environmental Impact Statement, 3-447.

16. Shiela Spores (silviculturist, Tongass National Forest), personal communication with author, May 17, 2010; USFS, Tongass National Forest, "Tongass Young Growth—Thinning and the Resource," fact sheet, ca. 2006; USFS, Tongass National Forest, "Evaluating Young-Growth Wood Quality in Southeast Alaska," fact sheet, ca. 2006; Susan J. Alexander, Eric B. Henderson and Randy Coleman, USFS, Alaska Region, *Economic Assessment of Southeast Alaska: Transitioning to a Young Growth Timber Industry and Stimulating Job Creation* (unpublished manuscript, 2010), 36–37.

APPENDIX A

1. Wyckoff, "History of Lumbering in Southeast Alaska," 38.
2. George T. Emmons, "The Woodlands of Alaska," (unpublished, 1902); Fred Eugene Wright and Charles Will Wright, *Ketchikan and Wrangell Mining Districts, Alaska*, USGS Bull. No. 347 (Washington: GPO, 1908), 96–97; Langille, "Report on Alexander Archipelago Forest Reserve, 1906"; Charles Will Wright, *Geology and Ore Deposits of Copper Mountain and Kasaan Peninsula, Alaska*, USGS Professional Paper No. 87 (Washington: GPO, 1915), 57.
3. "Fish Egg Booming," *Wrangell Sentinel*, January 18, 1912; "Fish Egg Property Sold," *Wrangell Sentinel*, February 1, 1912.
4. "Lumbering on Alaska Coast," *Timberman* (April 1917): 36; "Alaska Spruce Mill Ready in February, *West Coast Lumberman* (December 15, 1917): 45.
5. "Alaska Spruce Mill Ready in February," *West Coast Lumberman* (December 15, 1917): 45; Terrence Cole and Elmer E. Rasmuson, *Banking on Alaska: The Story of the National Bank of Alaska*, (Fairbanks: University of Alaska Press, 2001), 122; "Alaska Spruce for the Manufacture of Airplanes," *Wrangell Sentinel*, June 21, 1917; "Tromble Gets Big Spruce Contract," *Wrangell Sentinel*, January 17, 1918; "Spruce Sale on the Tongass," *Six Twenty-Six* (February 1, 1918): 15–16.
6. "Shattuck Has Taken a Lease Craig Sawmill," *Wrangell Sentinel*, May 30, 1918; Terrence Cole and Elmer E. Rasmuson, *Banking on Alaska: The Story of the National Bank of Alaska* (Fairbanks: University of Alaska Press, 2001), 122; "Craig Lumber Man Tells of Season's Work," *Wrangell Sentinel*, September 15, 1918 (reprinted from Ketchikan *Progressive Miner*); Rakestraw, *U.S. Forest Service in Alaska*, 77.
7. Cole and Rasmuson, *Banking on Alaska*, 122–123; "Local Interest," *Petersburg Weekly Report*, February 20, 1920; "Craig Lumber Company Case Now Decided," *Wrangell Sentinel*, May 19, 1921; "McDonald Logging Co. Wins in Suit Against Craig Lumber Co.," *Petersburg Weekly Report*, March 3, 1922.
8. "Craig Men Buy Big Stand of Timber," *Ketchikan Alaska Chronicle*, June 17, 1920; Rakestraw, *U.S. Forest Service in Alaska*, 77.
9. Advertisement, *Ketchikan Alaska Chronicle*, July 15, 1920; "Lumber Magazine Prints Story of Alaska Business," *Petersburg Weekly Report*, January 19, 1923; "Craig Sawmill Will Open Up First of June," *Ketchikan Alaska Chronicle*, May 14, 1920; Roy H. Shotwell, "Lumbering in Alaska," *Timberman* (December 1922): 110–112; "Lumber Shipped Out to England," *Ketchikan Alaska Chronicle*, July 18, 1922.
10. "Busy Season Expected on West Coast," *Ketchikan Alaska Chronicle*, December 9, 1924; "Craig Sawmill Has Been Sold," *Ketchikan Alaska Chronicle*, July 26, 1924; Wyckoff, "Lumber Industry on Tongass National Forest," 5; "Craig Saw-Mill," *Petersburg Herald*, February 12, 1925; Cole and Rasmuson, *Banking on Alaska*, 125; "Industries at Craig Prepare for Business," *Ketchikan Alaska Chronicle*, June 5, 1925; B. Frank Heintzleman, "Quarterly Report, History of Defense and War Activities of the Department of Agriculture Forest Service-Alaska Region," January 18, 1945, RG 95, NARA—Pacific Alaska Region, Anchorage.
11. E. E. Carter to Milton Briggs (Commodity Credit Corporation, USDA), March 6, 1945, RG 95, NARA—Pacific Alaska Region (Anchorage); Charles G. Burdick to Chief, USFS, February 26, 1945, RG 95, NARA—Pacific Alaska Region, Anchorage; *Tewkesbury's Who's Who in Alaska and Alaska Business Index* (Juneau: Tewkesbury Publishers, 1947), 294; *Sourdough Notes*, February 1, 1949; Fred Hamilton (long-time resident, Craig, Alaska) personal communication with author, May 26, 2006.
12. "Story of the First Manufacturing Industry Ever Established in Alaska," *West Coast Lumberman* (November 1, 1913): 29–30.
13. De Armond, "Early Sawmills, III"; John Daly (former part-owner, Ketchikan Spruce Mills), personal communication with author, September 24, 2004.
14. Albert Parker (son of A. L. Parker), personal communication with author, various dates; Howard M. Kutchin, *Report on the Salmon Fisheries of Alaska in 1907*, Bureau of Fisheries (Washington: GPO, 1908), 19; "Chronological History of Salmon Canneries in Southeastern Alaska," in *1949 Annual Report, Alaska Fisheries Board and Alaska Department of Fisheries*, Rept. No. 1 (Juneau: Alaska Fisheries Board and Alaska Dept. of Fisheries, 1949), 32; Frank T. Been, "Inspection of Glacier Bay National Monument, Alaska, August 1 to August 27, 1939," 32. Report on file at Glacier Bay National Park and Preserve, Gustavus, Alaska; Adam Greenwald, October 11, 2004.
15. Norman DesRosiers (former logger/sawmill operator, Excursion Inlet, Alaska), personal communication with author, December 12, 2004; Frank T. Been, "Field Notes and Interviews, Inspection of Glacier Bay and Sitka National Monuments, August 12 to September 4, 1942," 24–25. Report on file at Glacier Bay National Park and Preserve, Gustavus, Alaska.
16. Norman DesRosiers, various dates.

17. "Lumbering on Alaska Coast," *Timberman* (April 1917): 36.
18. Albert Parker and Glen Parker (former loggers/sawmill operators), personal communication with author, various dates.
19. "May in Alaska's History," *Alaska Sportsman* (May 1965): 32; "Unlocking Alaska," *American Forestry* (July 1914): 485; D. F. Houston (Secretary of Agriculture), "Forestry and the Paper Industry," *American Forestry* (April 1917): 205; "Alaska Spruce in California," *Morning Mail* (Ketchikan), October 24, 1913; "Alaska," *Timberman* (April 1914): 42; "Mill Makes Big Showing," *Alaska Sentinel* (Wrangell), December 18, 1913; "Alaska Lumber Revives Town," *Alaska Sentinel* (Wrangell), September 18, 1913; "Hadley Mill Coming Up," *Alaska Sentinel* (Wrangell), December 25, 1913; advertisement, *Morning Mail* (Ketchikan), September 22, 1913.
20. "New Company at Hadley, Alaska," *American Lumberman* (May 20, 1914): 53; "May in Alaska's History," *Alaska Sportsman* (May 1965): 32; "Hadley—Deserted Village," *Pathfinder* (Valdez, Alaska) (January 1921): 26.
21. John Schnabel (former sawmill operator, Haines, Alaska), personal communication with author, December 20, 2004; "Chronological History of Salmon Canneries in Southeastern Alaska," 30; "Musings from the Sheldon Museum," *Chilkat Valley News*, December 11, 1975; advertisement, Combs Lumber Company, *Haines Pioneer Press*, May 15, 1909; "Sawmill Burns, Destroying 15,000 Feet of Lumber," *Haines Pioneer Press*, November 16, 1912.
22. John Schnabel, interview by Ed and Milinda May, "Alaskana: Lumber for the Last Frontier," *Anchorage Daily News*, October 8, 2006.
23. "Alaska's Sawmills Produce," *Lumberman* (March 1956): 81.
24. Chuck Kleeschulte, "Chilkoot Lumber Aims to Light Up Haines," *Alaska Business Monthly* (February 1, 1989): 20; "Timber," *Historical Vignettes*, Sheldon Museum, Haines, Alaska, (http://sheldonmuseum.org/timber.htm).
25. "Alaska's Sawmills Produce," *Lumberman* (March 1956): 81; Fred Hosford (former sawmill owner, Dyea, Alaska), personal communication with author, April 19, 2005; "News on Haines, Alaska, Mill," *Lumberman* (April 1956): 144.
26. Alaska Resources Committee, *Alaska*, 100; Robert De Armond (Alaska historian), personal communication with author, February 2, 2005.
27. "First Raft of Alaska Spruce Arrives in Puget Sound," 32d; Joseph Smith (former sawmill owner, Juneau), personal communication with author, May 5, 2006.
28. Langille, "Report on Alexander Archipelago Forest Reserve, 1906"; De Armond, "Early Sawmills, III."
29. Hinkley, *Alaskan John G. Brady*, 123; Sheldon Jackson, *Education in Alaska*, Sen. Ex. Doc. No. 85, 49th Cong., 1st sess. (Washington: GPO, 1886),17.
30. Adam Greenwald, October 11, 2004; Alf Skaflestad (former logger), personal communication with author, October 11, 2004.
31. Robert De Armond, "Sitka Names & Places," *Sitka Sentinel*, August 12, 1999; *Report of the Governor of Alaska for the Fiscal Year 1889* (Washington, GPO, 10; U.S. Coast and Geodetic Survey, *Pacific Coast Pilot: Alaska*, Part I, 3rd ed. (Washington: GPO, 1891), 83.
32. George T. Emmons, "The Woodlands of Alaska," (unpublished, 1902); Langille, "Report on Alexander Archipelago Forest Reserve, 1906."
33. Howard M. Kutchin, *Report on the Salmon Fisheries of Alaska, 1900*, Sen. Doc. No. 168, 56th Cong. 2d. sess. (Washington: GPO, 1901), 79; Langille, "Report on Alexander Archipelago Forest Reserve, 1906."
34. "Lumbering on Alaska Coast," *Timberman* (April 1917): 36; "Hydaburg to Have Electric Plant," *Wrangell Sentinel*, April 17, 1919.
35. "Hyder—the Newest Alaskan Mining Camp," *Pathfinder* (Valdez, Alaska) (November 1921): 3; "News of Ketchikan," *Ketchikan Alaska Chronicle*, March 14, 1923.
36. Untitled, *Juneau City Mining Record*, April 19, 1888; advertisement, *Juneau City Mining Record*, April 12, 1888.
37. *Alaskan* (Sitka), May 30, 1894.
38. *Report of the Governor of Alaska for the Fiscal Year 1889* (Washington: GPO, 1889), 10; *Report of the Governor of Alaska* (Washington: GPO, 1890), 492.
39. Advertisement, *Douglas Island News* (Juneau), June 7, 1899; De Armond, "Early Sawmills, III"; "Russians Operated First Sawmill In Alaska With A Market In California," 32C.
40. "A New Saw Mill," *Wrangell Sentinel*, November 10, 1910; Wyckoff, "History of Lumbering in Southeast Alaska," 38; De Armond, "Early Sawmills, III"; Olmsted, "Inspection Report on the Alexander Archipelago Forest Reserve, 1906," 8; "Russians Operated First Sawmill In Alaska With A Market In California," 32C; "Lumbering on Alaska Coast," *Timberman* (April 1917): 36.
41. "Puget Sound Lumberman Joins Shattuck," *Daily Alaska Dispatch* (Juneau), February 24, 1913; "Lumbering on Alaska Coast"; "Will Take Mill Man's Body East," *Daily Alaska Dispatch* (Juneau), June 3, 1918.
42. "Eight Million Feet of Logs Cut by Mills," *Ketchikan Alaska Chronicle*, November 20, 1923; "Progress in Alaska Lumber Industry," *Timberman* (February 1925): 208; "Alaska Mills Enjoy Good Business," *Timberman* (August 1925): 161; "Largest Alaska Sawmill," *West Coast Lumberman* (September 1941): 12–13, 44; "Building Sawmills Around the World," *West Coast Lumberman* (May 1, 1928): 67; "Juneau Lumber Makes Progress Over 10 Years," *Daily Alaska Empire*, December 7, 1937; "Alaska Mills Plan for New Year," *American Lumberman* (December 24, 1927): 46; Alva W. Blackerby, "Opportunities for

Minor Wood Product Industries in Alaska," USFS, Alaska Region, January 1, 1945, 9; "Completes Biggest Year in Its History," *American Lumberman* (December 2, 1927): 53; advertisement, *Daily Alaska Empire* (Juneau), May 4, 1927; House Committee on Merchant Marine and Fisheries, *Fish Traps in Alaskan Waters* (hearing), 74th Cong., 2d sess., 1936, 213; advertisement, *Daily Alaska Empire* (Juneau), June 18, 1927; "Sawmilling in Alaska," *Timberman* (January 1947): 202.

43. Minutes of special meeting of stockholders of the Juneau Spruce Corp., Portland, Oregon, May 25, 1948, Juneau Spruce Corp. Records, 1947–1956, MS 191, Historical Collections, Alaska State Library, Juneau; "Oregon Men Buy Juneau, Alaska, Sawmill," *West Coast Lumberman* (June 1947): 117; Alaska State Library, Juneau Spruce Corporation, Historical Background, 2, Juneau Spruce Corp. Records, 1947–1956, MS 191, Historical Collections, Alaska State Library, Juneau; "Alaska Lumber Shipped East via Prince Rupert," *West Coast Lumberman* (May 1948): 114; "Forest Industries News," *Lumberman* (November 1949): 32.

44. George W. Rogers (Alaska economist), personal communication with author, March 15, 2006; Joseph Smith (former sawmill owner, Juneau), personal communication with author, May 4, 2006; "Forest Industries News," *Lumberman* (March 1951): 6, 52; "Alaska Plant Burns," *Lumberman* (March 1951): 52.

45. "Alaska Forest Products Manufactures Face Numerous Problems," *Lumberman* (November 1960): 56–57, 72; "Major Mill Cuts," *Lumberman* (March 25, 1954): 81; "Alaska Lumber Outlook," *West Coast Lumberman* (January 1941): 60; "Juneau Mill Closed for Season," *Ketchikan Daily News*, November 3, 1953.

46. Joseph Smith (former sawmill owner, Juneau), personal communication with author, May 4, 2006; "Lumber Firm Opening Yard," *Daily Alaska Empire* (Juneau), May 11, 1954; "Alaska Forest Products Manufacturers Face Numerous Problems," *Lumberman* (November 1960): 56–57, 72.

47. Joseph Smith, January 17, 2007, and various; *Sourdough Notes*, February 1, 1950.

48. Ralph Browne, *The Sitka District* (Juneau: Alaska Development Board, 1950), 41; *Sourdough Notes*, January 3, 1950.

49. Juan Munoz, *Juneau, Alaska: A Study of the Gastineau Channel Area* (Juneau: Alaska Resource Development Board, 1956), 84; George W. Rogers, *Alaska in Transition: The Southeast Region* (Baltimore: Johns Hopkins Press, 1960), 81; "Forest Industries News," *Lumberman* (October 1951): 12.

50. "Alaska Plywood Plant Planned at Juneau," *Timberman* (December 1951): 108; George W. Rogers (Alaska economist), personal communication with author, February 21, 2005.

51. "Columbia Lumber's Plywood Plant Finds Ready Market," *Ketchikan Alaska Chronicle*, May 15, 1954; Rogers, *Alaska in Transition*, 79; George W. Rogers, February 21, 2005.

52. "Alaska Plywood Plant Planned at Juneau," *Timberman* (December 1951): 108; "Plant Produces Spruce," *Lumberman* (January 1956): 82–83, 136; Timber sale advertisement, *Wrangell Sentinel*, September 21, 1951; Rogers, *Alaska in Transition*, 82.

53. "Tapping of Rich Forest Breeds a Batch of New Wood-Product Plants," *Wall Street Journal*, June 20, 1953; "Forest Industries News," *Lumberman* (September 1952): 20; "Forest Industries News," *Lumberman* (July 1953): 16.

54. George W. Rogers, February 21, 2005.

55. "Columbia Lumber's Plywood Plant Finds Ready Market," *Ketchikan Alaska Chronicle*, May 15, 1954; George W. Rogers, February 21, 2005.

56. George W. Rogers, February 21, 2005.

57. "Alaska Forest Products Manufactures Face Numerous Problems," *Lumberman* (November 1960): 56–57; Rogers, *Alaska in Transition*, 82.

58. "Late Forest Industries News," *Lumberman* (September 1956): 13; "Million Dollar Fire Hits Juneau Plant, *Daily Sitka Alaska Sentinel*, August 14, 1959.

59. "General News from Kake," *Petersburg Weekly Report*, January 14, 1920.

60. Charles Martinson, *Atlas of the Ketchikan Region: A Basis for Planning* (Ketchikan: Ketchikan Borough Planning Dept., 1979?); Howard M. Kutchin, *Report on the Salmon Fisheries of Alaska, 1903*, Dept. of Commerce and Labor (Washington: GPO, 1904), 16; Langille, "Report on Alexander Archipelago Forest Reserve, 1906"; De Armond, "Early Sawmills, III"; George T. Emmons, "The Woodlands of Alaska," (unpublished, 1902); "Alaska," *Pacific Fisherman* (June, 1904): 13; Wyckoff, "History of Lumbering in Southeast Alaska," 38.

61. Wyckoff, "History of Lumbering in Southeast Alaska," 38; J. M. Wyckoff, "Lumbering in Alaska," *Pathfinder of Alaska* (August 1923): 8; Advertisement, *Ketchikan Alaska Chronicle*, May 17, 1922.

62. "The First Native Enterprise on Business Principles," *Alaskan* (Sitka), January 21, 1893; George R. Tingle (inspector of salmon fisheries), *Report of the Salmon Fisheries in Alaska, 1896*, Treasury Department Doc. No. 1925 (Washington: GPO, 1897): 21; Jefferson F. Moser, *Salmon and Salmon Fisheries of Alaska* (Washington: GPO, 1899), 34; De Armond, "Early Sawmills, III"; Hinkley, *Alaskan John G. Brady*, 126; Patricia Roppel, *An Historical Guide to Revillagigedo and Gravina Islands, Alaska* (Wrangell: Farwest Research, 1995), 260.

63. *Mining Journal* (Ketchikan), January 5, 1901; Roppel, *An Historical Guide to Revillagigedo and Gravina Islands*, 252; C. Keith Stump, "75-Year-Old Sawmill is Still Part of City," *Ketchikan Daily News*, August 29, 1975.

64. Roppel, *An Historical Guide to Revillagigedo and Gravina Islands*, 14–15; De Armond, "Early Sawmills, III."

65. Howard M. Kutchin, *Report on the Salmon Fisheries of Alaska, 1903*, Dept. of Commerce and Labor (Washington: GPO, 1904), 16–17.
66. *Alaska Gazetteer and Directory (1911–1912)* (Seattle: R. L. Polk & Company, 1911), 325; advertisement, *Morning Mail* (Ketchikan), October 16, 1913; photograph, CD 80, KTN Shingle Mill, Tongass Historical Museum, Ketchikan; Stump, "75-Year-Old Sawmill is Still Part of City"; "Fleming Plant Will Start Its Season's Work," *Ketchikan Alaska Chronicle*, February 8, 1923; "Fleming Declares Mill Business Will Be Good in Alaska," *Ketchikan Alaska Chronicle*, December 1, 1923.
67. Ketchikan Power Company stationary, c. 1922, Ketchikan Spruce Mills Collection, Tongass Historical Museum, Ketchikan.
68. Stump, "75-Year-Old Sawmill is Still Part of City"; John Daly, September 25, 2004; advertisements, *Daily Miner* (Ketchikan), December 26, 1906, and March 14, 1907.
69. "Great Home Industry," *Mining Journal* (Ketchikan), Illustrated Annual, January 1907; "Story of the First Manufacturing Industry Ever Established in Alaska," *West Coast Lumberman* (November 1, 1913): 29–30; C. M. Archbold to M. J. Daly, July 14, 1959, Ketchikan Spruce Mills Collection, Tongass Historical Museum, Ketchikan; "Story of the First Manufacturing Industry Ever Established in Alaska," *West Coast Lumberman* (November 1, 1913): 29–30.
70. Unknown author, "For 60 Years Ketchikan Spruce Mills has Kept Pace with Alaska's Growth," 1963, Ketchikan Spruce Mills Collection, Tongass Historical Museum, Ketchikan; "Ketchikan Spruce Mills," *Ketchikan Alaska Chronicle*, ("New Era" edition), September 30, 1953; "Spruce Mills 50 Years Young; Is Major Industry," *Ketchikan Alaska Chronicle*, May 21, 1953.
71. "Big Sawmill Will Be Built in Ketchikan; Plans are Complete," *Ketchikan Alaska Chronicle*, September 4, 1924; "Most Modern Machinery in New Mill," *Ketchikan Alaska Chronicle*, October 17, 1924; "For 60 Years Ketchikan Spruce Mills has Kept Pace"; "Fire at Mill Extinguished Early Today," *Ketchikan Alaska Chronicle*, March 19, 1925; "Spruce Mill Made Record for Season," *Ketchikan Alaska Chronicle*, December 8, 1925.
72. "Box Factory of Sawmill Burns," *Ketchikan Alaska Chronicle*, October 8, 1925; "Daly Announces Plans for Box and Planing Factory," *Ketchikan Alaska Chronicle*, November 3, 1925; "Spruce Mills Begin Annual Season Today," *Ketchikan Alaska Chronicle*, March 1, 1926; "Ketchikan Mill Plans a Big Year," *Timberman* (February 1940): 39.
73. "Weyerhaeuser Pleased With Alaska Trip," *Ketchikan Alaska Chronicle*, June 26, 1926.
74. Advertisement, *Ketchikan Alaska Chronicle*, October 3, 1927; "Ketchikan Spruce Mills Will Operate Until Thanksgiving," *Ketchikan Alaska Chronicle*, August 28, 1928.
75. "Ketchikan Spruce Mills," *Ketchikan Alaska Chronicle*, ("New Era" edition), September 30, 1953; John Daly, September 25, 2004.
76. John Daly, September 25, 2004.
77. "History of Ketchikan Spruce Mill," Tongass Historical Museum files; "Ketchikan Mill Goes Unsold," *Anchorage Daily News*, October 19, 1984
78. "Lumbering on Alaska Coast," *Timberman* (April 1917): 36; J. M. Walley, "Report on Market for Cedar Poles at Prince Rupert or Elsewhere and Alaskan Market for Shingles," USFS (unpublished), 1925, RG 95, NARA—Pacific Alaska Region, Anchorage.
79. Advertisements, *Ketchikan Alaska Chronicle*, August 18, 1919, and July 29, 1921; "Progress of Alaska Lumber Industry," *Timberman* (February 1925): 208; "Operation of North End Mill Soon to Begin," *Ketchikan Alaska Chronicle*, March 24, 1932; Stump, "75-Year-Old Sawmill is Still Part of City."
80. Roppel, *An Historical Guide to Revillagigedo and Gravina Islands*, 272.
81. "Wide Demand for Alaska Shingles," *Alaska Fishing News* (Ketchikan), October 1, 1943; *Tewkesbury's Who's Who in Alaska and Alaska Business Index*, 296.
82. "Fire Destroys Sawmill, Two Boats," *Ketchikan Daily News*, March 19, 1954.
83. USFS, *KPC Pulp Mill Shutdown and 1997 Timber Demand Projections*, in *Land and Resource Management Plan – Tongass National Forest*, 1997, R10-MB 338-CD, Final Environmental Impact Statement, Supplement Evaluation, M-2.
84. *Ketchikan Pulp Company*, 14–15.
85. Greeley, "What is Wrong with Alaska," 205.
86. Tingle, *Report of the Salmon Fisheries in Alaska, 1896*, 17; Wyckoff, "History of Lumbering in Southeast Alaska," 38; "Russians Operated First Sawmill In Alaska With A Market In California," 32C; Moser, *Salmon and Salmon Fisheries of Alaska*, 110.
87. *Report of the Governor of Alaska for the Fiscal Year 1889* (Washington: GPO, 1890), 10.
88. Emmons, "The Woodlands of Alaska."
89. *Reid Brothers Logging Company v. Ketchikan Pulp Company* (June 8, 1981), "Timber Firms File Chapter 11," *Anchorage Daily News*, October 19, 1984; "Pact May Reopen Sawmill," *Anchorage Daily News*, December 19, 1985; "Timber Firm Will Start Work Again on Prince of Wales Island," *Anchorage Daily News*, March 18, 1988; Hal Bernton, "Tall Order the Future of the Tongass National Forest Depends On the Little Guy Finding Opportunities the Giants Missed," *Anchorage Daily News*, November 13, 1984; USDA Forest Service, Alaska Region, Timber Supply and Demand reports, 1983, 1984, 1985, 1987, 1991; Greg Head (son of Edward Head), personal communication with author, February 20, 2010; Kirk Dahlstrom (owner, Viking Lumber), personal communication with author, February 18, 2010.
90. William Duncan, "A Day at Metlakahtla," *Metlakahtlan*, November 1888.

91. *Metlakahtlan*, November 1889; "A Thriving Alaskan Town," *New York Times*, February 2, 1890; Howard M. Kutchin, *Report on the Salmon Fisheries of Alaska, 1900*, 56th Cong. 2d. sess., Sen. Doc. No. 168 (Washington: GPO, 1901), 69; De Armond, "Early Sawmills, III."

92. Bill of sale, December 31, 1917, Edward Marsden papers, AX 69, Div. of Spec. Collections and University Archives, University of Oregon, Eugene, Oregon 97403-1299.

93. Alaska Resources Committee, *Alaska*, 94; "Alaska Lumber Cut," *Timberman* (July 1941): 55; George Sundborg, *Opportunities in Alaska* (New York: Macmillan Company, 1945), 72; *Tewkesbury's Who's Who in Alaska and Alaska Business Index*, 298.

94. "Forest Industry News," *West Coast Lumberman* (March 1949): 36; "Greeley Greets Growth," *Ketchikan Pulp: A $52,500,000 Ultra-Modern Industry*, unpaged.

95. "Fire Destroys Metlakatla's New Sawmill," *Ketchikan Alaska Chronicle*, January 20, 1954; "Annette Island Spruce Plant Being Operated by 3 Everett Millmen," *1955 Yearbook of Alaska Timber Industries* (Ketchikan: *Ketchikan Alaska Chronicle*, 1955), unpaged; "Alaska Forest Products Manufactures Face Numerous Problems," *Lumberman* (November 1960): 56–57, 72; Roppel, *An Historical Guide to Revillagigedo and Gravina Islands*, 66; *Ketchikan Pulp Company*, 15; *Reid Brothers Logging Company v. Ketchikan Pulp Company* (9th Cir. 1983); USFS, *KPC Pulp Mill Shutdown and 1997 Timber Demand Projections*, 1997 Tongass Land Management Plan EIS Supplement Evaluation, M-2.

96. Wyckoff, "History of Lumbering in Southeast Alaska," 38.

97. Robert De Armond, February 2, 2005 and January 13, 2006; "Pelican—Alaska's New Fish Freezing Center," *Pacific Fisherman* (May 1944): 49; "9 Plants, 19 Logging Camps North of Petersburg, Alaska," *1955 Yearbook of Alaska Timber Industries* (Ketchikan: *Ketchikan Alaska Chronicle*, 1955), unpaged; John Clausen (long-time Pelican resident), personal communication with author, January 19, 2006; Ester Moy (long-time Pelican resident), personal communication with author, January 20, 2006. *Tewkesbury's Who's Who in Alaska and Alaska Business Index*, 298.

98. Donald Nelson, (Petersburg historian), e-mail message to author, January 14, 2006.

99. "New Saw Mill," *Daily Alaska Dispatch* (Juneau), August 19, 1903; Donald Nelson, January 18, 2005; Pat Ellis, *From Fish Camps to Cold Storages: A Brief History of the Petersburg Area to 1927* (Petersburg: Clausen Memorial Museum, 1998), 42; "Russians Operated First Sawmill In Alaska With A Market In California," 32C; "Brings Petersburg Logs," *Wrangell Sentinel*, June 5, 1913; "Sawmill Delivering First Shooks," *Wrangell Sentinel*, April 3, 1913; Donald Nelson, *The Story of Petersburg* (Petersburg: Pilot Publishing, 2001), 20.

100. "Big Saw-Mill Will Again Operate," *Petersburg Weekly Report*, June 22, 1915; *Petersburg Weekly Report*, January 31, 1919; "Mill to Start Soon," *Petersburg Weekly Report*, March 12, 1920; J. L. Mackenchuil (USFS ranger at Petersburg) to Forest Supervisor, March 27, 1920, RG 95, NARA—Pacific Alaska Region, Anchorage; "Will Double Mill Capacity," *Petersburg Weekly Report*, May 7, 1920; "To Ship Lumber," *Petersburg Weekly Report*, August 6, 1920; "Mill Finishes the Season," *Petersburg Weekly Report*, September 24, 1920; *Petersburg Weekly Report*, December 31, 1920; Louis McDonald, "Logging and Lumbering in Alaska," *Timberman* (January 1921): 41; *Petersburg Weekly Report*, January 20, 1922; "Petersburg Sawmill May Be Operated," *Petersburg Weekly Report*, February 9, 1923; Public Notice, *Petersburg Weekly Report*, February 1, 1924; "Progress in Alaska Lumber Industry," *Timberman* (February 1925): 208; "Big Sawmill Will Be Rebuilt At Once," *Petersburg Weekly Report*, May 29, 1925.

101. "Old Petersburg Mill Sold To Outside Party," *Petersburg Herald*, May 9, 1924; "Local Mill to Supply All of Lumber Needs," *Petersburg Herald*, September 2, 1924; advertisement, *Petersburg Herald*, October 31, 1924; "Big Sawmill Started Work for Summer," *Petersburg Herald*, May 1, 1925; "Timber for Airplanes is Turned Out," *Petersburg Herald*, May 22, 1925; "Sawmill To Be Rebuilt At Once," *Petersburg Herald*, May 29, 1925.

102. "Boiler Explodes In Knutsen's Saw Mill," *Petersburg Weekly Report*, October 27, 1922; "Barrel Plant May Be Started Again," *Petersburg Herald*, August 11, 1925; "Alaska Notes," *Timberman* (July 1936): 26; Donald Nelson, January 18, 2005.

103. *Petersburg Weekly Report*, March 13, 1915; "Alaska Notes," *Timberman* (July 1936): 26; "Petersburg, Alaska, Mill A Progressive Operation," *West Coast Lumberman* (May 1949): 118; various authors, Pioneer Profiles (Petersburg: Pioneers of Alaska, 2004), 17, 55; Donald Nelson, January 18, 2005; Arlene Peterson (daughter of Octor Arness), personal communication with author, January 12, 2006.

104. "Lumbering on Alaska Coast," *Timberman* (April 1917): 36.

105. The *(Petersburg) Alaskan* reported in August 1926 that the Alaska Spruce Mills (of Petersburg) had received an order for street lumber for Petersburg. This was possibly another name for the Northwest Mill Company, the Knutson Brothers Sawmill, or Olaf Arness's sawmill. See: Ellis, *From Fish Camps to Cold Storages*, 34; Eigil Buschmann, "List of most important men working at Petersburg, 1898–1900," Robert Thorstenson Sr. collection, Seattle, Wash.; untitled, *(Petersburg) Alaskan*, August 13, 1926.

106. Donald Nelson, January 18, 2005.

107. Advertisement, *Alaskan* (Petersburg), November 5, 1926; "Alaska Notes," *Timberman* (July 1936): 26; Donald Nelson, January 18, 2005.

108. "Power Wood Sawing Plant is Started Up," *Petersburg Weekly Report*, January 26, 1923; "Portable Mill is Installed by West," *Petersburg Weekly Report*, July 27, 1923.

109. Langille, "Report on Alexander Archipelago Forest Reserve, 1906."

110. "Speel River Plant Shut Down for Time," *Ketchikan Alaska Chronicle*, February 6, 1923; photograph of sawmill with boiler, Patrick Eugene Kennedy collection, PCA 444, Historical Collections, Alaska State Library Juneau.
111. Patricia Roppel, "The Gold of Rodman Bay," *Alaska History* (Fall 1933): 21.
112. *Alaskan* (Sitka), May 14, 1887; *Report of the Governor of Alaska for the Fiscal Year 1889* (Washington: GPO, 1889), 10; *Report of the Governor of Alaska, 1891* (Washington: GPO, 1891), 492; untitled, *Juneau City Mining Record*, July 24, 1891; U.S. Coast and Geodetic Survey, *Pacific Coast Pilot: Alaska*, Part I , 3rd ed. (Washington: GPO, 1891), 126; Ronald J. Benice, *Alaska Tokens* (Lake Mary, Fla.: Token and Medal Society, Inc., 1994), 231; Robert De Armond, "Around & About Alaska," *Daily Sitka Sentinel*, August 10, 1994.
113. "Juneau Enterprise," *Alaska Mining Record* (Juneau), special edition, probably late 1900; De Armond, "Early Sawmills, III"; Prospectus, Alaska Fish & Lumber Company, vertical file, Historical Collections, Alaska State Library, Juneau; Kutchin, *Report on the Salmon Fisheries of Alaska, 1905*, 9.
114. Ronald J. Benice, *Alaska Tokens* (Lake Mary, Fla.: Token and Medal Society, Inc., 1994), 231; Langille, "Report on Alexander Archipelago Forest Reserve, 1906"; De Armond, "Early Sawmills, III"; untitled, *Alaska Sentinel* (Wrangell), December 4, 1902; "Chronological History of Salmon Canneries in Southeastern Alaska," in *1949 Annual Report, Alaska Fisheries Board and Alaska Department of Fisheries*, Rept. No. 1 (Juneau: Alaska Dept. of Fisheries, 1949), 32; Shakan Salmon Company to Secretary of Commerce and Labor, April 21, 1909, RG 22, NARA, College Park, Md.
115. "Russians Operated First Sawmill in Alaska with a Market in California," 32C; Wyckoff, "History of Lumbering in Southeast Alaska," 38; "Chronological History of Salmon Canneries in Southeastern Alaska," in *1949 Annual Report, Alaska Fisheries Board and Alaska Department of Fisheries*, Rept. No. 1 (Juneau: Alaska Dept. of Fisheries, 1949), 32.
116. *Alaska Times* (Sitka), October 9, 1869 (citation from Robert De Armond); De Armond, "Around & About Alaska," *Daily Sitka Sentinel*, August 4, 1994.
117. Robert De Armond, "From Sitka's Past—No. 32," *Daily Sitka Sentinel*, May 6, 1986; *Report of Major E. H. Ludington*, reprinted in *Letter from the Secretary of War, in relation to the Territory of Alaska*, 42d. Cong., 1st sess., Ho. Ex. Doc. No. 5, 1871, 27; "An inventory of the public property in the City of New Archangel [Sitka] delivered to the United States of America, General Lovell H. Rousseau, United States Commissioner; by His Imperial Majesty the Emperor of Russia, Captain Alexis Pestchouroff, Russian Commissioner, on the 18th day of October, 1867, at New Archangel [Sitka]" in Archie W. Shiels, "The Purchase of Alaska," (unpublished, 1965), Archie Shiels Collection, Center for Pacific Northwest Studies, Bellingham, Wash.
118. De Armond, "From Sitka's Past—No. 32"; *Census of Sitka, Alaska Territory, October 24, 1870*, printed in *Letter from the Secretary of War, in relation to the Territory of Alaska*, 42d. Cong, 1st sess., Ho. Ex. Doc. No. 5, 1871,14–25.
119. De Armond, "Around & About Alaska," August 4, 1994; Andrews, *Sitka, The Chief Factory of the Russian American Company*, 85; "Tells About First Cannery in Alaska," *Petersburg Herald*, May 15, 1925; De Armond, "Around & About Alaska," August 10, 1994; Robert De Armond, "Sitka Fails to Capitalize on Sawmill Industry," *Daily Sitka Sentinel*, May 6, 1948.
120. De Armond, "Early Sawmills, III."
121. "Another Sitka Industry," *Pathfinder* (Valdez, Alaska) (December 1920): 7.
122. "Russians Operated First Sawmill In Alaska With A Market In California," 32C; "The Old Saw Mill a Thing of the Past," *Verstovian* (Sitka), November 1914.
123. De Armond, "Around & About Alaska," September 28, 1994; De Armond, "Around & About Alaska," October 12, 1994; "Progress of Alaska Lumber Industry," *Timberman* (February 1925): 208; "Alaska Notes," *Timberman* (July 1936): 26; "Alaska Forest Products Manufactures Face Numerous Problems," *Lumberman* (November 1960): 56–57, 72.
124. "A Major Sitka Spruce Operation," *West Coast Lumberman* (June 1949): 66, 68; "Progressive Alaska Lumber Company," *West Coast Lumberman* (September 1941): 54; "New Alaska Plant to Run in July," *Timberman* (July 1944): 56; De Armond, "Around & About Alaska," October 20, 1994.
125. "Alaska Forest Products Manufactures Face Numerous Problems," *Lumberman* (November 1960): 56–57, 72; De Armond, "Around & About Alaska," October 20, 1994.
126. Hinkley, *Alaskan John G. Brady*, 122–123.
127. Ibid., 123.
128. De Armond, "Around & About Alaska," August 19, 1994.
129. Hinkley, *Alaskan John G. Brady*, 124; *Alaskan*, August 29, 1891; Leslie W. Yaw, *Sixty Years in Sitka* (Sitka: Sheldon Jackson College Press, 1985), 64–65.
130. Hinkley, *Alaskan John G. Brady*, 329; De Armond, "Around & About Alaska," September 7, 1994.
131. De Armond, "Around & About Alaska," September 7, 1994.
132. *Report of the Governor of Alaska for the Fiscal Year 1889* (Washington, GPO, 10.
133. Yaw, *Sixty Years in Sitka*, 58–69.
134. *Report of the Governor of the District of Alaska to the Secretary of the Interior, 1899* (Washington: GPO), 13; *Alaska Mining Record* (Juneau), December 18, 1897.
135. "Juneau Enterprise," *Alaska Mining Record* (Juneau), special edition, probably late 1900; Frank Norris, "Sawmill Site at Dyea," Klondike Gold Rush National Historical Park, Skagway, Alaska; unpublished

report, Klondike Gold Rush National Historical Park, Skagway, Alaska; Dyea Wood Company document, Klondike Gold Rush National Historical Park, Skagway, Alaska.

136. Robert L. Spude, "The White Pass Trail Across Block 24, 1888–1897," draft manuscript (2001) on file at Klondike Gold Rush National Historical Park, Skagway, Alaska; "Skagway Saw Mill," *Daily Alaskan*, 3rd Annual Edition, January, 1901; "Moulding Machine Started," *Daily Alaskan*, July 7, 1900; "Sawmill Starts Up," *Daily Alaskan*, April 11, 1900; "Magnificent Timber from Native Shores," *Daily Alaskan*, July 20, 1900; "Another Citizen Gone," *Daily Alaskan*, June 18, 1901; De Armond, "Early Sawmills, III"; "Conditions in Alaska," *Columbia River and Oregon Timberman* (April 1902): 37; "Fire In Skagway; Sawmill Burns," *Daily Alaskan*, November 5, 1902; "Sawmill Not To Be Reconstructed," *Daily Alaskan*, November 6, 1902; Wyckoff, "History of Lumbering in Southeast Alaska," 38.

137. Fred Hosford (former Dyea, Alaska, sawmill operator), personal communication with author, April 19, 2005.

138. "Hosford's Sawmill" (unpublished), Klondike Gold Rush National Historical Park, Skagway, Alaska; Fred Hosford, April 19, 2005.

139. De Armond, "Around & About Alaska," September 14, 1994.

140. J. L. Mackenchuil (USFS ranger at Petersburg) to Forest Supervisor, March 27, 1920, RG 95, NARA—Pacific Alaska Region, Anchorage; USFS photo No. 219791 "Sawmill and Falls at Warm Springs Bay, Alaska," dated September 7, 1927, USFS Forestry Sciences Laboratory, Juneau; M. J. Kirchoff, *Baranof Island: An Illustrated History* (Juneau: Alaska Cedar Press, 1990), 116, 118, 120.

141. Dennis Sperl, *In the Wake of an Alaskan Mailboat* (Petersburg, Alaska: Dennis Sperl, 2001), 165–167.

142. *Alaska Sentinel* (Wrangell), March 5, 1903; *Alaskan* (Sitka), February 9, 1889; Samuel Sylvester, "Life and Adventures of Rufus Sylvester," *Alaska Sentinel* (Wrangell), February 9, 1905; "Wrangell Mill Company," *Stickeen River Journal* (Wrangell), May 14, 1898; "Alaska's Modern Sawmills," *Timberman* (October 1930): 41; *Stickeen River Journal* (Wrangell), January 22, 1898; "Our Introduction," *Alaska Missionary Herald* (Wrangell), November 1901; Robert De Armond, "Abaft the Beam: Early Sawmills, III," *Ketchikan Daily Alaska Fishing News*, February 9, 1946.

143. "Lumbering on Alaska Coast," *Timberman* (April 1917): 36; "Alaska," *Timberman* (April 1914): 42–43.

144. "Alaska Mill Burned," *West Coast Lumberman* (April 1, 1918): 24; J. W. Pritchett, "The Astounding Ignorance of J. J. Underwood Concerning the Timber Resources of Alaska," *Wrangell Sentinel*, July 25, 1921; "Modern Electric Plant Installed in Connection with Sawmill of Wrangell Lumber & Power Co.," *Wrangell Sentinel*, February 11, 1926; "Progress of Alaska Lumber Industry," *Timberman* (February 1925): 208.

145. "Alaska Mills Enjoy Good Business," *Timberman* (August 1925): 161; "Wrangell's New Power Plant Put Into Operation," *Ketchikan Alaska Chronicle*, August 8, 1925; "This Modern Alaska Mill Ranks With Best Coast Plants," *West Coast Lumberman* (April 15, 1926): 25.

146. "Alaska Timber Sales," *Timberman* (March 1926): 186; Ray Taylor, "Early Forest Research in Alaska," Unpublished report on file at USFS Forestry Sciences Laboratory, Juneau.

147. "This Modern Alaska Mill Ranks With Best Coast Plants," *West Coast Lumberman* (April 15, 1926): 25; "Wrangell, Alaska Mill Leased For Five Years," *West Coast Lumberman* (February 1929): 56.

148. George M. Cornwall, "Alaska and Its Lumber Folk," *Timberman* (July 1936): 20, 22; "Wrangell, Alaska Mill Leased For Five Years," *West Coast Lumberman* (February 1929): 56; "This Modern Alaska Mill Ranks With Best Coast Plants," 25.

149. Advertisement, *Wrangell Sentinel*, April 11, 1941; "Wrangell Mill Will Start Work Coming Week," *Wrangell Sentinel*, September 17, 1943; "Wrangell Mill To Start Operation Soon; $50,000 To Be Spent Rebuilding," *Wrangell Sentinel*, September 28, 1945.

150. "Wrangell Mill Gets Up Steam Preparatory To Opening," *Wrangell Sentinel*, March 15, 1946; "Wagners Take Over Ownership of Local Mill," *Wrangell Sentinel*, June 21, 1946; "Seattle Company Submits Bid on Local Sawmill," *Wrangell Sentinel*, February 4, 1949; "Large Alaskan Mill Resumes Operation," *West Coast Lumberman* (June 1946): 106; "First Load of Lumber Left Here Today for Puget Sound From Mill," *Wrangell Sentinel*, November 1, 1946; "Olson May Have Operator for Local Mill," *Wrangell Sentinel*, September 26, 1947.

151. "Seattle Company Submits Bid on Local Sawmill," *Wrangell Sentinel*, February 4, 1949; "Modern Mill for Alaska," *West Coast Lumberman* (August 1949): 96b; Leonard Campbell (Wrangell, Alaska tugboat business owner), personal communication with author, March 20, 2006; "New Sales To Cause Big Jump in Alaskan Timber Activities," *Lumberman* (July 1954): 111.

152. "Alaska's Sawmills Produce," *Lumberman* (March 1956): 78–79; *Biennial Report of the Alaska Resource Development Board, 1955–1957* (February 29, 1957), 11; "Alaskan Invasion," *Wall Street Journal*.

153. "Southeast Alaska Hopes Timber Will Help Lure New Industries," *New York Times*, July 4, 1965; masthead, *Wrangell Sentinel*, December 25, 1964.

154. Recorder's Office Notice of Location, February 21, 1898, Stickeen Lumber Mills folder, Irene Ingle Public Library (Wrangell); "Local News," Supplement to *Stickeen River Journal*, May 7, 1898; "Stickeen Lumber Mills," *Stickeen River Journal*, May 14, 1898; Moser, *Alaska Salmon Investigations in 1900*, 280.

155. *Wrangell Sentinel*, July 23, 1943; B. Frank Heintzleman, "Quarterly Report, History of Defense and War Activities of the Department of Agriculture Forest Service-Alaska Region," January 18, 1945, RG 95, NARA—Pacific Alaska Region.

156. Knox Marshall to Regional Forester, March 4, 1949, RG 95, NARA—Pacific Alaska Region, Anchorage; "Wrangell Lumber for Defense Work," *Wrangell Sentinel*, July 20, 1951.

157. "Wrangell Lumber for Ketchikan," *Wrangell Sentinel*, December 25, 1952; "Fire Destroys Mill Creek Lumber Co. Plant," *Wrangell Sentinel*, January 22, 1954.

158. "New Wrangell Sawmill," *1955 Yearbook of Alaska Timber Industries* (Ketchikan: *Ketchikan Alaska Chronicle*, 1955), unpaged; "Oral History Interview with Daniel L. Goldy," Truman Presidential Library and Museum (http://www.trumanlibrary.org/oralhist/goldyd.htm#transcript).

159. "Timber!" *Wrangell Sentinel*, April 9, *1954; 1954 Annual Report for the Fiscal Year Ended June 30,* Governor of Alaska to the Secretary of the Interior (Washington: GPO, 1954), 42; "Forest Industries Move Ahead," *Forest Industries Yearbook of Alaska* (Ketchikan: *Ketchikan Alaska Chronicle*, July 1955), unpaged; "Alaska Timber Sale Requires Pulp Mill," *Timberman* (May 1954): 30; "Gold Rush is Shaping Up," *Lumberman* (April 1956): 79; USFS, Draft Environmental Impact Statement, Pacific Northern Timber Company Timber Sale, 1977–1981 Operating Period, (1976), vii.

160. "Gold Rush is Shaping Up," 79; "Alaska Forest Products Manufactures Face Numerous Problems," *Lumberman* (November 1960): 56–57, 72; "Significant Industry News," *Lumberman* (January 1959): 11; Leonard Campbell, April 5, 2006; USFS, Draft Environmental Statement, Pacific Northern Timber Company Timber Sale, 1977–1981 Operating Period, (1976), vii.

161. In and around 1965, the PNTC sawmill may have operated as the Alaska Pacific Lumber Company. USFS, Draft Environmental Statement, Pacific Northern Timber Company Timber Sale, 1977–1981 Operating Period, (1976), vii; "First Ship Loads at Alaska Pacific Mill," *Marine Digest* (May 15, 1965): 18; Harris and Farr, *Forest Ecosystem of Southeast Alaska*, 9; Ralph Thomas, "The Best of Times and the Best of a Bad Situation," *Anchorage Daily News*, August 13, 1995; "Wrangell Sawmill Restarts—33 Jobs," *Juneau Empire*, April 26, 1998; "Logging Company Acts on Plan to Repay Creditors," *Anchorage Daily News*, January 9, 2004; "Wrangell Lumber Mill Set to Close," *Juneau Empire*, June 20, 2004; Leonard Campbell, March 10, 2006.

162. Advertisement, *Wrangell Sentinel*, February 11, 1949.

163. "A Boost for Our Shingles," *Wrangell Sentinel*, April 15, 1915.

164. "Repairing the Engine," *Wrangell Sentinel*, December 2, 1909; "Will Make Shingles," *Wrangell Sentinel*, March 16, 1911; "M. McKinney Buys the Shingle Mill," *Wrangell Sentinel*, September 23, 1915; "Shingle Mill is Working Full Time," *Wrangell Sentinel*, November 11, 1915; J. M. Walley (USFS), "Report on Market for Cedar Poles at Prince Rupert or Elsewhere and Alaskan Market for Shingles," (unpublished), 1925, RG 95, NARA—Pacific Alaska Region, Anchorage; McDonald, Louis, "Logging and Lumbering in Alaska," *Timberman* (January 1921): 41.

165. "Two New Local Projects Start in Wrangell; Bank, Mill," *Wrangell Sentinel*, February 15, 1946; *Wrangell Sentinel*, March 14, 1947; advertisement, Wrangell Cedar Products, *Wrangell Sentinel*, August 8, 1947; Leonard Angerman, (son of Fritz Angerman), personal communication with author, February 17, 2010.

166. Kutchin, *Report of the Salmon Fisheries of Alaska, 1904*, 15–16; Gary H. Gillette, *The Yakutat Fish Train* (Juneau: Alaska Dept. of Transportation, 1998), 5; William A. Langille to Forester, August 19, 1908, RG 95, NARA—Pacific Alaska Region, Anchorage.

167. *Tewkesbury's Who's Who in Alaska and Alaska Business Index*, 298.

APPENDIX B

1. E. T. Allen, "America's Transition from Old Forests to New" (Part II, "The Past"), *American Forestry* (March 1923): 163.

2. Khlebnikov, *Notes on Russian America*, 269–271.

3. Drushka, *Working in the Woods*, 32–33.

4. "Ketchikan Spruce Mills Buys Out J. R. Reynolds," *Ketchikan Daily News*, August 28, 1954; Adam Greenwald, January 26, 2005, and February 23, 2005.

5. Brixie Crabtree (former logger), personal communication with author, December 8, 2004.

6. Richard VanOrder (former logger), personal communication with author, September 29, 2009; Wes Tyler, Icy Straits Lumber Company (Hoonah), personal communication with author, February 17, 2010.

7. *Report of the Governor of the District of Alaska to the Secretary of the Interior* (Washington: GPO, 1899), 13.

8. Wyckoff, "Lumbering in Alaska," 8.

9. "Alaskan Forest Reserve," *Forestry and Irrigation* (January 1904): 6.

10. "Wrangell Is In The Very Front Rank," *Wrangell Sentinel*, March 23, 1911; W. H. Jackson, with Ethel Dassow, *Handloggers* (Anchorage: Alaska Northwest Publishing Company, 1974), 19.

11. Wyckoff, "Airplane Spruce Chances"; "Forest Service Has Busy Program in S.E. Alaska," *Wrangell Sentinel*, October 14, 1949.

12. W. H. Jackson, with Ethel Dassow, *Handloggers* (Anchorage: Alaska Northwest Publishing Company, 1974), 21.

13. S. H. Cooke, "Meet Up With the Handlogger, *Pacific Pulp & Paper Industry* (August 1928): 37.

14. R. E. Kan Smith, "Logging Chances on the West Cost of Prince of Wales Island" (1918), RG 95, NARA, College Park, Md.

15. Langille, "Report on Alexander Archipelago Forest Reserve, 1906," 6–7.

16. William B. Greeley to Donald Sutherland (Alaska's non-voting representative in Congress), February 28, 1927, RG 95, NARA, College Park, Md.

17. Wyckoff, "Lumbering in Alaska," 8; August Buschmann to Robert De Armond, December 2, 1959, Robert Thorstenson Collection, Seattle, Wash.

18. "Ketchikan Spruce Mills Buys Out J. R. Reynolds," *Ketchikan Daily News*, August 28, 1954; Wyckoff, "Lumbering in Alaska," 8; Tongass-Sales, William Paddock, September 5, 1913, RG 95, NARA—Pacific Alaska Region, Anchorage; Allen, *Spirit!*, 80; Adam Greenwald, July 19, 2005; Judith Brakel, "Historical and Cultural Aspects of the Falls Creek (Kahtaheena) Proposed Hydroelectric Project," in *Researchers Annual Reports, Preliminary Draft of Appendix 1 to the Application for a Hydropower License at Falls Creek Near Gustavus, Alaska* (February 2001), 17–19.
19. "Problems Confront Loggers," *Lumberman* (December 1955): 95, 96, 98, 134.
20. Rogers, *Alaska in Transition*, 152; "Hand-Boat Logging is All Arranged," *Ketchikan Alaska Chronicle*, October 20, 1953; "Boat Logging Program is Expanding," *Ketchikan Alaska Chronicle*, November 5, 1953.
21. Editor and S. Williams Asa, "Logging by Steam," *Forestry Quarterly* (March 1908): 2, 5.
22. J. H. Dickinson, "Safe Methods of Operating Mechanical Skidder," *American Lumberman* (January 2, 1926): 56.
23. De Armond, "Around & About Alaska," September 7, 1994; Hinkley, *Alaskan John G. Brady*, 329; Wyckoff, "Lumbering in Alaska," 8; Louis McDonald, "Logging and Lumbering in Alaska," *Timberman* (January 1921): 40; "Purchase New Type 'Donkey,'" *Ketchikan Alaska Chronicle*, May 6, 1924.
24. Roppel, *An Historical Guide to Revillagigedo and Gravina Islands*, 236; Wyckoff, "Lumbering in Alaska," 8.
25. "Shovels to Replace Tractors on Ketchikan Road Project," *Timberman* (October 1961): 41; "Logging in Alaska is Tougher," *Lumberman* (November 1955): 134.
26. "Problems Confront Loggers," *Lumberman* (December 1955): 95, 96, 98, 134.
27. "King-Size Skidder and Slackline Built for Alaska," *Lumberman* (September 1954): 107; caption on Leland J. Pratner photo, USFS Collection, PCA 207, Historical Collections, Alaska State Library, Juneau; Ketchikan Pulp Company Annual Report, 1955; "Logging in Alaska is Tougher," *Lumberman* (November 1955): 134, 135; "Problems Confront Loggers," *Lumberman* (December 1955): 98; John Hagen (former logger), personal communication with author, January 18, 2007.
28. "Alaska's Newest Pulp Mill Begins Operation," *Western Conservation Journal* (November–December 1959): 25; "Alaska Logging Methods," *Pulp & Paper* (February 1960): 106, 107; "Big Machines Solve Big Problems in Alaska," *Timberman* (November 1960): 47. Note: includes diagram of float.
29. Harris and Farr, *Forest Ecosystem of Southeast Alaska*, 56.
30. Frost, incomplete report of timber cruise in southeast Alaska, 3.
31. Kutchin, *Report of the Salmon Fisheries of Alaska, 1904*, 15–16; Gillette, *Yakutat Fish Train*, 5; Cobb, *Pacific Salmon Fisheries*, 448.
32. "Advantages and Disadvantages of Tractors in the Logging Industry," *American Lumberman* (August 7, 1920): 88.
33. Wyckoff, "History of Lumbering in Southeast Alaska," 39.
34. *Logging Systems Guide*, Series No. R10-21 (Juneau: Alaska Region, Division of Timber Management, USFS, April 1978), 32.
35. "Alaska Spruce," *Timberman* (February 1943): 12; "Progressive Alaska Lumber Company," *West Coast Lumberman* (September 1941): 54; "Shovels to Replace Tractors on Ketchikan Road Project," 40; *1953 Annual Report, Governor of Alaska to the Secretary of the Interior* (Washington: GPO, 1953), 38.
36. R. R. Robinson, "Alaska Goes Truck Logging," *West Coast Lumberman* (January 1938): 14.
37. Ibid.; "Truck Logging in Alaska," *Timberman* (April 1938): 56; "Ketchikan Spruce Mills," *Timberman* (August 1941): 34; James W. Girard and J. M. Wyckoff, "Report on the Suitability of Spruce in Southeastern Alaska for the Production of Airplane Lumber," 1942, USFS (unpublished, 1942), RG 95, NARA—Pacific Alaska Region, Anchorage.
38. Harris and Farr, *Forest Ecosystem of Southeast Alaska*, 54.
39. "Alaska Cut Continues to Develop," *Timberman* (May 29, 1961): 76; *1958 Annual Report of the Governor of Alaska* (Washington: GPO, 1958), 55.
40. National Forest Roads and Trails Act of 1964, Public Law 88-657 (October 13, 1964).
41. "Alaska Loggers Find Uses for Surplus Equipment," *Timberman* (April 1962): 38–40.

GLOSSARY

1. Clinton G. Smith, *Regional Development of Pulpwood Resources of the Tongass National Forest, Alaska*, USDA Bulletin No. 950 (Washington: GPO, 1921), 34–35; Heintzleman, "Newsprint from the North," 12; Timber Sale Agreement between USFS and Ketchikan Pulp & Paper Company, July 26, 1951, 9; Heintzleman, *Pulp-Timber Resources of Southeastern Alaska*, 18.
2. *Terms of the Trade* (Eugene, Oreg.: Random Lengths Publications, Inc., 2000), 91.
3. Brown, *American Lumber Industry*, 78.
4. "Forest Terminology," *Journal of Forestry* (January 1918): 18.
5. USFS, *Tongass National Forest Land and Resource Management Plan*, 2008, 2–7.
6. *Terms of the Trade*, 116.
7. O. Keith Hutchison and Vernon J. LaBau, *Forest Ecosystem of Southeast Alaska: No. 9, Timber Inventory, Harvesting, Marketing, and Trends*, Gen. Tech. Rept. PNW-34 (Portland, Oreg.: Pacific Northwest Forest and Range Experiment Station, 1975), 6; James Russell, "Mt. Verstiovia Russian-Era Harvest Young Single Cohort Stand," April 15, 1996, USFS, Sitka, Alaska files.

8. Whitey Lueck (University of Oregon, Eugene, Oreg.), "The Four Defining Features of Old-growth Forest," November 2006 (unpublished).

9. *Terms of the Trade*, 165.

10. "Forest Terminology," *Journal of Forestry* (January 1918): 41.

11. "Proposed Definition of Newsprint," *Paper Trade Journal* (April 9, 1925): 64.

12. *Terms of the Trade*, 262.

13. Ibid., 289.

14. In his *Woodsman's Handbook,* Henry Graves listed 51 different log rules. See: Graves, *Woodsman's Handbook*, Part 1 (Washington: GPO, 1903), 5–6; "Log Rules of Scribner and Doyle," *American Lumberman* (December 12, 1942): 34–35.

15. "Forest Terminology," 64.

16. E. P. Stamm, A. J. F. Brandstrom, and P. S. Bonney, "Confidential Report to U.S. War Production Board on Plans and Feasibility of Logging and Transporting High Grade Sitka Spruce From Alaska to Puget Sound for Airplane Stock," February–March 1942, RG 95, NARA—Pacific Alaska Region, Anchorage.

17. Richard A. Horn, *Morphology of Wood Pulp Fiber from Softwoods and Influence on Paper Strength*, FPL 242 (Madison, Wisc.: Forest Products Laboratory, 1974), 7; Ralph Dale, March 22, 2005.

Index

ABOUT THE AUTHOR

James Mackovjak was born in Cleveland, Ohio, and moved to Alaska in 1970. He worked mostly in the seafood industry and became involved in forest issues during the 1980s. During the late 1990s, he was a member of the governor's Southeast Alaska Timber Task Force. Mr. Mackovjak lives with his wife, Ann, in Gustavus, near Glacier Bay, in a home constructed almost completely of wood from the Tongass National Forest. His local history, Hope & Hard Work: The Early Settlers at Gustavus, Alaska, was published in 1988. More recently, he wrote the forthcoming Navigating Troubled Waters: A History of Commercial Fishing in Glacier Bay, Alaska, for the National Park Service. Current projects include a history of maritime freight service to the Aleutian Islands area and an administrative history of Lake Clark National Park.